Communications and Control Engineering

More information about this series at http://www.springer.com/series/61

Hendra I. Nurdin · Naoki Yamamoto

Linear Dynamical Quantum Systems

Analysis, Synthesis, and Control

Springer

Hendra I. Nurdin
School of Electrical Engineering
and Telecommunications
UNSW Australia
Sydney, NSW
Australia

Naoki Yamamoto
Department of Applied Physics
and Physico-Informatics
Keio University
Yokohama
Japan

ISSN 0178-5354 ISSN 2197-7119 (electronic)
Communications and Control Engineering
ISBN 978-3-319-55199-9 ISBN 978-3-319-55201-9 (eBook)
DOI 10.1007/978-3-319-55201-9

Library of Congress Control Number: 2017934052

MATLAB® is a registered trademark of The MathWorks, Inc., 3 Apple Hill Drive, Natick, MA 01760-2098, USA, http://www.mathworks.com.

Printed on acid-free paper

This Springer imprint is published by Springer Nature
The registered company is Springer International Publishing AG
The registered company address is: Gewerbestrasse 11, 6330 Cham, Switzerland

Preface

The main theme of this research monograph is modeling and control of a special class of open quantum systems that shall be referred to as linear dynamical quantum systems. They are also referred to in the literature as linear quantum stochastic systems with the qualifier "stochastic" added because these systems are coupled to and driven by quantum stochastic processes (i.e., "quantum noise"). For brevity, we will often refer to them simply as linear quantum systems. Such systems are frequently encountered in fields such as linear quantum optics and opto-mechanics, to name a few, with applications in areas including continuous-variable quantum communication and quantum precision sensing. They are essentially a quantum analogue of the distinguished class of classical (non-quantum) linear systems that have played a foundational role in the development of modern and post-modern systems and control theory.

The reader may wonder, is there a place for a research monograph dedicated to this class of quantum systems? Certainly, this monograph would not be the first text to discuss linear dynamical quantum systems. Indeed, the now standard reference text "Quantum measurement and Control" by Wiseman and Milburn [1] already has sections devoted to linear dynamical quantum systems (see Sects. 5.6 and 6.6 therein). The present monograph complements existing textbooks that treat the subject of quantum feedback control, as the text of Wiseman and Milburn and the more recent "Quantum Measurement Theory and its Applications" by Jacobs [2], in terms of its emphasis and approach. It focuses on the class of linear quantum systems and provides an in-depth treatment of the system-theoretic foundations of such systems, and the control theoretic formulation of quantum versions of familiar problems from the classical (non-quantum) setting, including estimation and filtering, realization theory, and linear quadratic control. Some of these topics may be unfamiliar to researchers in quantum control coming from outside of the control theory discipline. It is our hope that the monograph can at least partly bridge this gap and serve as good introduction to these topics as a stepping stone towards understanding more general nonlinear quantum stochastic systems.

The monograph is aimed at graduate students and researchers with a background in control theory, physics, or mathematics who have an interest in the topic of quantum feedback control theory, and linear quantum systems in particular. It assumes that the reader has working knowledge of quantum mechanics, at least at the undergraduate level, such as in D.J. Griffith's text "Introduction to Quantum Mechanics" (Pearson Education). It also assumes that the reader from a non-control theory background has working knowledge of modern control theory, in particular with the state-space representation of linear dynamical systems. Some exposure to continuous-time stochastic filtering and control theory would also be valuable.

With the 2012 Nobel Prize in physics being awarded to experimental physicists Serge Haroche and David Wineland for their groundbreaking contributions to experimental quantum control techniques, and continuously improving experimental capabilities in controlling quantum systems, it would not be an exaggeration to say that it is currently exciting times for quantum control. Linear quantum systems model a wide range of quantum systems that are taking part in the broader ongoing quantum revolution, which promises to significantly increase technological capabilities in sensing, communication, and computing. The quantum future looks bright, and it is our hope that this monograph can play a part in that future.

HN thanks Clare Donald for proofreading the language in some parts of the monograph, and Zhan Shi for her assistance in providing some figures for Sect. 6.2. HN and NY also thank Oliver Jackson from Springer, for inviting them to write this monograph and facilitating the writing process, and Karin de Bie, for technical assistance in preparing the monograph prior to her retirement.

Sydney, Australia — Hendra I. Nurdin
Yokohama, Japan — Naoki Yamamoto
November 2016

References

[1] H.M. Wiseman, G.J. Milburn, *Quantum Measurement and Control* (Cambridge University Press, Cambridge, 2010)
[2] K. Jacobs, *Quantum Measurement Theory and its Applications* (Cambridge University Press, Cambridge, 2014)

Contents

Notation

$\imath$	$\sqrt{-1}$
$\mathbb{R}$	The set of real numbers
$\mathbb{R}_+$	The set of nonnegative real numbers
$\mathbb{C}$	The set of complex numbers
$\mathcal{H}$	A Hilbert space
$L^2(\mathcal{I};\mathcal{H})$	The set of $\mathcal{H}$-valued square integrable functions on a Lebesque measurable set $\mathcal{I}$
$B(\mathcal{H})$	The space of bounded linear operators mapping $\mathcal{H}$ to itself
ker(X)	The kernel or null space of an operator X
range(X)	The range of an operator X
$[a_1 \quad a_2 \quad \ldots \quad a_n]$	A row vector with elements $a_1, a_2, \ldots, a_n$
$(a_1, a_2, \ldots, a_n)$	A row vector with elements $a_1, a_2, \ldots, a_n$
$(\cdot)^*$	The conjugate of a complex number, and the adjoint of a Hilbert space operator or a matrix of Hilbert space operators
$(\cdot)^\#$	For a matrix of complex number or Hilbert space operator X, denotes the elementwise conjugation operation $X^\# = [X_{ij}^*]$
$(\cdot)^\top$	For a matrix of complex number or Hilbert space operators X, denotes the transposition operation $X^\top = [X_{ji}]$
$\Re\{\cdot\}$	For a matrix of complex number or Hilbert space operators X, denotes the real part, $\Re\{X^\#\} = 1/2(X + X^\#)$
$\Im\{\cdot\}$	For a matrix of complex number or Hilbert space operators X, denotes the imaginary part, $\Im\{X^\#\} = 1/2(-\imath X + \imath X^\#)$
$[A, B]$	Commutator $AB - BA$ of two operators acting on the same Hilbert space
$[x, y^*]$	Commutator matrix $xy^* - (x^\# y^\top)^\top$ between two column vector x and y of operators acting on the same Hilbert space. It becomes $[x, y^\top] = xy^\top - (xy^\top)^\top$ when x and y have self-adjoint entries

$\|\psi\rangle$	A ket, a normalized vector on a Hilbert space
$\langle\psi\|$	A bra, the conjugate transpose of a normalized vector $\|\psi\rangle$ on a finite dimensional Hilbert space or the continuous linear functional on the Hilbert space associated with $\|\psi\rangle$
$\langle\psi\| \cdot \|\psi\rangle$	Quantum expectation with respect to the pure state $\|\psi\rangle$
$\langle\cdot\rangle$	Quantum expectation
Tr	Trace of a trace-class operator. If $M = [M_{ij}]$ is matrix of (trace-class) operators, then $\mathrm{Tr}(M)$ can mean either $\mathrm{Tr}(M) = \sum_{i=1}^{n} \mathrm{Tr}(M_{jj})$ (only if M is square) or $\mathrm{Tr}(M) = [\mathrm{Tr}(M_{ij})]$, depending on the context
$\mathrm{Tr}_{\mathcal{H}}$	Partial trace over $\mathcal{H}$
I_n	An $n \times n$ identity matrix, subscript n can be dropped if dimension of matrix can be inferred from context
$0_n, 0_{m\times n}$	An $n \times n$ and $m \times n$ zero matrix, respectively, subscript n or $m \times n$ can be dropped if dimension of matrix can be inferred from context
$\mathrm{diag}(M_1, \ldots, M_n)$	A block diagonal matrix with $M_1, \ldots, M_n$ on its diagonal blocks
$\mathrm{diag}_n(M)$	$I_n \otimes M$
$\mathbb{J}$	The skew-symmetric matrix $\begin{bmatrix} 0 & 1 \\ -1 & 0 \end{bmatrix}$
$\mathbb{J}_n$	The skew-symmetric matrix $I_n \otimes \mathbb{J}_n$
$\Gamma_s(\mathcal{H})$	Boson/symmetric Fock space over $\mathcal{H}$
$e(f)$	Coherent/exponential vector in $\Gamma_s(\mathcal{H})$
$W(f)$	Weyl operator on a Boson Fock space
q	Amplitude/position quadrature of a single mode resonator
p	Phase/momentum quadrature of a single mode resonator
$\xi_i(t)$	Quantum white noise for incoming field i
$\eta_i(t)$	Quantum white noise for outgoing field i
$\mathcal{A}_i(t)$	Input annihilation operator for field i, $\mathcal{A}_i(t) = \int_0^t \xi_i(s)ds$
$\mathcal{Y}_i(t)$	Output annihilation operator for field i
$\xi(t)$	$\xi(t) = (\xi_1(t), \xi_2(t), \ldots, \xi_m(t))^\top$, m denotes number of fields
$\eta(t)$	$\eta(t) = (\eta_1(t), \eta_2(t), \ldots, \eta_m(t))^\top$, m denotes number of fields
$\mathcal{A}(t)$	$\mathcal{A}(t) = (\mathcal{A}_1(t), \mathcal{A}_2(t), \ldots, \mathcal{A}_m(t))^\top$, m denotes number of fields
$\mathcal{Y}(t)$	$\mathcal{Y}(t) = (\mathcal{Y}_1(t), \mathcal{Y}_2(t), \ldots, \mathcal{Y}_{m'}(t))^\top$, m' denotes number of output fields ($m \le m'$)
$\Lambda_{ij}(t)$	Gauge process from channel j to channel i
$\Lambda(t)$	$\Lambda(t) = [\Lambda_{ij}(t)]_{i,j=1,\ldots,m}$, m denotes number of fields
$\triangleleft$	Series product of Markov input–output systems
$\boxplus$	Concatenation product of Markov input–output systems

$\mathrm{vN}\{y(s), 0 \leq s \leq t\}$	von Neumann algebra generated by the spectral projections of a self-adjoint process $\{y(s), 0 < s < t\}$
$\mathscr{Y}_t$	von Neumann algebra generated by the spectral projections of a self-adjoint process $\{y(s), 0 < s < t\}$

Chapter 1
Introduction

Abstract This chapter gives a brief overview of and introduction to quantum feedback control and linear quantum systems. It begins with a brief history of quantum feedback control, in particular the development of quantum analogues of ideas from stochastic control theory, and discusses the notion of closed and open quantum systems, the Markov property, and quantum Langevin equations. This chapter then proceeds with an introduction to linear quantum systems and several physical examples thereof, including optical cavities, non-degenerate and degenerate parametric amplifiers, and large atomic ensembles.

This monograph is concerned with the modeling and control of a special class of quantum systems that shall be referred to as *linear dynamical quantum systems* or simply *linear quantum systems* for short. It is also referred to in the literature as linear quantum stochastic systems, with the qualifier stochastic added to emphasize the role of quantum stochastic processes in the dynamics of these systems. This class [1–4] represents multiple distinct open quantum harmonic oscillators that are coupled linearly to one another and also to external fields such as Gaussian coherent laser beams and single/multiple photon fields. Indeed, many important devices can be modeled by a linear quantum system, ranging from optical cavities, mesoscopic mechanical resonators, optical and superconducting parametric amplifiers, to linear atomic quantum memories; see, e.g., [5, Chaps. 7 and 10] and [6–14]. In analogy with the distinguished classical linear state-space models from modern control theory, see, e.g., [15, 16], the dynamics of a linear quantum system in the Heisenberg picture of quantum mechanics can be completely described in terms of a quartet of matrices A, B, C, D. Note that throughout the monograph, we will use the qualifier "classical" to refer to objects that are not quantum mechanical (i.e., not subject to the laws of quantum mechanics). However, the matrices A, B, C, D in a linear

Sections 1.5.2 and 1.5.3 contain some materials reprinted, with permission, from [44] © 2012 IEEE.

H.I. Nurdin and N. Yamamoto, *Linear Dynamical Quantum Systems*,
Communications and Control Engineering, DOI 10.1007/978-3-319-55201-9_1

quantum system cannot be arbitrary but must satisfy a set of constraints imposed by quantum mechanics for them to represent a physical system, a restriction not encountered in the classical setting. As first demonstrated in [2], for the case where D is of the form $D = [\, I \;\; 0 \,]$, with I denoting an identity matrix, A and B must satisfy a certain nonlinear equality constraint, while B and C satisfy a linear equality constraint. These constraints on the A, B, C, D matrices are referred to as *physical realizability* constraints [2]. In analogy with classical linear systems, linear quantum systems present themselves as rather nice and relatively tractable class of quantum systems with which to explore and potentially develop fundamental ideas and principles of quantum control, possibly emulating the fundamental role of classical linear systems in the early development of systems and control theory.

A growing number applications of linear quantum systems have been proposed or experimentally demonstrated in the literature. From a theoretical point of view, they have been studied as sensors for extremely weak signals [17, 18] such as gravitational-wave forces [19–23] and have been proposed as coherent feedback controllers [2, 24] (treated in Chap. 5 of this monograph) for cooling of opto-mechanical systems [7], for back-action evasion in linear quantum sensors [25, 26] and realizing quantum non-demolition measurements [25], and for creating decoherence-free subsystems [25, 27]. They have also been proposed as optical filters for modifying the wave packet shape of single- [28, 29] and multi-photon [30] sources, and as a means for dissipatively generating Gaussian cluster states [31, 32], which are of interest for continuous-variable one-way quantum computers [33]. Applications of measurement-based feedback control strategies to linear quantum systems (see Chap. 5 of this monograph) have also been investigated for the purpose of cooling the motion of a particle [1, 34] and enhancing entanglement between separated nodes in a linear network [35, 36]. More recently, linear quantum models have been shown to be highly relevant for analyzing linear quantum memories [9, 12–14, 37]. Quantum memory schemes using switched linear quantum systems have been proposed based on atomic ensembles [38] and by combining passive linear optics with coherent feedback [27]. Experimentally, linear quantum systems have been demonstrated for broadband disturbance attenuation in bulk quantum optics [39] as well as silicon photonics [40]. They have also been experimentally shown to be able to modify the characteristics of squeezed light produced by an optical parametric oscillator [41–45], and to reshape the dynamics of an electromechanical circuit [8]. In [8], it is highlighted that multiple-input multiple-output linear state-space-like modeling of linear quantum networks, which are extensively employed in modern control theory for classical linear systems, provides a powerful tool for analyzing these networks when they involve many components and inputs and outputs. Venturing beyond quantum information technologies, linear quantum systems are also of interest as a component for classical signal processing on quantum devices, for instance for processing light when photons, rather than electrons, are employed to transport information between cores on a chip and between chips [46]. Together with proposed nonlinear ultra-low power optical logic gates, such as developed in [47], they form the building blocks for envisaged ultra-efficient all-optical classical information processing circuits of the future.

The rest of this introductory chapter is organized as follows. Section 1.1 provides a brief historical overview of the development of quantum feedback control theory from the 1970s up to the present, while Sect. 1.2 gives an overview of classical linear systems and control theory and the prospects for a linear quantum systems and control theory. This is then followed by a discussion of closed linear quantum systems in Sect. 1.3 and open quantum systems, in particular the special class of Markov open quantum systems, in Sect. 1.4. Finally, Sect. 1.5 introduces linear dynamical quantum systems as a special type of Markov open quantum systems and illustrates some concrete physical examples of such systems along with the applications in which they are employed.

1.1 Quantum Feedback Control: A Brief History

We begin with a brief historical overview of the field of quantum feedback control, especially from the systems and control theory perspective. Of course, this will only be a part of the whole story, as relevant for the theme of this monograph, see [48] for an additional overview. We should also remark that in the 1980s seminal work by Tarn, Huang, and Clark on control of *closed* quantum systems, represented by a controlled (time-dependent) Schrödinger equation, appeared, see, e.g., [49, 50]. However, due to the closed nature of the system, the type of control for these systems is *open-loop* control rather than feedback control. That is, the control signal is predesigned and does not rely on extracting (classical) information from the system by observing it, or exchanging quantum information with another quantum system in a feedback interconnection. A comprehensive treatment of open-loop control of closed quantum systems can be found in D'Alessandro's text "Introduction to Quantum Control and Dynamics" [51]. However, the concern of this monograph is linear quantum systems and feedback control on such systems; therefore, open-loop control schemes are not discussed further in the monograph.

Returning to our brief historical overview, it would be fair to say that the late Viacheslav "Slava" Belavkin pioneered the development of quantum mechanical analogues of notions and tools from stochastic systems and control theory. We find his seminal works in this area going back to the 1970s.[1] Belavkin was a student of the renowned Russian probabilist and stochastic control theorist Ruslan Stratonovich and the extension of ideas from classical stochastic systems and control theory (recall that classical refers to being non-quantum) to the so-called observable quantum systems (in the terminology of Belavkin) was a natural progression of the research themes being pursued in the Stratonovich school. One of the main questions that Belavkin aimed to address was how to describe the time evolution of the quantum state of an open system under continuous-time measurements/observations, starting with a study of quantum harmonic oscillators attached to a quantum transmission line. Belavkin's work in this direction is a seminal contribution toward a mathematical quantum

[1]Most, if not all, of Belavkin's works have have been archived on his memorial homepage at the University of Nottingham, https://www.maths.nottingham.ac.uk/personal/vpb/.

systems and control theory. Using conceptual tools from the theory of open quantum systems that were available at the time, such as the notion of a "quantum instrument" to describe quantum-state evolution [52], Belavkin succeeded in deriving what is now called the *Belavkin quantum filtering equation* or the *stochastic master equation* (SME) in the quantum optics community. This equation describes the evolution of a special class of open quantum systems (including a large class of quantum optical systems) under continuous observations. Here, quantum filtering, as in the classical setting of nonlinear filtering theory, is fundamentally connected to a quantum notion of conditional expectations. The derivation of the quantum filtering equation marks the first step toward establishing what is now referred to as *measurement-based quantum feedback control*, a problem which Belavkin also subsequently solved, see, e.g., [53–56], using techniques that can be viewed as essentially the quantum analogue of approaches for solving partially observable stochastic control problems, albeit in a more abstract form that relies heavily on concepts from mathematical physics and the theory of operator algebras.

Around the mid-1980s physical modeling of quantum systems interacting with propagating optical fields was undertaken by Gardiner and Collett. The starting point was to obtain an accurate dynamical model of several *open* optical devices, such as a cavity (resonator) containing a nonlinear crystal [57, 58] (to be elaborated further in Sects. 1.5.1–1.5.3). In the case of the optical cavity, the system is open since the *internal* optical field inside the cavity couples to *external* incoming and outgoing optical fields. The major significance of their work is to derive a mathematical model for the cavity as a dynamical system having inputs and outputs. Input–output models have been central in systems and control theory, and this point in time marks the emergence of analogous quantum *input–output* models. Notably, it later turned out that this specific input–output modeling method for quantum optics is applicable to many physical devices interacting with external fields, such as an atomic ensemble and a nano-mechanical oscillator [5]. The key tool enabling such accurate modeling is *quantum stochastic calculus*. Though such a calculus was obtained independently by Gardiner and Collett in [58] through physical intuition and arguments, a more general and mathematically rigorous version of quantum stochastic calculus had in fact been established earlier by Hudson and Parthasarathy in their landmark paper [59]. However, although Hudson and Parthasarathy devised a more complete calculus, they did not formulate the important notion of input–output quantum systems with this calculus, as Gardiner and Collett had done in their work. The input–output formulation of Gardiner and Collett and its subsequent developments and related research are summarized in the now classic Gardiner and Zoller quantum optics text "Quantum Noise" [5]. This book comprehensively describes several quantum systems having inputs and outputs with dynamical equations, described using *quantum stochastic differential equations* (QSDEs) as a quantum analogue of ordinary stochastic differential equations. Thus, the notion of a "quantum input–output system" could be established, paving the way for a systems and control theory for quantum systems. Of course, this input–output formulation can be formulated in the rigorous framework of the Hudson–Parthasarathy quantum stochastic calculus, see, e.g., [60].

In the early 1990s, a version of Belavkin's filtering theory was independently rediscovered simultaneously by several researchers in quantum optics [61–63], being obtained from a completely different point of view, using a different approach, and derived for the purpose of stochastic simulation of open quantum systems rather than for estimation. In the quantum optics literature, the theory goes by the name of "quantum trajectory theory" and a comprehensive treatment of this theory can be found, e.g., in the text [63] of Carmichael, one of its originators. As mentioned above, Belavkin's filtering equation in the latter is referred to as the SME. Roughly speaking, the SME describes the dynamics of the quantum state of a system conditioned on measurement results, which are obtained by detecting the system's output field. In other words, it models the evolution of the system as it is continuously monitored in time. Since the measurement outcomes can be processed and fed back to influence the system, the introduction of the SME spurred the development of quantum feedback control for optical systems in the physics literature in the 1990s, in particular through the influential contributions of Wiseman and Milburn; see their text [48] for a discussion of various important developments in this direction. This body of work had quite a distinctive flavor from the work of Belavkin in that system analysis and derivation of feedback laws relied on rigorous and extensive use of physical principles rather than mathematical ones. At around this period, the discipline of quantum information science [64] was in its early days, and it would later be widely recognized that some form of quantum control would be indispensable for the realization of any practical quantum information processing systems [48, 65]. This is simply because such systems tend to be very fragile and can easily interact with their environment (acting as a disturbance to the system).

The advent of the Hudson–Parthasarathy quantum stochastic calculus opened a path toward a modern formulation of Belavkin's work in a more accessible form and in a much closer analogy to classical stochastic control theory, including through the contributions of Bouten and van Handel, see, e.g., [60, 66–68]. Beside their own important original contributions to the subject, they translated major ideas in Belavkin's more abstract work into a language that is familiar to control theorists, probabilists, and engineers with a working knowledge of stochastic control theory. This brings out much more explicitly the role of the quantum filtering equations as a quantum analogue of classical filtering equations. In the classical case, the filtering equations represent an algorithm for real-time mean-square optimal estimation of a system variable. More precisely, it is a real-time Bayesian update rule based on the measurement record. This clarity helps in formulating quantum feedback control problems fully within the framework of quantum probability theory (the non-commutative extension of classical probability theory), and one can pose many important questions, such as, how can we utilize estimates from a quantum filter to engineer a feedback controller? Is it possible to formulate and solve optimal feedback control problems? If the system contains some uncertainty, can we design a robust estimator/controller? Can we replace a classical feedback controller composed of an estimator and actuator by a fully quantum device?

Many of the questions mentioned above have been fully or partly resolved. More recent developments have witnessed the emergence of quantum versions of electrical network synthesis theory, measurement-based/coherent feedback LQG control,

H^∞ control, robust estimation and control, stability analysis, system identification, to name a few. They are not mere analogues of classical theories, but have a role to play in applications to quantum mechanical systems, particularly in quantum information science, some of which will be highlighted in Chap. 6. Moreover, experimental demonstrations unimaginable a decade ago are within the scope of current technologies. The long-time dream of being able to control and manipulate quantum systems has now become a reality. Indeed, the Nobel Prize in Physics in 2012 was jointly awarded to experimental physicists Serge Haroche and David Wineland with the citation "for ground-breaking experimental methods that enable measuring and manipulation of individual quantum systems."

Due to its practical importance for quantum information processing, quantum feedback control theory has attracted extensive attention from various research disciplines. To bring together the interdisciplinary communities working in quantum control theory, and to accelerate the development of this emerging field, the workshop series "Principles and applications of control in quantum systems" (PRACQSYS) was established and held for the first time on the Caltech campus in 2004 [69]. Since the successful first meeting, the PRACQSYS series of workshops has been successively held almost every year in various locations in the USA, Europe, Australia, and Asia, playing the role of a venue to stimulate exchange of ideas among researchers working in the field.

To conclude, we emphasize that, from both theoretical and experimental viewpoints, the special class of linear quantum systems as the focus of this monograph can play a role in further accelerating developments in the field of quantum control theory for reasons that will be elaborated in Sect. 1.2.3 and illustrated in Chap. 6.

1.2 Classical Linear Systems and Control Theory

Here, we briefly describe some important results developed in systems and control theory for classical linear systems. Indeed, they remind us of the strength and merits of the theory, thereby serving as a motivation to study linear quantum systems encountered in the quantum regime.

1.2.1 Classical Linear Systems

There are several reasons why linear systems are of particular importance. The first is that many practical systems of interest have a dynamical time evolution that can be modeled quite faithfully by a linear dynamical system described by linear (deterministic or stochastic) differential equations. Sometimes the system contains a nonlinear term that can be well approximated by a local linear term via linearization.

Let us examine here a typical example of such classical linear systems: a classical mechanical oscillator. The basis of the physical modeling is, of course, the Newtonian equation of motion, or more generally the Hamilton equation

$$\frac{dX}{dt} = \{H, X\} = \frac{\partial H}{\partial p}\frac{\partial X}{\partial q} - \frac{\partial X}{\partial p}\frac{\partial H}{\partial q}, \tag{1.1}$$

where H is the Hamiltonian and (q, p) is a pair of canonical variables having the meaning of position and momentum of a single mechanical degree of freedom. In particular, let us consider the following quadratic Hamiltonian:

$$H = \frac{m\omega^2}{2}q^2 + \frac{1}{2m}p^2 - qu.$$

This governs the dynamics of an object in a harmonic potential with resonance frequency ω. The second and third terms represent the kinetic energy, where m is the mass of the object and $u(t)$ is a time-varying driving (control) force, i.e., an *input* to the system. Then, (1.1) immediately yields the following linear equation:

$$\frac{d}{dt}\begin{bmatrix} q \\ p \end{bmatrix} = \begin{bmatrix} 0 & 1/m \\ -m\omega^2 & 0 \end{bmatrix}\begin{bmatrix} q \\ p \end{bmatrix} + \begin{bmatrix} 0 \\ 1 \end{bmatrix}u, \quad y = [1\ 0]\begin{bmatrix} q \\ p \end{bmatrix}. \tag{1.2}$$

Here, we have added an *output* equation $y(t) = q(t)$, under the assumption that $q(t)$ can be observed directly. Note that even in the case where the system contains some nonlinear effects, the model (1.2) could be still effective; for instance, the dynamical equation $\dot{q} = p/m + q^3$ can be well approximated using $\dot{q} = p/m$ if the position q is confined in a small region satisfying $|q| \ll 1$.

Other than the above example, there are numerous physical systems that can be modeled using linear equations, such as the motion of an aircraft at equilibrium and an atomic force microscope; see [70]. They usually contain multiple variables, inputs, and outputs, which are sometimes of the order of hundreds, thousands, or more. Nonetheless, even large-scale multiple-input and multiple-output (MIMO) systems can often still be modeled by a linear equation, which can be of the following general form:

$$\dot{x} = Ax + B_1u + B_2w_1, \quad y = Cx + D_1u + D_2w_2. \tag{1.3}$$

Here, $x \in \mathbb{R}^n$ denotes the vector of the system's state variables, u is the vector of control input, and y is the vector of observed output. Further, in the above equation, additional noise variables w_1 and w_2 are included to represent a more general *open system* under environmental effects (w_1 and w_2 are external signals from the environment that disturb the system). Note however that, as we will see later, adding a noise term is essential in the quantum case if one wishes to do feedback control. $(A, B_1, B_2, C, D_1, D_2)$ are matrices representing the system dynamics. The strength of systems and control theory is that it can deal with a general model of the above form as long as matrices A, B_1, B_2, C, D_1, D_2 satisfy the conditions required by the particular technique being employed. Lastly, we remark that the above differential equation can be analytically solved as follows:

$$x(t) = e^{At}x(0) + e^{At}\int_0^t e^{-As}\Big(B_1u(s) + B_2w_1(s)\Big)ds.$$

This simple fact makes the analysis and controller design for linear systems more tractable and underlies some of the success of linear systems and control theory.

1.2.2 Linear Systems and Control Theory

Once a linear model of the system of interest is obtained, then linear systems and control theory provides the solution to several important problems. Let us return to the oscillator system (1.2). For this system, clearly we cannot directly control $q(t)$ with the input $u(t)$. But can we do that indirectly through $p(t)$? Also, we cannot directly observe $p(t)$. However, can we observe it indirectly through the output $y(t) = q(t)$? One may also think about the important problem of designing optimal control signals. For example, by appropriately designing the input signal $u(t)$ as a function of $y(t)$, we want to drive $q(0) \neq 0$ to $q(T) = 0$ at a specific time $t = T$, under a certain optimality criterion; but how can we formulate this problem and solve it? Moreover, it is often the case that some parameters contained in the system are unknown; for instance, if the driving force $u(t)$ is not a controllable force but an unknown signal, then how can we use the output $y(t)$ to estimate $u(t)$? Linear systems and control theory provides rigorous and systematic methodologies to solve these problems, analytically or numerically. Note that, for a nonlinear system, the problems can become significantly more difficult and in general no analytical solutions are available. Below we briefly highlight some important results.

(i) Let us consider the *controllability problem* of determining which components of $x(t)$ can be directly/indirectly controlled by the input $u(t)$ and the *observability problem* of determining which components of $x(t)$ can be observed directly/indirectly from the output $y(t)$. These notions are completely characterized by the following *controllability matrix* $\mathcal{C}$ and the *observability matrix* $\mathcal{O}$:

$$\mathcal{C} = [B_1,\ AB_1,\ A^2B_1, \dots, A^{n-1}B_1],$$
$$\mathcal{O} = [C^\top,\ A^\top C^\top,\ (A^\top)^2C^\top, \dots, (A^\top)^{n-1}C^\top]^\top.$$

First consider the controllability problem. Let us assume $\mathrm{rank}(\mathcal{C}) = m < n$. Then, there exists a coordinate transformation $x \to x' = [x_1', x_2']^\top$, where the dynamics part of (1.3) can be represented as

$$\frac{d}{dt}\begin{bmatrix} x_1' \\ x_2' \end{bmatrix} = \begin{bmatrix} A_{11}' & A_{12}' \\ O & A_{22}' \end{bmatrix}\begin{bmatrix} x_1' \\ x_2' \end{bmatrix} + \begin{bmatrix} B_{11}' \\ O \end{bmatrix} u + \begin{bmatrix} B_{21}' \\ B_{22}' \end{bmatrix} w_1, \qquad (1.4)$$

where $x_1'(t) \in \mathbb{R}^m$. Thus $x_1'(t)$ constitutes a *controllable subsystem* with respect to (w.r.t.) $u(t)$, while $x_2'(t)$ cannot be controlled by $u(t)$.

Next let us consider the observability problem and assume $\mathrm{rank}(\mathcal{O}) = \ell < n$. Then, as in the above case, the system dynamics has the following representation in a certain coordinate:

$$\frac{d}{dt}\begin{bmatrix} x_1'' \\ x_2'' \end{bmatrix} = \begin{bmatrix} A_{11}'' & O \\ A_{21}'' & A_{22}'' \end{bmatrix}\begin{bmatrix} x_1'' \\ x_2'' \end{bmatrix} + B_1''u + B_2''w_1,$$

$$y = [C_1'', \ O]\begin{bmatrix} x_1'' \\ x_2'' \end{bmatrix} + D_1u + D_2w_2. \tag{1.5}$$

This clearly shows that $x_2''(t)$ cannot be observed from the output $y(t)$, while $x_1''(t)$ can; hence $x_1''(t)$ constitutes an *observable subsystem* while $x_2''(t)$ is not.

(ii) A particularly important technique related to the observability property is that of *filtering* [71]; that is, we wish to *estimate* the system variable $x(t)$ only through the output $y(t)$, in the presence of the noisy environment. Let us denote such an estimator by $z(t)$, which should be a function of the set of outputs, $\mathscr{Y}_t = \{y(s) \mid 0 \leq s \leq t\}$. Then, if the noises w_1 and w_2 are independent Gaussian white noises, the most celebrated and powerful result of linear stochastic systems theory is that the optimal estimate

$$\hat{x}(t) = \mathrm{argmin}_z \mathbf{E}\Big[\big(x(t) - z(t)\big)\big(x(t) - z(t)\big)^\top \ \Big| \ \mathcal{Y}_t\Big]$$

can be explicitly calculated by the following *Kalman filter*:

$$\frac{d}{dt}\hat{x}(t) = A\hat{x}(t) + B_1u(t) + V(t)C^\top\big[y(t) - C\hat{x}(t)\big], \tag{1.6}$$

$$\frac{d}{dt}V(t) = AV(t) + V(t)A^\top - V(t)C^\top CV(t) + B_2B_2^\top, \tag{1.7}$$

where $V(t)$ represents the estimation error covariance matrix:

$$V(t) := \mathbf{E}\Big[\big(x(t) - \hat{x}(t)\big)\big(x(t) - \hat{x}(t)\big)^\top\Big]. \tag{1.8}$$

In the above formula, $\mathbf{E}$ denotes the mean with respect to the Gaussian distribution. Also for simplicity, we have assumed $D_1 = 0$ and $D_2 = I$, and the covariance matrix of w_1 and w_2 are both the identity matrices. Equation (1.7) is called a *Riccati differential equation*. The initial conditions $\hat{x}(0)$ and $V(0)$ are determined according to an "initial guess," which of course should be carefully considered.

(iii) In the linear case, several control problems can be well formulated and, further, some of them have analytical or partly analytical solutions. A particularly beautiful and useful theory is the theory of *Linear Quadratic Gaussian* (LQG) control [72, 73]. The problem is, for the linear system (1.3), to find a feedback control law $u(t)$ that is a function of $\mathscr{Y}_t$ and that minimizes the following quadratic-type cost function:

$$J[u(t)] = \mathbf{E}\Big[\int_0^T \big(x(t)^\top Px(t) + u(t)^\top Qu(t)\big)dt + x(T)^\top Rx(T)\Big], \tag{1.9}$$

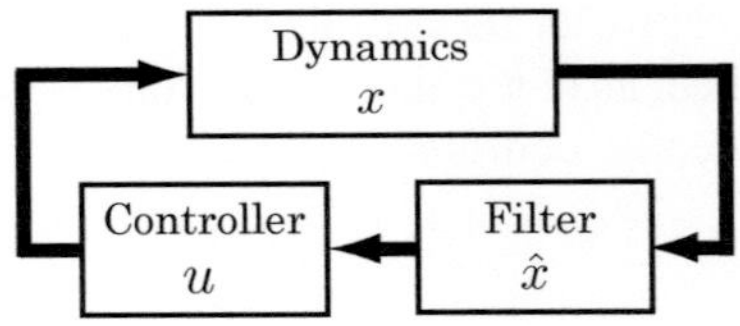

Fig. 1.1 Diagram of classical LQG control

where (P, Q, R) are the weighting matrices of the appropriate dimensions (Q is positive definite, and (P, R) are positive semidefinite). Let us assume again that w_1 and w_2 are independent Gaussian white noises, and $D_1 = 0$, $D_2 = I$. Then, the optimal feedback input minimizing $J[u(t)]$ is given by $u^{\mathrm{opt}}(t) = -Q^{-1}B_1^\top K(t)\hat{x}(t)$, where the time-varying matrix $K(t)$ satisfies

$$\dot{K}(t) + K(t)A + A^\top K(t) - K(t)B_1 Q^{-1} B_1^\top K(t) + P = O, \tag{1.10}$$

with terminal condition $K(T) = R$. A notable fact is that the optimal feedback control input $u^{\mathrm{opt}}(t)$ is a function of the optimal estimate $\hat{x}(t)$, the latter being generated recursively from the Kalman filter equations (1.6) and (1.7). This means that $u^{\mathrm{opt}}(t)$ is obtained by computing $K(t)$ and $\hat{x}(t)$ *independently* of each other. This fact is known as the *separation principle*; see Fig. 1.1.

1.2.3 Toward Systems and Control Theory for Linear Quantum Systems

In the previous sections, we have seen that, in the classical case, linear systems and control theory provides a number of powerful tools that can be applied in many practical situations. It is surprising that many quantum system models of interest, in particular arising in quantum optical and quantum opto-mechanical systems, can be described or well approximated by a linear dynamical equation, and, further, there exists a quantum version of Kalman filtering and LQG control. One of the main purposes of this monograph is, as mentioned at the beginning of this chapter, to detail this class of quantum systems, and describe the quantum linear systems and control theory. There is certainly a beautiful underlying theory that is still undergoing development and finding natural applications.

Classical linear systems have played a major role in the development of modern control theory, where many crucial systems and control concepts were first discovered for these classes of systems first before being extended to other classes of dynamical systems. Linear quantum systems could potentially also play a role as a class of quantum models with which to investigate and discover fundamental concepts for quantum systems and control theory. This will not only be of theoretical but also of practical interest. Various physical systems that can be modeled as linear quantum systems, as alluded to above, are relevant for applications such as nano-mechanical cooling [7, 74], high-precision sensing [19–23], and quantum memories [9, 12–14, 37, 38].

1.3 Closed Linear Quantum Systems

We will now describe linear quantum systems, starting with a quantum version of the classical harmonic oscillator (1.2). The Hamiltonian has exactly the same form:

$$H = \frac{m\omega^2}{2}q^2 + \frac{1}{2m}p^2 - qu. \tag{1.11}$$

But in quantum mechanics, the position q and the momentum p must be represented by self-adjoint operators satisfying the *canonical commutation relation* (CCR)

$$[q, p] = qp - pq = 2\imath. \tag{1.12}$$

For convenience, here we assume physical units in which the Planck constant is $\hbar = 2$, a commonly used convention in quantum optics and continuous-variable quantum information that we shall adopt for the remainder of the monograph without further comment; see, e.g., [75, 76]. The dynamical equation of (q, p) is described by the Heisenberg equation of motion rather than the Hamilton equation (1.1); i.e., given a Hamiltonian H, the dynamics of any quantum operator X evolves in time according to the following differential equation:

$$\frac{dX}{dt} = \imath[H, X]. \tag{1.13}$$

Hence, substituting the quadratic Hamiltonian (1.11) for (1.13), we find that $X = q$ and $X = p$ obey the following linear equation:

$$\frac{d}{dt}\begin{bmatrix} q \\ p \end{bmatrix} = \begin{bmatrix} 0 & 1/m \\ -m\omega^2 & 0 \end{bmatrix}\begin{bmatrix} q \\ p \end{bmatrix} + \begin{bmatrix} 0 \\ 1 \end{bmatrix} u. \tag{1.14}$$

Surprisingly, this has completely the same form as the dynamical equation of the classical harmonic oscillator, (1.2). But recall that q and p are operator-valued variables, which always satisfy the CCR (1.12). Actually, when $u \equiv 0$ for simplicity, the above dynamical equation has a solution of the following form:

$$q(t) = \cos(\omega t)q(0) - \sin(\omega t)p(0), \quad p(t) = \sin(\omega t)q(0) + \cos(\omega t)p(0),$$

thus, $[q(t), p(t)] = [q(0), p(0)] = 2\imath$ holds for all t.

Another important fact is that, although $(q(t), p(t))$ evolve in time in a deterministic way, they are non-commutative random variables or observables, with their spectra (eigenvalues) corresponding to possible measurement outcomes (if they are measured). The mean values of these observables are hence well defined; let $\langle q(t)\rangle$ and $\langle p(t)\rangle$ be those mean values, then they satisfy

$$\frac{d}{dt}\begin{bmatrix}\langle q\rangle\\ \langle p\rangle\end{bmatrix}=\begin{bmatrix}0 & 1/m\\ -m\omega^2 & 0\end{bmatrix}\begin{bmatrix}\langle q\rangle\\ \langle p\rangle\end{bmatrix}+\begin{bmatrix}0\\ 1\end{bmatrix}u. \tag{1.15}$$

As now the means $\langle q(t)\rangle$ and $\langle p(t)\rangle$ are scalar variables, this is a classical dynamical equation. This can be analyzed in the same way as with (1.2). For instance, it is possible to plot a trajectory of the dynamics (1.15) in the phase space of $(\langle q\rangle, \langle p\rangle)$.

We now highlight an important difference between the classical and quantum systems considered above. Unlike the classical case (1.2), there is no output equation accompanying the quantum dynamics (1.14). This is because the quantum harmonic oscillator is *closed* in that it is perfectly isolated from any external influence; that is, roughly speaking, it evolves "inside a shielded box." This means that neither the operator-valued dynamics (1.14) nor the mean dynamics (1.15) represent the real-time evolution of the system under a monitoring (observation) process. Hence, the input $u(t)$ appearing in the equations cannot depend on a quantity obtained from observation of the system (e.g., by performing continuous measurement), and feedback control is not possible. The input $u(t)$ can only be a signal that is designed independently of the real-time dynamics of the system, and the only type of control possible for a closed system is open-loop control. Therefore, if we want to manipulate the system via real-time feedback control, the system should be *open*, as described in the next section.

1.4 Open Quantum Systems, the Markov Approximation, and Quantum Langevin Equations

1.4.1 Open Quantum Systems

Real-world quantum systems are never completely decoupled from their environment. In reality, they will, to some degree, be interacting with their environment. Quantum systems interacting with an external environment are said to be *open quantum systems*. They are indispensable for modeling various important physical phenomena, for instance, to model the decay of the energy of an atom by spontaneous emission of photons (particles of light). The environment is often modeled to be another quantum system that can be viewed as a *heat bath* to which an open quantum system can dissipate energy or from which it can gain energy. To describe this interaction between an open quantum system and an external heat bat, often idealizations are introduced to derive approximate but tractable mathematical models. One such idealization is the so-called *Markov* assumption: The coupling of the system and bath are essentially "memoryless" in the sense that future evolution of the dynamics of the system given its present state is independent of its past states and those of the bath. Open quantum systems with such a property are said to be *Markov*. Markov open quantum systems are important as they provide effective and accurate approximate models for various practically important open quantum systems that are encountered in fields such as quantum optics, opto-mechanics, and microwave superconducting circuits. As models, their power derives from their relatively tractability for analysis as their dynamics can be written in terms of first-order operator differential equations.

In this section, we give an illustration of the Markov approximation in a concrete physical setting. This is followed by an introduction to the evolution equation (in the Heisenberg picture) for a special class of Markov open quantum dynamics, known as the quantum Langevin equations.

1.4.2 Illustration of the Markov Approximation and Markov Open Quantum System Dynamics

Here, we shall derive the *quantum Langevin equation*, which describes the evolution of a large class of Markov open quantum system, for a single quantum harmonic oscillator with mode c (satisfying the CCR $[c, c^*] = 1$) coupled to a single external electromagnetic bath; for a more detailed discussion, see, e.g., [5, Chaps. 5 and 7]. The essential ideas are summarized as follows:

(i) First compute the Heisenberg equation of *both* the system and the bath.
(ii) Make a Markov approximation on the system and bath coupling, which leads to a dynamical equation for the system observables.

Let us begin with a description of the bath. It is composed of a continuum of modes with annihilation operator b_ω, where ω denotes the frequency for that mode and which is also used to label it. As will be shown later, the continuum of frequencies will range from $-\infty$ and ∞. Technically, negative frequencies are unphysical, but in the model that we discuss here, typically arising in quantum optics and related fields, negative frequencies arise through the application of a *rotating-wave approximation* (RWA), which we assume has already been applied (we refer to [5] for the details). The fields at distinct frequencies do not couple with one another and commute with each other, satisfying the CCR

$$[b_\omega, b_{\omega'}] = 0, \quad [b_\omega, b_{\omega'}^*] = \delta(\omega - \omega'), \tag{1.16}$$

where $\delta(\cdot)$ denotes the Dirac delta function. The Hamiltonian for the oscillator is $H_{\rm sys} = \omega_o c^* c$, where ω_o is the oscillator's resonance frequency. The idealized bath Hamiltonian is given by

$$H_{\rm bath} = \int_{-\infty}^{\infty} \omega b_\omega^* b_\omega d\omega,$$

indicating explicitly the modeling of the bath as a collection of independent infinitesimal harmonic oscillators oscillating at distinct frequencies in a continuum of values. Next, the system and the bath are assumed to be coupled through the following interaction Hamiltonian:

$$H_{\rm int} = \imath \int_{-\infty}^{\infty} \kappa(\omega)\Big[b_\omega^* c - c^* b_\omega\Big] d\omega.$$

Let us now write the Heisenberg equations of both the system and the bath. The bath equation is given by

$$\frac{db_\omega(t)}{dt} = \imath[H_{\rm sys} + H_{\rm bath} + H_{\rm int}, b_\omega(t)] = -\imath\omega b_\omega(t) + \kappa(\omega)c(t),$$

where (1.16) was used and $b_\omega(t)$ and $c(t)$ are the Heisenberg picture evolution of b_ω and c, respectively. This immediately yields the solution

$$b_\omega(t) = e^{-\imath\omega t} b_\omega + e^{-\imath\omega t}\int_0^t e^{\imath\omega s}\kappa(\omega)c(s)ds. \tag{1.17}$$

Likewise, an arbitrary system variable $X(t)$ obeys

$$\begin{aligned}\frac{dX(t)}{dt} &= \imath[H_{\rm sys} + H_{\rm int}, X(t)]\\ &= \imath[H_{\rm sys}, X(t)] + \int_{-\infty}^{\infty}\kappa(\omega)\Big\{[c(t)^*, X(t)]b_\omega(t) - b_\omega(t)^*[c(t), X(t)]\Big\}d\omega.\end{aligned}$$

Then substituting (1.17) into this equation, we obtain the dynamical equation of the system observables. But as readily seen, the dynamics is not Markov since the time evolution of $X(t)$ in the future $[t, \infty)$ depends on the past history in $[0, t]$. Nevertheless, in several typical situations such as the case where the bath is a coherent laser beam, we are allowed to make a valid approximation that converts the system to a Markov one. In our case, this means that $\kappa(\omega)$ can be approximated to be constant, say $\kappa(\omega) = \sqrt{\gamma/2\pi}$; i.e., the system-bath coupling does not depend on the frequency. Then, due to the identity $\int_{-\infty}^{\infty} e^{-\imath\omega t}d\omega = 2\pi\delta(t)$, the above system equation reduces to

$$\begin{aligned}\frac{dX(t)}{dt} &= \imath[H_{\rm sys}, X(t)] + \sqrt{\gamma}\Big\{[c(t)^*, X(t)]b_{\rm in}(t) - b_{\rm in}(t)^*[c(t), X(t)]\Big\}\\ &\quad + \gamma\Big(c(t)^*X(t)c(t) - \frac{1}{2}c(t)^*c(t)X(t) - \frac{1}{2}X(t)c(t)^*c(t)\Big).\end{aligned} \tag{1.18}$$

This is called a quantum Langevin equation, describing the (approximate) Markov time evolution of the system. Here, we have defined the *input field*:

$$b_{\rm in}(t) = \frac{1}{\sqrt{2\pi}} \int_{-\infty}^{\infty} e^{-\imath\omega t} b_\omega d\omega. \tag{1.19}$$

That is, $b_{\rm in}(t)$ is the Fourier transform of b_ω. Note the following important commutation relations:

$$[b_{\rm in}(s), b_{\rm in}(t)] = 0, \quad [b_{\rm in}(s), b_{\rm in}(t)^*] = \delta(s-t) \ \ \forall s, t \geq 0.$$

Thus, $b_{\rm in}(t)$ can be regarded as the quantum analogue of classical white noise. Hence the input to the system (1.18) is precisely this operator-valued *quantum white noise process*. In particular, under the Markov approximation, the interaction Hamiltonian takes the form of a time-varying Hamiltonian, $H_{\rm int}(t) = \imath\sqrt{\gamma}(b_{\rm in}(t)^* c - c^* b_{\rm in}(t))$. Moreover, we can also immediately see that the optical field

$$b_{\rm out}(t) = \frac{1}{\sqrt{2\pi}} \int_{-\infty}^{\infty} b_\omega(t) d\omega \tag{1.20}$$

satisfies the following equation (interpreted as a boundary condition):

$$b_{\rm out}(t) = \sqrt{\gamma} c(t) + b_{\rm in}(t). \tag{1.21}$$

This is the portion of the optical light field $b_{\rm in}(t)$ *immediately after* its interaction with the system, which contains some information about the system as indicated by the first term on the right-hand side of (1.21). Also, as will be shown in the examples, the phase or intensity of $b_{\rm out}(t)$ can be measured by using a photo detector. That is, $b_{\rm out}(t)$ represents the *output field* of the system, and thus, (1.21) is called the *output equation*. For instance, if the input is a laser beam, then the output is the part of the beam that departs from the oscillator after the former's interaction with the latter.

We can write down a particular form of the Langevin equation when the system has m input–output fields:

$$\begin{aligned} \frac{dX}{dt} &= \imath[H_{\rm sys}, X] + \sum_{j=1}^{m} \Big(L_j^* X L_j - \frac{1}{2} L_j^* L_j X - \frac{1}{2} X L_j^* L_j \Big) \\ &\quad + \sum_{j=1}^{m} \Big\{ [L_j^*, X]\xi_j - \xi_j^* [L_j, X] \Big\}, \end{aligned} \tag{1.22}$$

$$\eta_j = L_j + \xi_j, \quad (j = 1, \ldots, m), \tag{1.23}$$

where $\sqrt{\gamma} c$, $b_{\rm in}$, and $b_{\rm out}$ are now replaced by L, ξ_j, and η_j, respectively. The multiple-input white noise processes ξ_j and multiple-output white noise processes η_j satisfy $[\xi_j(s), \xi_k(t)^*] = \delta_{jk}\delta(t-s)$ and $[\xi_j(s), \xi_k(t)] = 0$, and $[\eta_j(s), \eta_k(t)^*] = \delta_{jk}\delta(t-s)$ and $[\eta_j(s), \eta_k(t)] = 0$, respectively. Also note that the interaction Hamiltonian between the system and the jth bath is now given by

$$H_{\text{int}}^{(j)}(t) = \imath(\xi_j(t)^* L - L^* \xi_j(t)). \tag{1.24}$$

Note that the quantum Langevin equation above does not represent the most general Markov open quantum dynamics as it does not include the so-called gauge process which will be introduced and discussed in Chap. 2.

Finally, we remark that justifying the Markov approximation and Langevin equation for general open quantum systems and coupling operators L is in general subtle. A different set of physical assumptions from the ones we have described above for the single model oscillator with $L = \sqrt{\gamma}c$, and a different approximating procedure, may be required to justify employing a Markov approximation. We refer the reader to [77, 78] and the references therein for the details.

1.5 Linear Dynamical Quantum Systems: Description and Physical Examples

Armed with the background materials from the preceding section, we are now in a position to introduce the class of linear dynamical quantum systems, with their precise description and detailed mathematical modeling being deferred until the next chapter. To make the exposition of this section more concrete, the introduction will be followed by a discussion of several examples of physical systems that can be modeled within this class: optical cavities, non-degenerate parametric amplifiers, degenerate parametric amplifiers, opto-mechanical systems, and large atomic ensembles.

Linear dynamical quantum systems arise in practice as idealized models of a collection of coupled distinct open quantum harmonic oscillators that are in a Markov interaction with their surrounding heat baths (acting as a quantum noise source) with the following properties:

(i) The system Hamiltonian is of a *quadratic* form, quadratic here being in the position and momentum operators of the oscillators, such as $H_{\text{sys}} = (q^2 + p^2)/2$.
(ii) The system's canonical position and momentum operators are *linearly* coupled to one or more external (quantum) heat baths, e.g., via the coupling operator $L = (q + \imath p)/2$.

This results in a linear quantum Langevin equation for the position and momentum operators of the oscillators driven by stochastic heat baths. For instance, consider the treatment in [34] of an atom trapped in an optical cavity. The atomic dipole is strongly coupled to light inside the cavity that results in random mechanical forces acting on the atom as it absorbs and emits light. When the parameter values are in the right regime, the details of the dynamics of the optical and atomic dipole become unimportant, and the optical field can be viewed as an environment for the atomic motion. The observables relating to the motion of the trapped atom, its position and momentum operators, can then be treated like those of an open quantum harmonic oscillator linearly coupled to a bosonic bath. Linear Markov open quantum models are particularly prominent in quantum optics. They are used to describe various linear quantum optical devices and are also employed to describe counterparts of these optical devices in other physical platforms such as superconducting microwave circuits. In the following, we will present several concrete physical examples of linear quantum systems, taken from quantum optics, nano-mechanics, and atom optics; these are very active research fields playing important roles for both a purely academic purpose, for understanding and testing quantum mechanics, as well as for an applied purpose, the engineering of quantum information processing systems. In each example, we write down the system equations without detailed derivation, but connections to the general Langevin equation described in the previous section will be discussed so that the physical background of the dynamics can be reasonably conceived.

1.5.1 Optical Cavities

A simple open linear system is an empty optical cavity as illustrated in Fig. 1.2. The dynamical behavior of this optical device is represented by only one variable; that is, a standing optical wave with fixed resonance frequency ω_o is generated inside the cavity, and its quantized complex amplitude, i.e., the annihilation operator, $a_{\omega_o}(t)$ with "mode" ω_o is the variable characterizing the system dynamics. We call this single variable the *cavity mode* and in what follows will be denoted by $a(t)$ (in the Heisenberg picture). This mode satisfies the CCR $[a(t), a(t)^*] = 1$ for all $t \geq 0$. The oscillation dynamics of this optical field is described by the system Hamiltonian $H_{\rm sys} = \omega_o a^* a$, with ω_o as before. Now, the meaning of "open" is as follows. As shown in the figure, the cavity is composed of two partially reflecting mirrors and one perfectly reflecting mirror. Through the partially reflecting mirrors, the intracavity mode $a(t)$ interacts with two external fields. Let $\xi_1(t)$ and $\xi_2(t)$ be the continuous-mode annihilation process of those input optical fields, which satisfy $[\xi_i(s), \xi_j(t)^*] = \delta_{ij}\delta(s-t)$. As described later, in this setting, ξ_1 serves as a coherent laser field with amplitude $\langle \xi_1(t) \rangle = \beta(t)$, while ξ_2 is a vacuum field (i.e., a field containing no photons, just vacuum fluctuations). The interaction Hamiltonian of a and ξ_j at time t, which corresponds to (1.24), is now given by $H_{\rm int}^{(j)} = \imath\sqrt{\kappa_j}(a^*\xi_j - a\xi_j^*)$. Therefore, $L_j = \sqrt{\kappa_j} a$. Here, κ_j represents the interaction strength of these two optical fields; more precisely, it is given by $\kappa_j = cT_j/\ell$ with c the speed of light, T_j

Fig. 1.2 Mode-cleaning cavity

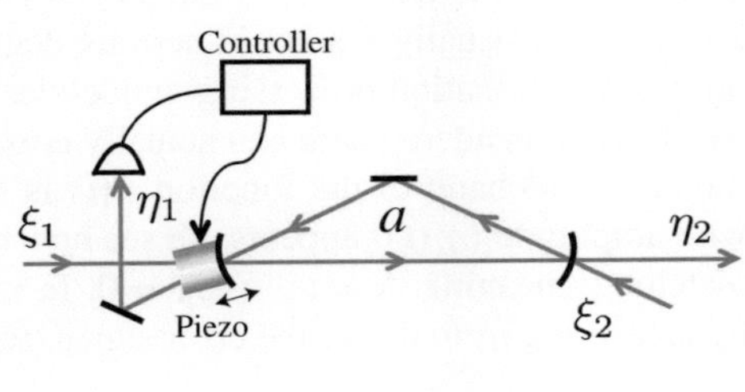

the dimensionless transmissivity of the jth mirror, and ℓ the optical path length in the cavity (hence, κ_j has dimension of [1/s]). In this setting, the quantum Langevin equation for a is obtained from (1.22) as

$$\frac{da}{dt} = -\imath\omega_o a - \frac{\kappa_1+\kappa_2}{2}a - \sqrt{\kappa_1}\xi_1 - \sqrt{\kappa_2}\xi_2.$$

It is often convenient to move to a "rotating frame" where the new variables describing the dynamics are taken to be $\tilde{a}(t) = e^{\imath\omega_p t}a(t)$ and $\tilde{\xi}_j(t) = e^{\imath\omega_p t}\xi_j(t)$, with ω_p the center frequency of the driving laser field. With this transformation, where we again denote the new rotating frame variables by $a(t)$ and $\xi_j(t)$ rather than $\tilde{a}(t)$ and $\tilde{\xi}_j(t)$, the above Langevin equation can be rewritten as

$$\frac{da}{dt} = -\imath\Delta a - \frac{\kappa_1+\kappa_2}{2}a - \sqrt{\kappa_1}\xi_1 - \sqrt{\kappa_2}\xi_2. \tag{1.25}$$

The frequency difference $\Delta = \omega_o - \omega_p$ is called the *detuning*. For high-quality quantum optical experiments, Δ should be tuned to be zero.

The external field operators also change due to the interaction; as shown in (1.23), under the Markov approximation, the output fields are given by

$$\eta_1 = \sqrt{\kappa_1}a + \xi_1, \quad \eta_2 = \sqrt{\kappa_2}a + \xi_2. \tag{1.26}$$

As a consequence, it becomes apparent that the optical cavity can be modeled as a linear dynamical system with input ξ_j and output η_j. Note that the above equations are more rigorously described by a quantum stochastic differential equation (QSDE), which in the *Itō form* is given by

$$da = -\Big(\imath\Delta + \frac{\kappa_1+\kappa_2}{2}\Big)adt - \sqrt{\kappa_1}d\mathcal{A}_1 - \sqrt{\kappa_2}d\mathcal{A}_2,$$
$$d\mathcal{Y}_1 = \sqrt{\kappa_1}adt + d\mathcal{A}_1, \quad d\mathcal{Y}_2 = \sqrt{\kappa_2}adt + d\mathcal{A}_2.$$

The QSDE will be discussed in more detail in the next chapter.

Typically, this system functions as an optical low-pass filter, see [79]. In reality, any single-mode coherent laser field is not an ideal monochromatic source that only contains a single center frequency, but it is composed of several (continuous) modes of frequency; in our case, the amplitude of the coherent input field ξ_1, i.e., $\langle\xi_1(t)\rangle = \beta(t)$, is not an ideal (sinusoidal) function with frequency ω_p. Since this imperfection usually results in serious degradation of the laser's performance in quantum information processing, mode-cleaning of the field is important. The cavity system considered here can actually erase unwanted optical frequencies; that is, the noisy sideband of the function $\beta(t)$ is suppressed and a mode-cleaned output with amplitude $\langle\eta_2(t)\rangle$ appears. To see how this works, we first assume exact mode-matching, meaning $\Delta = \omega_o - \omega_p = 0$. In practice, this condition can be achieved by measuring η_1 to detect the error signal, which is then fed back to a piezoactuator mounted at one of the mirrors for locking the optical path length in the cavity; see Fig. 1.2. Then, the resulting system dynamics is given by

$$\frac{da}{dt} = -\kappa a - \sqrt{\kappa}\xi_1 - \sqrt{\kappa}\xi_2, \quad \eta_2 = \sqrt{\kappa}a + \xi_2,$$

where we set $\kappa = \kappa_1 = \kappa_2$. Now, as in the classical case, for a linear quantum system we can formally define the transfer function to examine the input–output relationship. The transfer function from ξ_1 to η_2 is given by $\Xi_{1\to 2}[s] = -\kappa/(s+\kappa)$, and its gain in the Fourier domain $s = \imath\omega$ (i.e., essentially, the power spectral density) is obtained as

$$|\Xi_{1\to 2}[\imath\omega]|^2 = \frac{\kappa^2}{\omega^2 + \kappa^2}.$$

Thus, input components with higher frequencies are attenuated in the output field (note that $\omega = 0$ corresponds to the laser center frequency ω_p since we are working in the rotating frame). This specific type of cavity is known as a *mode-cleaning cavity* (MCC) and is often used in quantum optics experiments [6, 79].

1.5.2 Non-degenerate Optical Parametric Amplifiers

We would now like to describe a type of linear optical parametric amplifier, called a *non-degenerate optical parametric amplifier* (NOPA). This device is of significant importance in the area of continuous-variable quantum information where the NOPA is used to generate entangled beams of light. We here follow the standard treatment of the NOPA, see, e.g., [80]. As depicted in Fig. 1.3, one realization of a NOPA is as an optical device consisting of mirrors, forming an optical ring cavity, and a nonlinear $\chi^{(2)}$ optical crystal. The three mirrors M_2, M_3, and M_4 are fully reflecting while the mirror M_1 is partially transmissive. An external coherent light is used as a pump beam to supply quanta to the NOPA. Inside the nonlinear crystal, the pump beam interacts with vacuum fields in a so-called parametric down conversion process to generate two photons of the same frequency but of orthogonal polarization, known as signal and idler photons (in other implementations, the two photons generated can have the same polarization but different resonance frequencies). We will work in the scenario where the pump beam is strongly coherent and assumed to remain more or less undepleted during this interaction, so that it can be treated as a classical field. Through the mirror M_1, the orthogonally polarized photons circulating inside the cavity can escape to the outside. The escaping photons form a continuous-mode Gaussian beam with correlations in the polarization modes that can be regarded as Einstein-Podolsky-Rosen (EPR)-like entanglement [80].

The NOPA can be mathematically modeled as follows. We basically have two cavity modes a_1 and a_2 with the same resonance frequency ω_c but are orthogonally polarized with respect to one another, so that they are distinguishable. They satisfy the

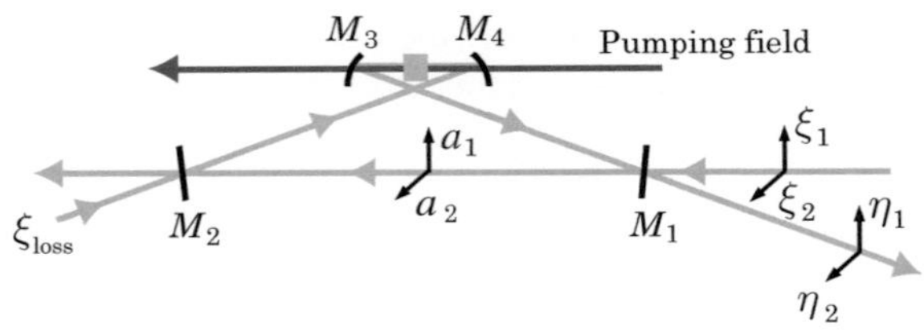

Fig. 1.3 A non-degenerate optical parametric amplifier composed of a bow-tie type optical cavity and a $\chi^{(2)}$ nonlinear crystal implementing the parametric amplifier. The mirror M_1 is partially transmissive, while the other mirrors are perfectly reflective for the input–output field. Now the input field is set to the vacuum. The internal optical loss is modeled by the fictitious vacuum input with mode ξ_{loss}. The crystal is pumped by a strong classical field, which is perfectly transmissive for the mirrors M_3 and M_4. Reprinted with permission from [44] © 2012 IEEE

CCR $[a_i, a_j^*] = \delta_{ij}$ and $[a_i, a_j] = 0$ and are coupled by an interaction Hamiltonian $H_{\text{sys}} = \imath(\epsilon/2)\left(e^{-\imath\omega_{\text{p}}t}a_1^*a_2^* - e^{\imath\omega_{\text{p}}t}a_1a_2\right)$, where ϵ is an effective amplitude, related to the amplitude of the pump beam and the $\chi^{(2)}$ coefficient of the nonlinear crystal, and ω_p is the frequency of the pump beam. A *stand-alone* NOPA has four inputs (continuous-mode or broadband) fields, all in the vacuum state [5]. In terms of the Langevin equation (1.22), the mode a_1 is coupled to the incoming fields ξ_1 and ξ_2 by the coupling operators $L_1 = \sqrt{\gamma}a_1$ and $L_2 = \sqrt{\gamma}a_2$, respectively, where γ denotes the coupling strength. Furthermore, to model optical losses, the mode is also coupled to additional vacuum input fields $\xi_{\text{loss},1}$ and $\xi_{\text{loss},2}$ with coupling operators $L_3 = \sqrt{\kappa}a_1$ and $L_4 = \sqrt{\kappa}a_2$, where κ denotes the loss rate. These fields carry photons away which then become completely lost. We consider the matched case where $\omega_{\text{c}} = \omega_{\text{p}}/2$. From (1.22), the quantum Langevin equation of the NOPA modes a_1 and a_2 in the Heisenberg picture is given by

$$\begin{aligned}\frac{da_1}{dt} &= -\imath\omega_{\text{c}}a_1 - \frac{\gamma+\kappa}{2}a_1 + \frac{\epsilon}{2}e^{-2i\omega_pt}a_2^* - \sqrt{\gamma}\xi_1 - \sqrt{\kappa}\xi_{\text{loss},1},\\ \frac{da_2}{dt} &= -\imath\omega_{\text{c}}a_2 - \frac{\gamma+\kappa}{2}a_2 + \frac{\epsilon}{2}a_1^* - \sqrt{\gamma}\xi_2 - \sqrt{\kappa}\xi_{\text{loss},2}.\end{aligned} \tag{1.27}$$

Also from (1.23), we obtain the output equations:

$$\eta_1 = \sqrt{\gamma}a_1 + \xi_1, \quad \eta_2 = \sqrt{\gamma}a_2 + \xi_2. \tag{1.28}$$

The equations above are time-dependent due to the dependence on the oscillating term $e^{\imath\omega_{\text{p}}t}$. Typically, the equation is transformed to a time-independent one by moving to rotating frame at half the pump frequency $\omega_{\text{p}}/2$. In this rotating frame, one makes the substitutions $a_j(t)e^{-\imath\omega_pt/2} \to a_j(t)$, $\xi_j(t)e^{\imath\omega_pt/2} \to \xi_j(t)$, and $\eta_j(t)e^{\imath\omega_pt/2} \to \eta_j(t)$ for $j = 1, 2$ and all t. After the substitution is made, the equations become time-independent,

$$\frac{da_1}{dt} = -\frac{\gamma+\kappa}{2}a_1 + \frac{\epsilon}{2}a_2^* - \sqrt{\gamma}\xi_1 - \sqrt{\kappa}\xi_{\text{loss},1},$$
$$\frac{da_2}{dt} = -\frac{\gamma+\kappa}{2}a_2 + \frac{\epsilon}{2}a_1^* - \sqrt{\gamma}\xi_2 - \sqrt{\kappa}\xi_{\text{loss},2}, \tag{1.29}$$

$$\eta_1 = \sqrt{\gamma}a_1 + \xi_1, \quad \eta_2 = \sqrt{\gamma}a_2 + \xi_2. \tag{1.30}$$

It must be kept in mind that all quantities are now rotating frame quantities although we have retained the same notation a_j, ξ_j, η_j as in the original (non-rotating frame) equation. Hence, as with the MCC from the previous example, the NOPA can be modeled as a linear time-invariant dynamical quantum system. But an important difference from the case of the MCC is that the equations are now expressed in terms of both annihilation and creation operators, a_j and a_j^*. To analyze the NOPA, it is convenient to transform it into quadrature form (to be discussed in more detail in Chap. 2) by defining the following vector of operators:

$$z = (q_1, p_1, q_2, p_2)^\top,$$
$$\xi = (\xi_1^q, \xi_1^p, \xi_2^q, \xi_2^p, \xi_{\text{loss},1}^q, \xi_{\text{loss},1}^p, \xi_{\text{loss},2}^q, \xi_{\text{loss},2}^p)^\top,$$
$$y = (\eta_1^q, \eta_1^p, \eta_2^q, \eta_2^p)^\top,$$

where $q_j = a_j + a_j^*$ and $p_j = (a_j - a_j^*)/\imath$. Likewise, for the noise operators we have defined $\xi_j^q = \xi_j + \xi_j^*$ and $\xi_j^p = (\xi_j - \xi_j^*)/\imath$, etc. The dynamics of the NOPA can then be expressed as

$$\frac{dz}{dt} = Az + B\xi, \quad y = Cz + D\xi,$$

where

$$A = \begin{bmatrix} -(\gamma+\kappa)/2 & 0 & \epsilon/2 & 0 \\ 0 & -(\gamma+\kappa)/2 & 0 & -\epsilon/2 \\ \epsilon/2 & 0 & -(\gamma+\kappa)/2 & 0 \\ 0 & -\epsilon/2 & 0 & -(\gamma+\kappa)/2 \end{bmatrix},$$
$$B = -[\sqrt{\gamma}I_4 \; \sqrt{\kappa}I_4],$$
$$C = \sqrt{\gamma}I_4,$$
$$D = [I_4 \; 0_4].$$

In the operation of a NOPA, we require that the matrix A be Hurwitz (all its eigenvalues lie in the left half plane) so that the system is stable. Stable here is in the sense that the average number of photons contained in the modes a_1 and a_2 remains finite at all times, corresponding to the NOPA operating in what is known as the below threshold regime.

As alluded to earlier, the type of entanglement produced by a NOPA is a so-called EPR entanglement, which is characterized by two-mode squeezing between the pair of output fields η_1 and η_2. A two-mode squeezed state between a pair of continuous-mode output fields such as η_1 and η_2 is characterized by correlation and anticorrelation between quadratures of the two fields. As we will see below, in the model of the NOPA above this is correlation between quadratures η_1^p and η_1^p, and anticorrelation between η_1^q and η_1^q. The amount of correlations, or two-mode squeezing, between the two beams can be quantified in the frequency domain via some squeezing spectral densities that will be introduced next.

Let $\tilde{\eta}^q(t) = \eta_1^q(t) + \eta_2^q(t)$ and $\tilde{\eta}^p(t) = \eta_1^p(t) - \eta_2^p(t)$. Also, let $\tilde{\Xi}^q[\imath\omega]$, $\tilde{\Xi}^p[\imath\omega]$, $Z[\imath\omega]$, and $\Xi[\imath\omega]$, respectively, be the Fourier transforms of $\tilde{\eta}^q(t)$, $\tilde{\eta}^p(t)$, $z(t)$, and $\xi(t)$, defined via the transform relation $F[\imath\omega] = \frac{1}{\sqrt{2\pi}} \int_{-\infty}^{\infty} f(t) e^{-\imath\omega t} dt$, with F the Fourier transform of f. Therefore,

$$
\begin{aligned}
\tilde{\Xi}^q[\imath\omega] &= C_1 Z[\imath\omega] + D_1 \Xi[\imath\omega], \\
\tilde{\Xi}^p[\imath\omega] &= C_2 Z[\imath\omega] + D_2 \Xi[\imath\omega],
\end{aligned}
$$

where $C_1 = [1\,0\,1\,0]C$, $C_2 = [0\,1\,0\,-1]C$, $D_1 = [1\,0\,1\,0]D$, and $D_2 = [0\,1\,0\,-1]D$. The two-mode squeezing spectral densities V_+ and V_- can be defined through the following identities:

$$
\begin{aligned}
\langle \tilde{\Xi}^q[\imath\omega]^* \tilde{\Xi}^q[\imath\omega'] \rangle &= V_+(\imath\omega)\delta(\omega - \omega'), \\
\langle \tilde{\Xi}^p[\imath\omega]^* \tilde{\Xi}^p[\imath\omega'] \rangle &= V_-(\imath\omega)\delta(\omega - \omega'),
\end{aligned}
\tag{1.31}
$$

where $\langle \cdot \rangle$ denotes quantum expectation with respect to the initial state of the oscillator modes and the vacuum state of the incoming fields. The quantity V_+ is called the two-mode amplitude squeezing spectrum and V_- the two-mode phase squeezing spectrum. The two outgoing fields η_1 and η_2 are entangled at frequency ω rad/s if, see, e.g., [81],

$$
V = V_+(\imath\omega) + V_-(\imath\omega) < 4. \tag{1.32}
$$

A perfect maximally entangled continuous-mode EPR state would have $V_+(\imath\omega) = V_-(\imath\omega) = V_+(\imath\omega) + V_-(\imath\omega) = 0$ for all ω. However, this ideal EPR state is not physically achievable as it would require an infinite amount of energy to produce and no losses can be permitted. Thus, in practical devices the goal is to make $V_+(\imath\omega)$ and $V_-(\imath\omega)$ small over a frequency range (bandwidth) such that one has $V_+(\imath\omega) + V_-(\imath\omega) < 4$ in this frequency range. This yields EPR-like states of the outgoing fields that are practical approximations of the ideal continuous-mode EPR state. For the model considered here, the symmetry in the parameter values implies that $V_+(\imath\omega) = V_-(\imath\omega)$, so entanglement at frequency ω is verified by $V_\pm(\imath\omega) < 2$. Moreover, for low nonnegative values of ω (note that since we are working in a rotating frame, the actually frequency is around half the pump frequency, $\omega_p/2$) we have that, to an excellent approximation, $V_+(\imath\omega) \approx V_+(0)$ and $V_-(\imath\omega) \approx V_-(0)$. Therefore, $V_+(0) = V_-(0) = V_\pm(0)$ can be employed as a figure of merit for the degree of two-mode squeezing or entanglement between the output fields η_1 and η_2.

Define $H_j[\imath\omega] = C_j (\imath\omega I - A)^{-1} B + D_j$ ($j = 1, 2$), then it is straightforward to show that

$$V_+(\imath\omega) = \mathrm{Tr}\left(H_1[\imath\omega]^* H_1[\imath\omega]\right), \tag{1.33}$$
$$V_-(\imath\omega) = \mathrm{Tr}\left(H_2[\imath\omega]^* H_2[\imath\omega]\right). \tag{1.34}$$

When $\kappa = 0$, we have that [80]

$$V_\pm(0) = \frac{2(1-k)^2}{(1+k)^2}, \tag{1.35}$$

with $k = \epsilon/\gamma$ taking a value $0 \leq k < 1$ in the typical below threshold operation.[2]

Note that (1.32) is a sufficient condition for EPR entanglement, with the two fields squeezed in amplitude and phase quadratures. However, in general, they may instead be squeezed in other quadratures. Hence, we now give the following general notion of EPR entanglement. Let $\eta_1^{\psi_1} = e^{\imath\psi_1}\eta_1$ and $\eta_2^{\psi_2} = e^{\imath\psi_2}\eta_2$ with $\psi_1, \psi_2 \in (-\pi, \pi]$. Denote the corresponding two-mode squeezing spectra between $\eta_1^{\psi_1}$ and $\eta_2^{\psi_2}$ as $V_\pm^{\psi_1,\psi_2}(\imath\omega, \psi_1, \psi_2)$, following the definition of $V_\pm$ given above but making the substitutions $\eta_j^{\psi_j} \to \eta_j$, $j = 1, 2$.

Definition 1.1 The fields η_1 and η_2 are EPR entangled at the frequency ω rad/s if $\exists\, \psi_1, \psi_2 \in (-\pi, \pi]$ such that

$$V_+^{\psi_1,\psi_2}(\imath\omega, \psi_1, \psi_2) + V_-^{\psi_1,\psi_2}(\imath\omega, \psi_1, \psi_2) < 4. \tag{1.36}$$

EPR entanglement is said to vanish at ω if there are no values of ψ_1 and ψ_2 satisfying the above criterion. Unless otherwise specified, EPR entanglement will refer to the case with $\psi_1 = \psi_2 = 0$.

1.5.3 Degenerate Parametric Amplifiers/Optical Parametric Oscillators

Mathematically, the degenerate parametric amplifier (DPA) or optical parametric oscillator (OPO) is just a special case of the non-degenerate one from the preceding section, in the sense that the two modes a_1 and a_2 become "degenerate" as depicted in Fig. 1.4; see [5, 57]. In this case, the two modes are identical in all respects (frequency and polarization) and cannot be distinguished. As a result, again expressing all quantities in a rotating frame at half the pump frequency as in the NOPA, the system Hamiltonian inside the cavity becomes $H_{\mathrm{sys}} = \imath\epsilon\left(a^{*2} - a^2\right)/2$, where a is

---

[2] Note that [80] and the formula (1.35) take $\xi^q_{out,a} - \xi^q_{out,b}$ and $\xi^p_{out,a} + \xi^p_{out,b}$ as the squeezed two-mode quadratures. To obtain squeezing in the quadratures $\xi^q_{out,a} + \xi^q_{out,b}$ and $\xi^p_{out,a} - \xi^p_{out,b}$, one need only take $\epsilon < 0$ and the formula (1.35) is modified to be $V_\pm(0) = 2(1+k)^2/(1-k)^2$.

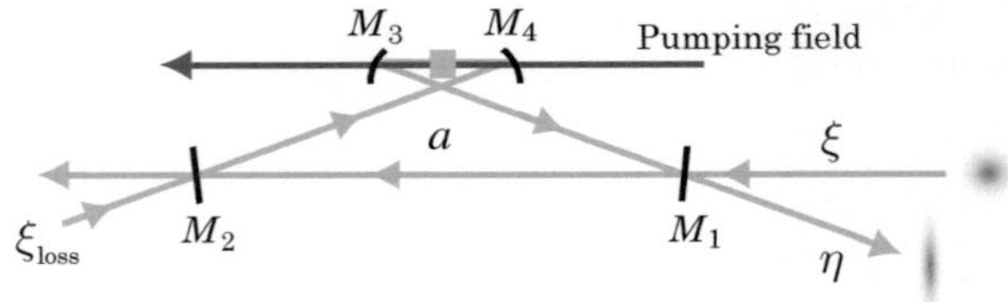

Fig. 1.4 Optical parametric oscillator. Reprinted with permission from [44] © 2012 IEEE

the cavity mode. As in the NOPA case, a couples with the external vacuum field ξ through the partially reflected mirror M_1 with coupling constant γ; thus, $L = \sqrt{\gamma}a$. Consequently, the system's dynamical equation is given by

$$\frac{da}{dt} = -\frac{\gamma+\kappa}{2}a + \frac{\epsilon}{2}a^* - \sqrt{\gamma}\xi - \sqrt{\kappa}\xi_{\rm loss}, \quad \eta = \sqrt{\gamma}a + \xi, \tag{1.37}$$

where again the fictitious vacuum field $\xi_{\rm loss}$ is added to model the internal optical loss. The above equation contains both a and a^*; hence, as before it is sometimes more convenient to work with the quadrature form equation in terms of the amplitude quadrature $q = a + a^*$ and phase quadrature $p = (a - a^*)/\imath$:

$$\frac{d}{dt}\begin{bmatrix} q \\ p \end{bmatrix} = \frac{1}{2}\begin{bmatrix} -\gamma+\epsilon & 0 \\ 0 & -\gamma-\epsilon \end{bmatrix}\begin{bmatrix} q \\ p \end{bmatrix} - \sqrt{\gamma}\begin{bmatrix} 1 & 0 \\ 0 & 1 \end{bmatrix}\begin{bmatrix} \xi^q \\ \xi^p \end{bmatrix},$$
$$\begin{bmatrix} \eta^q \\ \eta^p \end{bmatrix} = \begin{bmatrix} \sqrt{\gamma} & 0 \\ 0 & \sqrt{\gamma} \end{bmatrix}\begin{bmatrix} q \\ p \end{bmatrix} + \begin{bmatrix} \xi^q \\ \xi^p \end{bmatrix},$$

where $\xi_{\rm loss}$ is omitted for simplicity. Like the NOPA, the DPA also plays a very important role in quantum optics and information [6, 82]; in the laboratory, this device is utilized as a generator of a *squeezed Gaussian optical field.*

Let us briefly describe how the DPA produces a squeezed Gaussian optical field; a precise treatment of Gaussian fields will be given in the next chapter. First, we mention that although the input is vacuum, the output is not a vacuum field (but does have zero amplitude). The non-trivial property of the output field can be easily seen by examining the transfer function matrix from the input (ξ^q, ξ^p) to the output (η^q, η^p):

$$\Xi[s] = {\rm diag}\Big(\frac{s-(\gamma+\epsilon)/2}{s+(\gamma-\epsilon)/2}, \ \frac{s-(\gamma-\epsilon)/2}{s+(\gamma+\epsilon)/2}\Big).$$

Now, the input is simply the vacuum field; hence, the power in each quadrature component is $\langle|\xi^q[\imath\omega]|^2\rangle = \langle|\xi^p[\imath\omega]|^2\rangle = 1$; this is called the *quantum noise limit* (QNL), which corresponds to a scenario in which quantum fluctuations are equally distributed in the quadratures q and p, while keeping the minimum uncertainty relation, $\langle|\xi^q[\imath\omega]|^2\rangle\langle|\xi^p[\imath\omega]|^2\rangle = 1$. (Note that in the Fourier domain, the Heisenberg uncertainty relation is represented by $\langle|\xi^q[\imath\omega]|^2\rangle\langle|\xi^p[\imath\omega]|^2\rangle \geq 1$.) On the other hand, the power of the output fields can be calculated to be

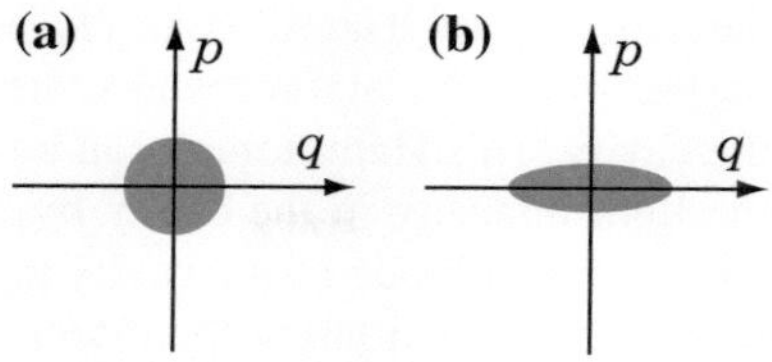

Fig. 1.5 Phase-space picture for **a** a vacuum state and **b** a squeezed vacuum state

$$\langle |\eta^q[\imath\omega]|^2\rangle = \frac{\omega^2+(\gamma+\epsilon)^2/4}{\omega^2+(\gamma-\epsilon)^2/4}, \quad \langle |\eta^p[\imath\omega]|^2\rangle = \frac{\omega^2+(\gamma-\epsilon)^2/4}{\omega^2+(\gamma+\epsilon)^2/4}.$$

They satisfy the relation $\langle |\eta^q[\imath\omega]|^2\rangle\langle |\eta^p[\imath\omega]|^2\rangle = 1$ for all ω, so that the output field is also in a minimum uncertainty state. But importantly, the quadratures have different fluctuation variances; for instance when $\epsilon = \gamma/2$, then at the center frequency $\omega = 0$ they are given by $\langle |\eta^q|^2\rangle = 9$ and $\langle |\eta^p|^2\rangle = 1/9$. That is, the fluctuation of the p-quadrature is reduced below the QNL, while the q-quadrature fluctuates above the QNL. Hence, the output field is "squeezed" in that there are non-uniform fluctuations in η^q and η^p, as depicted in Fig. 1.5b. Squeezed light fields are important in various situations in quantum mechanics and its applications; in particular, they are used for generating entangled light beams and for high-precision detection of a tiny force such as in gravitational-wave interferometers [19, 22].

1.5.4 Opto-mechanical Systems

Nano-mechanical devices are also promising platforms for testing quantum mechanics as well as carrying out quantum information processing. A typical one is an opto-mechanical oscillator as depicted in Fig. 1.6. This system is composed of a two-sided optical cavity wherein one of these mirrors is movable (not fixed); see [21, 23, 83, 84]. Thus, the moving mirror can serve as a mechanical oscillator. Let q_1 and p_1 be the oscillator's (i.e., the moving mirror's) position and momentum operators satisfying the CCR $[q_1, p_1] = 2\imath$. Also define $q_2 = a_2 + a_2^*$ and $p_2 = (a_2 - a_2^*)/\imath$ with a_2 the annihilation operator of the cavity mode. The oscillator is in a harmonic potential with Hamiltonian $H_{\rm har} = m\omega_o^2 q_1^2/2 + p_1^2/2m$ where m and ω_o represent the mass and the resonant frequency of the oscillator, respectively. The oscillator and the intracavity field interact with one another through a radiation pressure force,

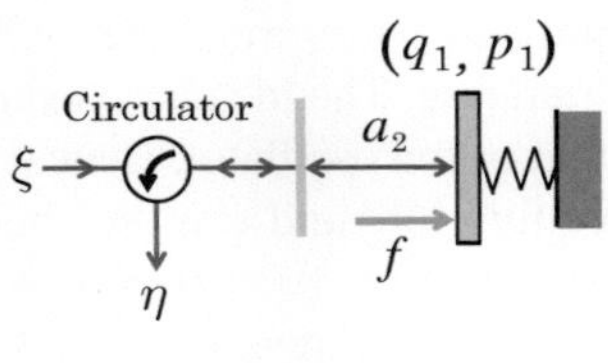

Fig. 1.6 Opto-mechanical oscillator. Figure adapted from [25]

described by the Hamiltonian $H_{\rm rp} = \gamma' q_1 a_2^* a_2$. Roughly speaking, the oscillator is pushed by an optical force whose strength is proportional to the photon number inside the cavity. This Hamiltonian can be approximated, around the equilibrium point of the light intensity in the cavity by $H'_{\rm rp} = \gamma' q_1(\alpha^* a_2 + \alpha a_2^*)$ where $\alpha \in \mathbb{C}$ denotes the mean amplitude of the cavity mode, i.e., $\alpha = \langle a_2 \rangle$. This approximation is valid when $|\alpha|$ is much larger than the motion of q_1. Moreover, the cavity field couples to the incoming optical field $\xi(t)$ through the other partially reflective fixed mirror, via the singular interaction Hamiltonian $H_{\rm int} = \imath\sqrt{\kappa}(a_2^*\xi - a_2\xi^*)$. The dynamical equation of the composite system is given by the quantum Langevin equation (1.22) with $H_{\rm sys} = H_{\rm har} + H'_{\rm rp}$ and $L = \sqrt{\kappa} a_2$, leading to

$$\frac{d}{dt}\left[\begin{array}{c} q_1 \\ p_1 \\ \hline q_2 \\ p_2 \end{array}\right] = \left[\begin{array}{cc|cc} 0 & 1/m & 0 & 0 \\ -m\omega^2 & 0 & \gamma & 0 \\ \hline 0 & 0 & -\kappa/2 & 0 \\ \gamma & 0 & 0 & -\kappa/2 \end{array}\right]\left[\begin{array}{c} q_1 \\ p_1 \\ \hline q_2 \\ p_2 \end{array}\right] - \sqrt{\kappa}\left[\begin{array}{cc} 0 & 0 \\ 0 & 0 \\ \hline 1 & 0 \\ 0 & 1 \end{array}\right]\begin{bmatrix} \xi^q \\ \xi^p \end{bmatrix} + \left[\begin{array}{c} 0 \\ 1 \\ \hline 0 \\ 0 \end{array}\right] f, \tag{1.38}$$

$$\begin{bmatrix} \eta^q \\ \eta^p \end{bmatrix} = \begin{bmatrix} \sqrt{\kappa} & 0 \\ 0 & \sqrt{\kappa} \end{bmatrix}\begin{bmatrix} q_2 \\ p_2 \end{bmatrix} + \begin{bmatrix} \xi^q \\ \xi^p \end{bmatrix} \tag{1.39}$$

where (ξ^q, ξ^p) and (η^q, η^p) are the quadratures of the input field mode ξ and the output field mode η, respectively. Also, we have assumed $\alpha \in \mathbb{R}$ and set $\gamma = \gamma'\alpha$. Note that, because of the uncertainty principle, both η^q and η^p cannot be measured simultaneously; quantum mechanics allows us to measure only one of them.

In the above system dynamics, we have added a classical force f that acts only on the oscillator. Typically, f is a very tiny unknown force such as a gravitational-wave force, and this opto-mechanical oscillator system can act as a sensor for estimating f. What should be emphasized here is the fact that systems and control theory offers a number of powerful tools for general dynamical estimation problems or, more broadly, *quantum metrology*. This fact is indeed applicable to the above sensing problem for f. In Sect. 6.3 we will see a feedback control method to construct an effective sensor.

Before closing this section, we want to demonstrate that a simplified model of Eqs. (1.38) and (1.39) can be obtained via a technique known in physics as *adiabatic elimination* and in mathematics and engineering as *singular perturbation*; this is indeed an important topic and will be encountered again in Chap. 3. Here, we only provide a very rough heuristic treatment that is commonly employed in the physics literature. The idea is based on the fact that the cavity dynamics is much faster than that of the oscillator. Hence, one imagines that the cavity state goes immediately to equilibrium and can thus be eliminated from the dynamical equation. To proceed, we set $\dot{q}_2 = \dot{p}_2 = 0$, which yields $q_2 = -2\xi^q/\sqrt{\kappa}$ and $p_2 = 2\gamma q_1/\kappa - 2\xi^p/\sqrt{\kappa}$. Substituting them into (1.38), we end up with

$$\frac{d}{dt}\begin{bmatrix} q_1 \\ p_1 \end{bmatrix} = \begin{bmatrix} 0 & 1/m \\ -m\omega^2 & 0 \end{bmatrix}\begin{bmatrix} q_1 \\ p_1 \end{bmatrix} - \sqrt{\lambda}\begin{bmatrix} 0 \\ 1 \end{bmatrix}\xi^q + \begin{bmatrix} 0 \\ 1 \end{bmatrix} f, \tag{1.40}$$

where $\lambda = 4\gamma^2/\kappa$. Also substitution into (1.39) yields

$$\begin{bmatrix} \eta^q \\ \eta^p \end{bmatrix} = \begin{bmatrix} 0 & 0 \\ \sqrt{\lambda} & 0 \end{bmatrix}\begin{bmatrix} q_1 \\ p_1 \end{bmatrix} - \begin{bmatrix} \xi^q \\ \xi^p \end{bmatrix}.$$

Since η^q clearly does not contain any information about the oscillator and, accordingly, the unknown force f, we should measure η^p. Hence, the output equation describing the informative measurement result is given by

$$y = [\sqrt{\lambda}\ \ 0]\begin{bmatrix} q_1 \\ p_1 \end{bmatrix} - \xi^p. \tag{1.41}$$

Equations (1.40) and (1.41) are quantum versions of the classical open harmonic oscillator (1.2). Note that (1.2) is valid even without disturbance noise. Also recall now that (1.14) has the same form, but it is a closed system having no associated real-time output signal. In the quantum case, the coupling of the cavity mode to the external optical field turns the system into an open one, in which information can be extracted from the system with the optical field serving as a quantum probe.

1.5.5 Large Atomic Ensemble

The last example is taken from the field of atom optics. Let us consider a large atomic ensemble trapped in a cavity that is coupled to an external optical field as illustrated in Fig. 1.7. In an ultimately cooled cavity, as a good approximation we can treat each atom as an ideal "qubit" composed of two energy levels: the ground and excited states; the observable representing the energy of the kth atom is given by

$$\sigma_z^{(k)} = \frac{1}{2}\begin{bmatrix} 1 & 0 \\ 0 & -1 \end{bmatrix}.$$

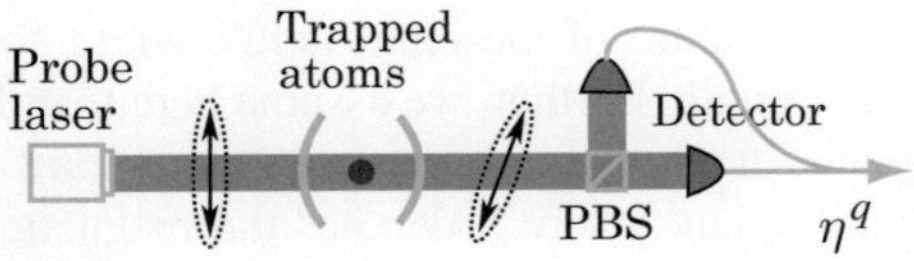

Fig. 1.7 Atomic ensemble trapped in a cavity. PBS indicates a polarized beam splitter, which is used for detecting the polarization of the output field. Figure adapted from [25]

Hence, measuring the energy gives the result $+1/2$ or $-1/2$, corresponding to "excited" and "ground" states, respectively. Besides $\sigma_z^{(k)}$ there are also two other especially important observables that do not commute with $\sigma_z^{(k)}$,

$$\sigma_x^{(k)} = \frac{1}{2}\begin{bmatrix} 0 & 1 \\ 1 & 0 \end{bmatrix}, \quad \sigma_y^{(k)} = \frac{1}{2}\begin{bmatrix} 0 & -\imath \\ \imath & 0 \end{bmatrix}.$$

Here, the observables of interest are those of a collection of the above two-state system, in particular the total energy of the atomic ensemble, which is mathematically represented by

$$J_z = \sum_{k=1}^{N} \sigma_z^{(k)} = (\sigma_z^{(1)} \otimes I \otimes \cdots I) + \cdots + (I \otimes I \otimes \cdots \sigma_z^{(N)}), \tag{1.42}$$

where I denotes the 2×2 identity matrix and N is the number of atoms. Likewise, we define

$$J_x = \sum_{k=1}^{N} \sigma_x^{(k)}, \quad J_y = \sum_{k=1}^{N} \sigma_y^{(k)}. \tag{1.43}$$

A notable fact is that they satisfy the spin CCR $[J_i, J_j] = \imath\epsilon_{ijk} J_k$, where ϵ_{ijk} denotes the Levi-Civita symbol.

The particular system configuration we are interested in here is one that facilitates *quantum non-demolition* (QND) measurement of J_z; see e.g., [85, 86]. This atomic ensemble couples with the optical field, having a polarization degree of freedom, via the *Faraday interaction*. Through this interaction, the polarization of the output field rotates depending on the energy level, but ideally the observable representing the energy level does not change; as a result, J_z can be measured without being disturbed, and this is the essence of QND measurement of J_z. In terms of the Langevin equation (1.22), this interaction is described by the operator $L = \sqrt{M} J_z$, from which the dynamics of the system's observables (J_x, J_y, J_z) is given by

$$\frac{d}{dt}\begin{bmatrix} J_x \\ J_y \\ J_z \end{bmatrix} = \begin{bmatrix} -M/2 & 0 & -b \\ 0 & -M/2 & 0 \\ b & 0 & 0 \end{bmatrix}\begin{bmatrix} J_x \\ J_y \\ J_z \end{bmatrix} + \sqrt{M}\begin{bmatrix} -J_y \\ J_x \\ 0 \end{bmatrix}\xi^p, \tag{1.44}$$

where ξ^p is the phase quadrature of the input field's white noise process, and M represents the coupling strength. Further, we assume here that the atomic ensemble is subjected to a magnetic field, which corresponds to a term of the form $H_{\text{sys}} = bJ_y$ in the system Hamiltonian, where b denotes the magnetic field strength. The output equation is also obtained, which now has a non-trivial component only in the amplitude quadrature:

$$y = \eta^q = 2\sqrt{M} J_z + \xi^q. \tag{1.45}$$

In particular, when $b = 0$, J_z does not change while J_z still appears in the output equation. This indicates that we can indeed extract some information about J_z without disturbing it; such an observable is in general referred to as a *QND observable*.

A typical problem setting is that the magnetic field is very tiny and its strength b is unknown; we are interested in estimating this tiny magnetic field [17]. This *magnetometry* problem is also an important parameter estimation problem, like the force estimation problem described in Sect. 1.5.4. To solve this problem, the system theoretic approach based on the linearized model is again very powerful. Now note that, when M is small, the system variables obey a skew-Hermitian dynamics, implying that they preserve $J_x^2 + J_y^2 + J_z^2$. Hence, in the large ensemble limit $N \gg 1$ and over a short time period, their dynamics are constrained in the tangent space of the manifold of the dynamics. This is the situation where we are allowed to linearize the system dynamics. In particular, let us assume that the system is nearly along the x-axis; thus, we set J_x to be a constant rather than a quantum observable. Then due to $J_x = J = N/2$, the only dynamics that is important to us is $\dot{J}_z = bJ$. In particular, if we know that b is a classical random variable obeying a certain stochastic dynamics, the problem boils down to one that can be attacked with a system theoretic approach. Let us assume that b obeys $\dot{b} = -\gamma b + \sqrt{\sigma}\xi^b$, where γ is the dissipation rate and ξ^b is a standard Gaussian white noise (ξ^q and ξ^b are independent); then, as a result the system's dynamics is described by the following linear equation:

$$\frac{d}{dt}\begin{bmatrix} J_z \\ b \end{bmatrix} = \begin{bmatrix} 0 & J \\ 0 & -\gamma \end{bmatrix}\begin{bmatrix} J_z \\ b \end{bmatrix} + \begin{bmatrix} 0 \\ \sqrt{\sigma} \end{bmatrix}\xi^b, \quad y = 2\sqrt{M}J_z + \xi^q. \tag{1.46}$$

Thus, clearly we can gain access to the unknown value b indirectly through the dynamics of J_z. A specific scheme for estimating b was established in [17].

Finally, we remark that, although the above discussion has been focused on an atomic ensemble system in a dispersive regime, another important linear approximation arises in a dissipative regime, where the coupling operator is given by $L = \sqrt{M}(J_y + \imath J_z)$. In this case, L is approximated by an annihilation operator $L = \sqrt{M}a$, which as a result yields a strictly stable linear system of the same form as the cavity dynamics discussed in Sects. 1.5.1–1.5.3. (Note that the linear approximated system considered here, (1.46), is marginally stable.) A typical application of such a strictly stable linear system is for quantum memories; that is, we are interested in transferring a quantum state encoded in an input optical field to the atomic state; see [9, 12–14, 37, 38].

References

1. V.P. Belavkin, S.C. Edwards, Quantum filtering and optimal control, in *Quantum Stochastics and Information: Statistics, Filtering and Control (University of Nottingham, UK, 15–22 July 2006)*, ed. by V.P. Belavkin, M. Guta (World Scientific, Singapore, 2008), pp. 143–205

2. M.R. James, H.I. Nurdin, I.R. Petersen, H^{∞} control of linear quantum stochastic systems. IEEE Trans. Autom. Control **53**(8), 1787–1803 (2008)
3. H.I. Nurdin, M.R. James, A.C. Doherty, Network synthesis of linear dynamical quantum stochastic systems. SIAM J. Control Optim. **48**(4), 2686–2718 (2009)
4. J.E. Gough, M.R. James, H.I. Nurdin, Squeezing components in linear quantum feedback networks. Phys. Rev. A **81**, 023804 (2010)
5. C.W. Gardiner, P. Zoller, *Quantum Noise: A Handbook of Markovian and Non-Markovian Quantum Stochastic Methods with Applications to Quantum Optics*, 3rd edn. (Springer, Berlin, 2004)
6. A. Furusawa, P. van Loock, *Quantum Teleportation and Entanglement: A Hybrid Approach to Optical Quantum Information Processing* (Wiley-VCH, Berlin, 2011)
7. R. Hamerly, H. Mabuchi, Advantages of coherent feedback for cooling quantum oscillators. Phys. Rev. Lett. **109**, 173602 (2012)
8. J. Kerckhoff, R.W. Andrews, H.S. Ku, W.F. Kindel, K. Cicak, R.W. Simmonds, K.W. Lehnert, Tunable coupling to a mechanical oscillator circuit using a coherent feedback network. Phys. Rev. X **3**, 021013 (2013)
9. M. Hush, A.R.R. Carvalho, M. Hedges, M.R. James, Analysis of the operation of gradient echo memories using a quantum input-output model. New J. Phys. **15**, 085020 (2013)
10. A.A. Clerk, M.H. Devoret, S.M. Girvin, F. Marquardt, R.J. Schoelkopf, Introduction to quantum noise, measurement, and amplification. Rev. Mod. Phys. **82**, 1155 (2010)
11. N. Bergeal, F. Schackert, M. Metcalfe, R. Vijay, V. Manucharyan, L. Frunzio, D.E. Prober, R.J. Schoelkopf, S.M. Girvin, M.H. Devoret, Phase-preserving amplification near the quantum limit with a Josephson ring modulator. Nature **465**, 64–68 (2010)
12. Q.Y. He, M.D. Reid, E. Giacobino, J. Cviklinski, P.D. Drummond, Dynamical oscillator-cavity model for quantum memories. Phys. Rev. A **79**, 022310 (2009)
13. I. Novikov, A.V. Gorshkov, D.F. Phillips, A.S. Sorensen, M.D. Lukin, R.L. Walsworth, Optimal control of light pulse storage and retrieval. Phys. Rev. Lett. **98**, 243602 (2007)
14. A.V. Gorshkov, A. Andre, M.D. Lukin, A.S. Sorensen, Photon storage in lambda-type optically dense atomic media I. Cavity model. Phys. Rev. A **76**, 033804 (2007)
15. W.L. Brogan, *Modern Control Theory*, 3rd edn. (Prentice-Hall, Upper Saddle River, 1991)
16. K. Zhou, J.C. Doyle, K. Glover, *Robust and Optimal Control* (Prentice-Hall, Upper Saddle River, 1995)
17. J.K. Stockton, J. Geremia, A.C. Doherty, H. Mabuchi, Robust quantum parameter estimation: coherent magnetometry with feedback. Phys. Rev. A **69**, 032109 (2004)
18. M. Guta, N. Yamamoto, System identification for passive linear quantum systems. IEEE Trans. Autom. Control **61**(4), 921–936 (2016)
19. V.B. Braginsky, F.Y. Khalili, *Quantum Measurement* (Cambridge University Press, Cambridge, 1992)
20. C.M. Caves, K.S. Thorne, R.W.P. Drever, V.D. Sandberg, M. Zimmermann, On the measurement of a weak classical force coupled to a quantum mechanical oscillator. I. Issues of principle. Rev. Mod. Phys. **52**, 341–392 (1980)
21. M. Tsang, C.M. Caves, Coherent quantum-noise cancellation for optomechanical sensors. Phys. Rev. Lett. **105**, 123601 (2010)
22. H. Miao, *Exploring Macroscopic Quantum Mechanics in Optomechanical Devices* (Springer, Berlin, 2012)
23. Y. Chen, Macroscopic quantum mechanics: theory and experimental concepts of optomechanics. J. Phys. B: At. Mol. Opt. Phys. **46**, 104001 (2013)
24. H.I. Nurdin, M.R. James, I.R. Petersen, Coherent quantum LQG control. Automatica **45**, 1837–1846 (2009)
25. N. Yamamoto, Coherent versus measurement feedback: linear systems theory for quantum information. Phys. Rev. X **4**, 041029 (2014)
26. Y. Yokotera, N. Yamamoto, Geometric control theory for quantum back-action evasion. EPJ Quantum Technol. **3**(15), 1–22 (2016)

27. H.I. Nurdin, J.E. Gough, Modular quantum memories using passive linear optics and coherent feedback. Quantum Inf. Comput. **15**(11–12), 1017–1040 (2015)
28. G.J. Milburn, Coherent control of single photon states. Eur. Phys. J. **159**, 113–117 (2008)
29. G. Zhang, M.R. James, On the response of quantum linear systems to single photon input fields. IEEE Trans. Autom. Control **58**(5), 1221–1235 (2013)
30. G. Zhang, Analysis of quantum linear systems' response to multi-photon states. Automatica **50**(2), 442–451 (2014)
31. K. Koga, N. Yamamoto, Dissipation-induced pure Gaussian state. Phys. Rev. A **85**, 022103 (2012)
32. N. Yamamoto, Pure Gaussian state generation via dissipation: a quantum stochastic differential equation approach. Philos. Trans. R. Soc. A **370**, 5324–5337 (2012)
33. N.C. Menicucci, P. van Loock, M. Gu, C. Weedbrook, T.C. Ralph, M.A. Nielsen, Universal quantum computation with continuous-variable cluster states. Phys. Rev. Lett. **97**, 110501 (2006)
34. A.C. Doherty, K. Jacobs, Feedback-control of quantum systems using continuous state-estimation. Phys. Rev. A **60**, 2700 (1999)
35. N. Yamamoto, H.I. Nurdin, M.R. James, I.R. Petersen, Avoiding entanglement sudden-death via feedback control in a quantum network. Phys. Rev. A **78**, 042339 (2008)
36. H.I. Nurdin, N. Yamamoto, Distributed entanglement generation between continuous-mode Gaussian fields with measurement-feedback enhancement. Phys. Rev. A **86**, 022337 (2012)
37. J. Yoshikawa, K. Makino, S. Kurata, P. van Loock, A. Furusawa, Creation, storage, and on-demand release of optical quantum states with a negative Wigner function. Phys. Rev. X **3**, 041028 (2013)
38. N. Yamamoto, M.R. James, Zero dynamics principle for perfect quantum memory in linear networks. New J. Phys. **16**, 073032 (2014)
39. H. Mabuchi, Coherent-feedback quantum control with a dynamic compensator. Phys. Rev. A **78**, 032323 (2008)
40. M. Sarovar, D.B.S. Soh, J. Cox, C. Brif, C.T. DeRose, R. Camacho, P. Davids, Silicon nanophotonics for scalable quantum coherent feedback networks. EPJ Quantum Technol. **3**(14), 1–18 (2016)
41. M. Yanagisawa, H. Kimura, Transfer function approach to quantum control-part i: dynamics of quantum feedback systems. IEEE Trans. Autom. Control **48**(12), 2107–2120 (2003)
42. M. Yanagisawa, H. Kimura, Transfer function approach to quantum control - part ii: control concepts and applications. IEEE Trans. Autom. Control **48**(12), 2121–2132 (2003)
43. J.E. Gough, S. Wildfeuer, Enhancement of field squeezing using coherent feedback. Phys. Rev. A **80**, 042107 (2009)
44. S. Iida, M. Yukawa, H. Yonezawa, N. Yamamoto, A. Furusawa, Experimental demonstration of coherent feedback control on optical field squeezing. IEEE Trans. Autom. Control **57**(8), 2045–2050. Reprinted, with permission, © 2012 IEEE (2012)
45. O. Crisafulli, N. Tezak, D.B.S. Soh, M.A. Armen, H. Mabuchi, Squeezed light in an optical parametric amplifier oscillator network with coherent feedback quantum control. Opt. Express **21**, 18371–18386 (2013)
46. R.G. Beausoleil, P.J. Keukes, G.S. Snider, S.-Y. Wang, R.S. Williams, Nanoelectronic and nanophotonic interconnect. Proc. IEEE **96**, 230–247 (2007)
47. H. Mabuchi, Nonlinear interferometry approach to photonic sequential logic. Appl. Phys. Lett. **99**, 153103 (2011)
48. H.M. Wiseman, G.J. Milburn, *Quantum Measurement and Control* (Cambridge University Press, Cambridge, 2010)
49. T.J. Tarn, G. Huang, J.W. Clark, Modelling of quantum mechanical control systems. Math. Model. **1**(1), 109–121 (1980)
50. G.M. Huang, T.J. Tarn, J.W. Clark, On the controllability of quantum-mechanical systems. J. Math. Phys. **24**(11), 2608–2618 (1983)
51. D. D'Alessandro, *Introduction to Quantum Dynamics and Control*. Applied Mathematics and Nonlinear Science Series (Chapman & Hall/CRC, London, 2008)

52. E. Davies, *Quantum Theory of Open Systems* (Academic Press, New York, 1976)
53. V.P. Belavkin, Optimal measurement and control in quantum dynamical systems. Institute of Physics, Nicolaus Copernicus University, Torun, preprint 411 (1979)
54. V. Belavkin, On the theory of controlling observable quantum systems. Autom. Remote Control **44**(2), 178–188 (1983)
55. V.P. Belavkin, Nondemolition measurements, nonlinear filtering, and dynamic programming of quantum stochastic processes, in *Modelling and Control of Systems in Engineering, Quantum Mechanics, Economics, and Biosciences*, ed. by A. Blaquiere (Springer, New York, 1988), pp. 245–265
56. V.P. Belavkin, Continuous non-demolition observation, quantum filtering and optimal estimation, *Quantum Aspects of Optical Communication*, vol. 45, Lecture Notes in Physics (Springer, Berlin, 1991), pp. 151–163
57. M.J. Collett, C.W. Gardiner, Squeezing of intracavity and traveling-wave light fields produced in parametric amplification. Phys. Rev. A **30**(3), 1386–1391 (1984)
58. C. Gardiner, M. Collett, Input and output in damped quantum systems: quantum stochastic differential equations and the master equation. Phys. Rev. A **31**, 3761–3774 (1985)
59. R.L. Hudson, K.R. Parthasarathy, Quantum Ito's formula and stochastic evolution. Commun. Math. Phys. **93**, 301–323 (1984)
60. L. Bouten, R. van Handel, M.R. James, An introduction to quantum filtering. SIAM J. Control Optim. **46**, 2199–2241 (2007)
61. J. Dalibard, Y. Castin, K. Mölmer, Wave-function approach to dissipative processes in quantum optics. Phys. Rev. Lett. **68**(5), 580–583 (1992)
62. R. Dum, P. Zoller, H. Ritsch, Monte Carlo simulation of the atomic master equation for spontaneous emission. Phys. Rev. A **45**(7), 4879–4887 (1992)
63. H. Carmichael, *An Open Systems Approach to Quantum Optics* (Springer, Berlin, 1993)
64. M.A. Nielsen, I.L. Chuang, *Quantum Computation and Quantum Information* (Cambridge University Press, Cambridge, 2000)
65. J.P. Dowling, G.J. Milburn, Quantum technology: the second quantum revolution. Philos. Trans. R. Soc. Lond. A **361**, 1655–1674 (2003)
66. L. Bouten, Filtering and control in quantum optics. Ph.D. dissertation, Catholic University of Nijmegen (2004)
67. L. Bouten, R. van Handel, Quantum filtering: a reference probability approach (2006). arXiv:math-ph/0508006 (arXiv preprint)
68. L. Bouten, R. van Handel, On the separation principle of quantum control, in *Quantum Stochastics and Information: Statistics, Filtering and Control (University of Nottingham, UK, 15–22 July 2006)*, ed. by V.P. Belavkin, M. Guta (World Scientific, Singapore, 2008), pp. 206–238
69. H. Mabuchi, N. Khaneja, Principles and applications of control in quantum systems. Int. J. Robust Nonlinear Control **15**(15), 647–667 (2005)
70. K.J. Astrom, R.M. Murray, *Feedback Systems: An Introduction for Scientists and Engineers* (Princeton University Press, Princeton, 2008)
71. B.D.O. Anderson, J.B. Moore, *Optimal Control: Linear Quadratic Methods* (Prentice-Hall, Englewood Cliffs, 1990)
72. O.L.R. Jacobs, *Introduction to Control Theory* (Oxford University Press, Oxford, 1993)
73. P. Whittle, *Optimal Control* (Wiley, Chichester, 1996)
74. D.J. Wilson, V. Sudhir, N. Piro, R. Schilling, A. Ghadimi, T.J. Kippenberg, Measurement-based control of a mechanical oscillator at its thermal decoherence rate. Nature **524**, 325–329 (2015)
75. S. Pirandola, A. Serafini, S. Lloyd, Correlation matrices of two-mode bosonic systems. Phys. Rev. A **79**, 052327 (2009)
76. J. Laurat, G. Keller, J.A. Oliveira-Huguenin, C. Fabre, T. Coudreau, A. Serafini, G. Adesso, F. Illuminati, Entanglement of two-mode Gaussian states: characterization and experimental production and manipulation. J. Opt. B: Quantum Semiclass. Opt. **7**, S577–S587 (2005)
77. L. Accardi, J. Gough, Y.G. Lu, On the stochastic limit for quantum theory. Rep. Math. Phys. **36**(2), 155–187 (1995)

78. L. Accardi, Y.G. Lu, I. Volovich, *Quantum Theory and Its Stochastic Limit*. Series, Physics and Astronomy (Springer, Berlin, 2002)
79. H. Bachor, T. Ralph, *A Guide to Experiments in Quantum Optics*, 2nd edn. (Wiley-VCH, Weinheim, 2004)
80. Z.Y. Ou, S.F. Pereira, H.J. Kimble, Realization of the Einstein–Podolski–Rosen paradox for continuous variables in nondegenerate parametric amplification. Appl. Phys. B **55**, 265–278 (1992)
81. D. Vitali, G. Morigi, J. Eschner, Single cold atom as efficient source of EPR-entangled light. Phys. Rev. A **74**, 053814 (2006)
82. S.L. Braunstein, P. van Loock, Quantum information with continuous variables. Rev. Mod. Phys. **77**, 513 (2005)
83. C.K. Law, Interaction between a moving mirror and radiation pressure: a Hamiltonian formulation. Phys. Rev. A **51**, 2537–2541 (1995)
84. G.J. Milburn, M.J. Woolley, An introduction to quantum optomechanics. Acta Physica Slovaca **61**, 483–601 (2011)
85. L.K. Thomsen, S. Mancini, H.M. Wiseman, Continuous quantum nondemolition feedback and unconditional atomic spin squeezing. J. Phys. B: At. Mol. Opt. Phys. **35**, 4937–4952 (2002)
86. R. Inoue, S. Tanaka, R. Namiki, T. Sagawa, Y. Takahashi, Unconditional quantum-noise supression via measurement-based quantum feedback. Phys. Rev. Lett. **110**, 163602 (2013)

Chapter 2
Mathematical Modeling of Linear Dynamical Quantum Systems

Abstract This chapter provides a review of the mathematical theory of linear quantum systems, which is based on the Hudson–Parthasarathy quantum stochastic calculus as a mathematical tool for describing Markov open quantum systems interacting with external propagating quantum fields. A precise definition of linear quantum systems is given as well as quantum stochastic differential equations representing their linear equation of motion in the Heisenberg picture. The important notion of physical realizability for linear quantum stochastic differential equations is introduced, and necessary and sufficient conditions for physical realizability reviewed. Complete parameterizations for linear quantum systems are given, and transfer functions defined. Also, the special class of completely passive linear quantum systems is introduced and the notion of stability for linear quantum systems is developed.

Our aim here is to give a brief exposition of the key ideas behind quantum stochastic calculus, as required for the purposes of this monograph. For a more detailed exposition, we refer the reader to two excellent and comprehensive texts on the subject, K.R. Parthasarathy's *An Introduction to Quantum Stochastic Calculus* [1] and P.-A. Meyer's *Quantum Probability for Probabilists* [2]. We also mention the remarkably well-written tutorial paper [3]. Once the basis for quantum stochastic calculus has been laid out, we proceed to use it to formulate linear dynamical quantum systems and their stochastic dynamics.

H.I. Nurdin and N. Yamamoto, *Linear Dynamical Quantum Systems*,
Communications and Control Engineering, DOI 10.1007/978-3-319-55201-9_2

2.1 Quantum Stochastic Calculus

2.1.1 *The Boson Fock Space, Exponential Vectors, and Fundamental Processes on the Fock Space*

The linear quantum systems that are the main subject of this monograph consist of linearly coupled quantum harmonic oscillators that are in turn coupled to one or more distinct freely propagating optical fields. The freely propagating fields are bosons and each field can contain an indefinite number of bosons. A bosonic field can be mathematically described by a special type of Hilbert space known as a (*symmetric*) *Fock space*. Let $\mathcal{H}$ denote a complex Hilbert space with complex inner product $\langle \cdot, \cdot \rangle_{\mathcal{H}}$ linear in the second slot and antilinear in the first, referred to as the one-particle Hilbert space. Then the symmetric Fock space over $\mathcal{H}$, denoted by $\Gamma_s(\mathcal{H})$, is defined as

$$\Gamma_s(\mathcal{H}) = \mathbb{C} \oplus \bigoplus_{j=1}^{\infty} \mathcal{H}^{\otimes_s j},$$

where $\mathcal{H}^{\otimes_s j} = \underbrace{\mathcal{H} \otimes_s \mathcal{H} \otimes_s \cdots \otimes_s \mathcal{H}}_{j\text{-times}}$, and $\otimes_s$ denotes the symmetric tensor product. For any j elements $f_1, f_2, \ldots, f_j \in \mathcal{H}$, the symmetric tensor product of these elements is given by $f_1 \otimes_s f_2 \otimes_s \cdots \otimes_s f_j = \frac{1}{j!} \sum_{\pi \in \mathcal{P}_j} f_{\pi(1)} \otimes f_{\pi(2)} \otimes \cdots \otimes_s f_{\pi(j)}$, where $\otimes$ is the ordinary tensor product on Hilbert spaces, and $\mathcal{P}_j$ denotes the set of all permutation maps π of $\{1, 2, \ldots, j\}$ to itself (there are $j!$ such maps). The Hilbert space $\mathcal{H}^{\otimes_s j}$ is referred to as the j-particle subspace of $\Gamma_s(\mathcal{H})$. Thus, the symmetric Fock space is the infinite direct sum of finite particle Hilbert spaces $\mathcal{H}^{\otimes_s j}$ (including the 0-particle space corresponding to $\mathbb{C}$), representing the fact that a bosonic field can contain an indefinite number of bosons. The symmetric nature of the Fock space, captured by the symmetric tensor product space on each finite particle subspace, reflects the fact that bosons are a class of indistinguishable particles (the other being fermions) with a wavefunction which is symmetric with respect to an interchange of any pair of its arguments (fermions, on the other hand, have wavefunctions that are antisymmetric with respect to an interchange of any pair of its arguments). Every vector $\psi \in \Gamma_s(\mathcal{H})$ can be expressed as an infinite-dimensional vector

$$\psi = (\psi_0, \psi_1, \psi_2, \ldots, \psi_j, \ldots),$$

with $\psi_0 \in \mathbb{C}$ and $\psi_j \in \mathcal{H}^{\otimes_s j}$ for $j \geq 1$. For any two elements $\psi, \phi \in \Gamma_s(\mathcal{H})$, we have the inner product

$$\langle \psi, \phi \rangle = \psi_0^* \phi_0 + \sum_{j=1}^{\infty} \langle \psi_j, \phi_j \rangle_{\otimes_s j},$$

with $\langle \cdot, \cdot \rangle_{\otimes_s j}$ denoting the inner product on $\mathcal{H}^{\otimes_s j}$. This inner product is defined via the identity

$$\langle u_1 \otimes_s \cdots \otimes_s u_n, v_1 \otimes_s \cdots \otimes_s v_n \rangle_{\otimes_s n} = \text{Perm}\left([\langle u_i, v_j \rangle_{\mathcal{H}}]_{i,j=1,2,\ldots,n} \right),$$

where Perm$(\cdot)$ denotes the permanent of a square matrix.

An important class of vectors in $\Gamma_s(\mathcal{H})$ is the class of exponential or coherent vectors $e(f)$ that is parametrized by $f \in \mathcal{H}$. It is defined as

$$e(f) = \left(1, f, \frac{1}{2!}f^{\otimes_s 2}, \ldots, \frac{1}{k!}f^{\otimes_s k}, \ldots\right),$$

(note that $f^{\otimes_s j} = f^{\otimes j}$) with inner product

$$\langle e(g), e(f)\rangle = \exp(\langle g, f\rangle_{\mathcal{H}}).$$

In particular, we have the norm

$$\|e(f)\| = \sqrt{\exp(\|f\|_{\mathcal{H}}^2)},$$

where $\|\cdot\|_{\mathcal{H}}$ is the norm on $\mathcal{H}$, $\|f\|_{\mathcal{H}} = \sqrt{\langle f, f\rangle_{\mathcal{H}}}$.

Let $\mathcal{H}_1, \mathcal{H}_2, \ldots, \mathcal{H}_r$ be Hilbert spaces. The symmetric Fock space has the property that $\Gamma_s(\oplus_{j=1}^r \mathcal{H}_j) \cong \otimes_{j=1}^r \Gamma_s(\mathcal{H}_j)$ for any integer $r \geq 1$, where $\cong$ denotes that the two spaces are unitarily equivalent. Therefore, $\Gamma_s(\oplus_{j=1}^r \mathcal{H}_j)$ can be identified with $\otimes_{j=1}^r \Gamma_s(\mathcal{H}_j)$. This identification can be made via the correspondence $e((f_1, f_2, \ldots, f_r)) \leftrightarrow e(f_1) \otimes e(f_2) \otimes \cdots \otimes e(f_r)$, for any $f_j \in \mathcal{H}_j$.

In this monograph, we will be working exclusively with a boson Fock space over $\mathcal{H} = L^2(\mathbb{R}_+; \mathbb{C}^m)$, with $\mathbb{R}_+$ denoting the set of nonnegative real numbers, which we denote throughout the rest of the monograph as $\mathcal{F}_m$, i.e., $\mathcal{F}_m = \Gamma_s(L^2(\mathbb{R}_+; \mathbb{C}^m))$. Let $\mathcal{F}_m(\mathcal{I}) = \Gamma_s(L^2(\mathcal{I}; \mathbb{C}^m))$ for any (Lebesque) measurable set $\mathcal{I} \subset \mathbb{R}_+$. If $\mathcal{I}_1, \mathcal{I}_2, \ldots, \mathcal{I}_r$ are disjoint subsets of $\mathbb{R}_+$, then $\mathcal{F}_m(\mathcal{I}_1 \cup \mathcal{I}_2 \cup \cdots \cup \mathcal{I}_r) = \otimes_{j=1}^r \mathcal{F}_m(\mathcal{I}_j)$. In the following, we will frequently use the shorthand notation $t] \equiv [0, t]$, $t) \equiv [0, t)$, $[t \equiv [t, \infty)$, and $(t \equiv (t, \infty)$.

2.1.2 Adapted Processes and Quantum Stochastic Integrals

Let $\mathcal{E}$ denote the complex linear space spanned by the exponential vectors on the Fock space $\mathcal{F}_m$, $\mathcal{E} = \text{span}\{e(f) \mid f \in L^2(\mathbb{R}_+; \mathbb{C}^m)\}$. For any $g \in L^2(\mathbb{R}_+; \mathbb{C}^m)$ and bounded self-adjoint operator $\Pi : L^2(\mathbb{R}_+; \mathbb{C}^m) \to L^2(\mathbb{R}_+; \mathbb{C}^m)$, define the operators $\mathcal{A}(f)$, $\mathcal{A}^*(f)$, and $\Lambda(\Pi)$ with domain $\mathcal{E}$ via their action on the coherent vectors:

$$\mathcal{A}(g)e(f) = \left(\int_0^\infty g(s)^* f(s) ds\right) e(f),$$

$$\mathcal{A}^*(g)e(f) = \frac{d}{dt} e(f + tg)\Big|_{t=0},$$

$$\Lambda(\Pi)e(f) = \mathcal{A}^*(\Pi f)e(f).$$

In particular, we have the relation $\langle \mathcal{A}^*(f)e(h), e(g)\rangle = \langle e(h), \mathcal{A}(f)e(g)\rangle$ and the commutation relations

$$\begin{aligned}
[\mathcal{A}(f), \mathcal{A}(g)] &= [\mathcal{A}^*(f), \mathcal{A}^*(g)] = 0, \\
[\mathcal{A}(f), \mathcal{A}^*(g)] &= \int_0^\infty f(s)^* g(s) ds, \\
[\Lambda(\Pi_1), \Lambda(\Pi_2)] &= \Lambda([\Pi_1, \Pi_2]), \\
[\mathcal{A}(f), \Lambda(\Pi)] &= \mathcal{A}(\Pi^* f), \\
[\mathcal{A}^*(f), \Lambda(\Pi)] &= -\mathcal{A}^*(\Pi f).
\end{aligned}$$

Also, $\mathcal{A}^*(f)$ is the adjoint of $\mathcal{A}(f)$ with domain $\mathrm{Dom}(A^*(f)) \supset \mathcal{E}$, and $\Lambda(\Pi)$ is self-adjoint on a domain $\mathrm{Dom}(\Lambda(\Pi)) \supset \mathcal{E}$.

Let e_i be a standard basis vector in $\mathbb{C}^m$ (as a column vector) that is 0 everywhere except at the i-th element which is 1, and let $1_{\mathcal{I}}$ be the indicator function on the set $\mathcal{I}$. Define the so-called fundamental processes $\mathcal{A}_j(t)$, $\mathcal{A}_j^*(t)$, and $\Lambda_{jk}(t)$ acting on $\mathcal{E}$ as

$$\begin{aligned}
\mathcal{A}_j(t) &= \mathcal{A}(e_j 1_{t]}), \\
\mathcal{A}_j^*(t) &= \mathcal{A}^*(e_j 1_{t]}), \\
\Lambda_{jk}(t) &= \Lambda(e_j e_k^\top 1_{t]}),
\end{aligned}$$

with the indices j and k ranging from 1 until m. The subscript j for $\mathcal{A}_j(t)$ and $\mathcal{A}_j^*(t)$ indicates that they are field *annihilation and creation processes* on the j-th bosonic field, respectively, while the subscript jk on $\Lambda_{jk}(t)$ indicates that it is a photon exchange operator from the k-th boson field to the j-th field. We may then *identify* (and we shall do so for the rest of the monograph without further comment) $\mathcal{A}(t)$ and $\mathcal{A}^*(t)$ as vectors of operators on $\mathcal{F}_m$:

$$\begin{aligned}
\mathcal{A}(t) &= [\, \mathcal{A}_1(t)\ \mathcal{A}_2(t)\ \ldots\ \mathcal{A}_m(t)\,]^\top, \\
\mathcal{A}^*(t) &= [\, \mathcal{A}_1^*(t)\ \mathcal{A}_2^*(t)\ \ldots\ \mathcal{A}_m^*(t)\,]^\top,
\end{aligned}$$

and identify $\Lambda(t)$ with the matrix of operators,

$$\Lambda(t) = [\Lambda_{jk}(t)]_{j,k=1,2,\ldots,m}.$$

From the properties of the creation process, we will freely make the identification $\mathcal{A}_j^*(t) = \mathcal{A}_j(t)^*$, and $\mathcal{A}^*(t) = \mathcal{A}(t)^*$ on $\mathcal{E}$. We also note the fact that $\Lambda(t) = \Lambda(t)^*$ on $\mathcal{E}$.

The process $\Lambda(t)$ is known as the *counting* or *gauge process*. Let $\mathfrak{h}$ denote the underlying Hilbert space of a quantum system that is coupled to bosonic fields on the Fock space $\mathcal{F}_m$. A process $\{X(t),\ t \geq 0\}$ defined on $\mathfrak{h} \otimes \mathcal{F}_m$ is said to be adapted if $X(t) \left(\psi(t) \otimes e(f 1_{[t})\right) = \phi(t) \otimes e(f 1_{[t})$ for any $f \in L^2(\mathbb{R}_+; \mathbb{C}^m)$, $\psi(t) \in \mathfrak{h} \otimes \mathcal{F}_m(t])$, and some $\phi(t) \in \mathfrak{h} \otimes \mathcal{F}_m(t])$. That is, an adapted process acts trivially (i.e., as the identity operator) on the future factor $\mathcal{F}_m([t)$ of $\mathcal{F}_m$.

Observe that $(\mathcal{A}(t) - \mathcal{A}(s))e(f) = e(f 1_{s]}) \otimes \mathcal{A}(1_{[s,t]}) e(f 1_{[s,t]}) \otimes e(f 1_{[t})$, and similarly, $(\mathcal{A}^*(t) - \mathcal{A}^*(s))e(f) = e(f 1_{s]}) \otimes \mathcal{A}^*(1_{[s,t]}) e(f 1_{[s,t]}) \otimes e(f 1_{[t})$, and

$(\Lambda(t) - \Lambda(s))e(f) = e(f1_{s]}) \otimes \Lambda(1_{[s,t]})e(f1_{[s,t]}) \otimes e(f1_{[t})$. With these properties, one can define quantum stochastic integrals with respect to adapted processes:

$$\int_0^t (X_1(s)d\mathcal{A}(s) + X_2(s)d\mathcal{A}(s) + X_3(s)d\Lambda(s))$$
$$= \int_0^t (X_1(s) \otimes d\mathcal{A}(s) + X_2(s) \otimes d\mathcal{A}(s) + X_3(s) \otimes d\Lambda(s)), \qquad (2.1)$$

where X_1, X_2, X_3 are adapted processes, and the quantum stochastic integral is defined on $\mathfrak{h} \otimes \mathcal{F}_m$. The quantum stochastic integral then also defines an adapted process. These quantum stochastic integrals can be constructed in a fashion that is similar to the construction of classical Itō stochastic integrals; see [1, 2, 4] for details. Also, note the important property that any adapted process $X(t)$ commutes with $\mathcal{A}_j(1_{[\tau_1,\tau_2)})$, $\mathcal{A}_j^*(1_{[\tau_1,\tau_2)})$, and $\Lambda_{jk}(1_{[\tau_1,\tau_2)})$ for any j, k and any $t \leq \tau_1 < \tau_2 \leq \infty$.

From a physical point of view, the processes $\mathcal{A}_j(t)$, $\mathcal{A}_j^*(t)$, and $\Lambda_{jk}(t)$ arise as integrated version of idealized quantum white noise processes $\xi_j(t), j = 1, 2, \ldots, m$, satisfying the singular commutation relations $[\xi_j(t), \xi_k(s)^*] = \delta_{jk}\delta(t-s)$ [5, 6], introduced earlier in Chap. 1. That is, we may formally write $\mathcal{A}_j(t) = \int_0^t \xi_j(s)ds$, $\mathcal{A}_j^*(t) = \int_0^t \xi_j(s)^*ds$, and $\Lambda_{jk}(t) = \int_0^t \xi_j(s)^*\xi_k(s)ds$. Due to the singular nature of the quantum white noise processes, it is mathematically simpler to work with the more regular integrated processes $\mathcal{A}_j(t)$, $\mathcal{A}_j^*(t)$, and $\Lambda_{jk}(t)$ as these can be rigorously and explicitly constructed as processes on the Fock space $\mathcal{F}_m$, as we have already briefly elaborated upon. However, from the perspective of describing the physics involved in the interaction between the system and the bosonic environment, the white noise picture is more meaningful, and we shall often employ this in the discussion that ensues. That is to say that the more fundamental processes from the underlying physics are the quantum white noise processes rather than the mathematically more convenient "derived" integrated processes. It should be noted that quantum white noise processes arise as a consequence of making a Markov assumption regarding the interaction between the system and the bosonic environment, and it is only within this approximation that the quantum white noise processes can be given a sensible physical interpretation. The modeling of a bosonic environment as a quantum white noise process without the Markov approximation can be problematic and gives rise to physical inconsistencies, see, e.g., [7] for a discussion.

2.1.3 *The Quantum Itō Table in Vacuum and the Quantum Itō Rule*

The fundamental processes $\mathcal{A}_j(t)$, $\mathcal{A}_j^*(t)$, and $\Lambda_{jk}(t)$ are quantum stochastic processes on the Fock space $\mathcal{F}_m$. The processes $\mathcal{A}_j(t) + \mathcal{A}_j^*(t)$ and $-\imath\mathcal{A}_j(t) + \imath\mathcal{A}_j^*(t)$ are non-commuting processes that are each isomorphic to a classical standard Wiener processes, that is, each can be viewed as Fock space representations of the standard

Wiener process. On the other hand, $\Lambda_{jj}(t)$ is isomorphic to a classical Poisson process, that is, the latter is a realization of the former on the Fock space. As a quantum mechanical system, the bosonic fields have a quantum state that determines their statistics under measurement. We now assume that the bosonic fields are in the vacuum state $|\Omega\rangle = e(0)$. This is a state in which the fields do not contain any photons. In this state, we have that $\mathcal{A}(f)e(0) = 0$ for all $f \in L^2(\mathbb{R}_+; \mathbb{C}^m)$, and the forward-pointing differentials $d\mathcal{A}_j(t) = \mathcal{A}_j(t+dt) - \mathcal{A}_j(t)$, $d\mathcal{A}_j^*(t) = \mathcal{A}_j^*(t+dt) - \mathcal{A}_j^*(t)$, and $d\Lambda_{jk}(t)$ satisfy the quantum Itō product rule, as a quantum adaptation of the classical Itō product rules: $d\mathcal{A}_j(t)d\mathcal{A}_k^*(t) = \delta_{jk}dt$; $d\mathcal{A}_j^*(t)d\mathcal{A}_k(t) = d\mathcal{A}_j(t)d\mathcal{A}_k(t) = d\mathcal{A}_j^*(t)d\mathcal{A}_k^*(t) = 0$, and

$$
\begin{aligned}
d\Lambda_{jk}(t)d\Lambda_{j'k'}(t) = \delta_{kj'}d\Lambda_{jk'}(t),\ d\mathcal{A}_j(t)d\Lambda_{kl}(t) = \delta_{jk}d\mathcal{A}_l(t),\\
d\Lambda_{jk}d\mathcal{A}_l^*(t) = \delta_{kl}d\mathcal{A}_j^*(t).
\end{aligned}
$$

Let $X(t)$ and $Y(t)$ be two adapted processes on $\mathfrak{h} \otimes \mathcal{F}_m$ that can be expressed as quantum stochastic integrals with respect to the fundamental processes $\mathcal{A}_j(t)$, $\mathcal{A}_j^*(t)$, and $\Lambda_{jk}(t)$, as in (2.1). The quantum Itō product rule holds for the forward-pointing differential of the product process $X(t)Y(t)$:

$$
d(X(t)Y(t)) = (dX(t))Y(t) + X(t)dY(t) + dX(t)dY(t).
$$

Note the third term that serves as a quantum analogue of the second-order correction term in the classical Itō stochastic calculus. Since X and Y are quantum stochastic integrals with respect to the fundamental processes, the correction term can be calculated using (2.1) together with the quantum Itō product rule for the fundamental processes given above.

2.1.4 The Hudson–Parthasarathy Quantum Stochastic Differential Equation

Using the quantum stochastic integrals and the quantum Itō rules, one can define adapted processes as solutions of quantum stochastic differential equations (QSDEs). A particularly important class of QSDEs that describe the physical scenario of Markov open quantum systems coupled to a bosonic environment is the Hudson–Parthasarathy QSDE of the form:

$$
\begin{aligned}
dU(t) = \Big(&\mathrm{Tr}\big((S - I)^\top d\Lambda(t)\big) + d\mathcal{A}(t)^* L - L^* S d\mathcal{A}(t)\\
&-(\imath H + 1/2 L^* L)dt\Big)U(t),
\end{aligned}
\tag{2.2}
$$

with initial condition $U(0) = I$. Here, H is the Hamiltonian of the system and is a self-adjoint operator on the system Hilbert space $\mathfrak{h}$, $L = [L_1\ L_2\ \ldots\ L_m]^\top$ is a vector of operators L_j on $\mathfrak{h}$, and $S \in \mathbb{C}^{m\times m} \otimes \mathfrak{h}$ is a unitary operator (i.e., $S^*S = SS^* = I$), with entries S_{jk} being operators on $\mathfrak{h}$, called a *scattering matrix*. They are the operator-valued "coefficients" or parameters of the QSDE. The form of the equation is such that the solution $U(t)$ of the QSDE is a unitary adapted process. That is, $U(t)$ is an adapted process and is unitary for each t: $U(t)^*U(t) = U(t)U(t)^* = I$.[1]

Following [9], we denote a quantum system G, whose dynamics are governed by the unitary solution of a Hudson–Parthasarahy QSDE (2.2) with parameters S, L, and H, by the shorthand $G = (S, L, H)$.

We will now elaborate on properties of the Hudson–Parthasarathy QSDE that will be crucial for the (measurement-based) feedback control theory of open quantum systems. Let $\theta_t : L^2(\mathbb{R}; \mathbb{C}^m) \to L^2(\mathbb{R}; \mathbb{C}^m)$ denote the one-particle left shift operator, defined by $\theta_t f(\cdot) = f(\cdot + t)\ \forall t \in \mathbb{R}_+$. Let Θ_t be the so-called second quantization of the one-particle left shift operator, defined as an operator on $\mathcal{F}_{m-} = \Gamma_s(L^2(\mathbb{R}; \mathbb{C}^m))$ by its action on the exponential vectors as $\Theta_t e(f) = e(\theta_t f)$. It is straightforward to verify that $\Theta_t^* e(f) = e(f(\cdot - t))$ and Θ_t can be uniquely extended to a unitary operator on $\mathcal{F}_{m-}$, $\Theta_t^*\Theta_t = \Theta_t\Theta_t^* = I$. The solution $U(t)$ of a Hudson–Parthasarathy QSDE, with $U(t)$ extended from $\mathfrak{h} \otimes F_m$ to $\mathfrak{h} \otimes F_{m-}$ as $I \otimes U(t)$, has the property that it is a left cocycle with respect to Θ_t, meaning that the following holds for all $0 \le s \le t$,

$$U(t) = \Theta_s^* U(t - s)\Theta_s U(s).$$

We note that $V(t, s) = \Theta_s^* U(t - s)\Theta_s$ acts non-trivially only on the portion $\mathfrak{h} \otimes \mathcal{F}_m([s, t])$ of $\mathfrak{h} \otimes \mathcal{F}_m$. Note that $U(t) = V(t, 0)$ and the cocycle property can be equivalently expressed as $V(t, s) = V(t, \tau)V(\tau, s)$ for all $0 \le s \le \tau \le t$. Indeed, we have that

$$\begin{aligned} V(t,\tau)V(\tau,s) &= \Theta_\tau^* U(t-\tau)\Theta_\tau\Theta_s^* U(\tau - s)\Theta_s \\ &= \Theta_\tau^* U(t-\tau)\Theta_{\tau-s} U(\tau - s)\Theta_s \\ &= \Theta_s^*\Theta_{\tau-s}^* U(t-\tau)\Theta_{\tau-s} U(\tau - s)\Theta_s \\ &= \Theta_s^* V(t-s, \tau-s) U(\tau - s)\Theta_s \\ &= \Theta_s^* U(t-s)\Theta_s \\ &= V(t,s), \end{aligned}$$

where in the second and third lines we have used the fact that $\Theta_\tau\Theta_s^* = \Theta_{\tau-s}$. A consequence of this property is that $\tilde{U}(t) = \Theta_t U(t)$ defines a strongly continuous one-parameter semigroup of unitary operators on $\mathfrak{h} \otimes \mathcal{F}_{m-}$, $\tilde{U}(t)\tilde{U}(s) = \tilde{U}(t+s)$ for all $s, t \ge 0$ with $\tilde{U}(0) = I$. Therefore there exists a densely defined essentially self-adjoint operator K such that $\tilde{U}(t) = e^{-\imath K t}$. The operator K is the Stone generator

[1] Technically, the existence of a unique solution of the Hudson–Parthasarathy QSDE that is unitary is guaranteed whenever the coefficients S, L, H are all bounded operators. When they are unbounded then additional technical assumptions need to be assumed, see, e.g., [8].

of the strongly continuous semigroup $\{\tilde{U}(t),\ 0 \leq s \leq t\}$. The difficult and long-standing problem of characterizing this generator was finally resolved by Gregoratti [10, 11] building on the work of Chebotarev [12].

Let us now turn to the Heisenberg picture of quantum mechanics, where states are fixed and operators evolve in time. This will be the natural setting for studying quantum filtering and control problems that will be considered later in the monograph. Let $j_t(\cdot) = U(t)^* \cdot U(t)$ and let $Z(t)$ denote any adapted process on the boson Fock space $\mathcal{F}_m$. Then by the cocycle property of $V(t,s)$ we find that, for any $t \geq s \geq 0$,

$$\begin{aligned} j_t(Z(s)) &= U(t)^* Z(s) U(t) \\ &= U(s)^* \Theta_s^* U(t-s)^* \Theta_s Z(s) \Theta_s^* U(t-s) \Theta_s U(s) \\ &= U(s)^* Z(s) U(s) \\ &= j_s(Z(s)). \end{aligned}$$

Using this property, it follows that for any (bounded linear) system operator X on $\mathfrak{h}$ and $0 \leq s \leq t$,

$$\begin{aligned} j_t(X) j_s(Z(s)) &= j_t(X) j_t(Z(s)) \\ &= U(t)^* X Z(s) U(t) \\ &= U(t)^* Z(s) X U(t) \\ &= j_t(Z(s)) j_t(X) \\ &= j_s(Z(s)) j_t(X). \end{aligned}$$

Therefore, $[j_t(X), j_s(Z(s))] = 0$ for any system operator X and any adapted process $Z(t)$ on $\mathcal{F}_m$. In other words, $\{j_t(X), j_s(Z(s)),\ 0 \leq s \leq t\}$ forms a commutative family of operators, a property of input–output Markov open quantum systems known as the *non-demolition* property. Now, let $Z(t)$ be an adapted process on $\mathcal{F}_m$ with the additional property that $[Z(t), Z(s)] = 0$ for all $s, t \geq 0$. For example, $Z(t) = e^{-\imath\phi}\mathcal{A}(t) + e^{\imath\phi}\mathcal{A}(t)^*$ ($\phi \in \mathbb{R}$) and $Z(t) = \Lambda(t)$, all have these properties. It follows that

$$\begin{aligned} j_t(Z(t)) j_s(Z(s)) &= j_t(Z(t)) j_t(Z(s)) \\ &= U(t)^* Z(t) Z(s) U(t) \\ &= U(t)^* Z(s) Z(t) U(t) \\ &= j_t(Z(s)) j_t(Z(t)) \\ &= j_s(Z(s)) j_t(Z(t)). \end{aligned}$$

Hence, when $Z(t)$ has the stipulated properties, $[j_t(Z(t)), j_s(Z(s))] = 0$ for all $s, t \geq 0$ and $\{j_t(Z(t)), j_s(Z(s)),\ 0 \leq s \leq t\}$ forms a commutative family of operators. This is a property of input–output Markov open quantum systems known as the *self non-demolition* property.

Now, let $Z(t)$ be an observable for each t, $Z(t) = Z(t)^*$, the self non-demolition property means that $\mathscr{Z}_t = \{j_s(Z(s)),\ 0 \leq s \leq t\}$ describes a continuous output process of the quantum fields that may be continuously monitored/measured (depending on what $Z(t)$ actually is), to yield a real-valued measurement record. When $Z(t)$ is self non-demolition, the non-demolition property means that information about $j_t(X)$ can be inferred from an observation record of $\mathscr{Z}_t$. This inference process is known as quantum filtering, the quantum analogue of classical nonlinear filtering theory [13], and will be treated in Chap. 4.

2.2 Linear Dynamical Quantum Systems: Joint Unitary Evolution of Oscillators and Boson Fields

We now begin to introduce and specialize to the class of linear quantum stochastic systems as the central theme of this monograph. This section describes the joint evolution of a set of coupled oscillators that are also coupled to external bosonic fields, via a certain form of interaction, which results in linear dynamics of the oscillators' position and momentum operators in the Heisenberg picture. The linear dynamics is a defining feature of linear quantum systems.

Let there be n independent quantum harmonic oscillators. The j-th quantum harmonic oscillator has position and momentum operators q_j and p_j acting on elements of the underlying Hilbert space $L^2(\mathbb{R};\mathbb{C})$, the space of all square integrable complex-valued function on $\mathbb{R}$. These operators satisfy the canonical commutation relations (CCR)[2]:

$$[q_j, p_k] = 2\imath\delta_{jk}\,,\ [q_j, q_k] = 0\,,\ [p_j, p_k] = 0. \tag{2.3}$$

The position and momentum operators of the n oscillators can be collected in a single column vector of operators x defined by $x = (q_1, p_1, q_2, p_2, \ldots, q_n, p_n)^\top$. We can then express the CCR more compactly as,

$$[x, x^\top] = xx^\top - (xx^\top)^\top = 2\imath\mathbb{J}_n,$$

with $\mathbb{J}_n = I_n \otimes \mathbb{J}$, with

$$\mathbb{J} = \begin{bmatrix} 0 & 1 \\ -1 & 0 \end{bmatrix}.$$

Also note in passing that $xx^\top \neq (xx^\top)^\top$ since some elements of x do not commute with one another. The composite system of n quantum harmonic oscillators has a *quadratic Hamiltonian* H given by $H = (1/2)x^\top Rx$, with R a real symmetric $2n \times 2n$

[2]We take the common convention that units are taken such that $\hbar = 1$. Also, we will take (2.3) as the default CCR for the position and momentum operators of multiple distinct oscillators. However, it is easy to adapt the results of this chapter for a different definition of these operators that satisfy a different set of commutation relations; see Remark 2.2.

matrix. The quantum harmonic oscillators are also coupled to m distinct external quantum bosonic fields. They are coupled to the k-th quantum field via a singular interaction of the form Hamiltonian $H_k = \imath(L_k\xi_k^*(t) - L_k^*\xi_k(t))$ [5, 6], where $L_k = K_k x$ (with $K_k \in \mathbb{C}^{1\times 2n}$) is a linear coupling operator describing the coupling of the position and momentum operators to $\xi_k(t)$. Here $\xi_k(t)$ is a quantum white noise process as discussed in Sect. 2.1.2. We can collect the coupling operators $L_1, L_2, \ldots, L_m$ together in one *linear coupling vector* $L = [\,L_1\; L_2\; \ldots\; L_m\,]^\top = Kx$, with $K = [\,K_1^\top\; K_2^\top\; \ldots\; K_m^\top\,]^\top$. The *joint* evolution of the oscillators and quantum fields is then given by a unitary adapted process $U(t)$ satisfying the Hudson–Parthasarathy QSDE (2.2), with S being a fixed unitary matrix in $\mathbb{C}^{m\times m}$, i.e., the entries of S in this case are complex numbers rather than operators on $\mathfrak{h}$.

Using the shorthand notation introduced earlier, a linear quantum stochastic system with parameters $S, L = Kx, H = (1/2)x^\top Rx$ as described above can be expressed as $G = (S, Kx, (1/2)x^\top Rx)$. In a later section, we will slightly generalize the notion of a linear quantum system by allowing so-called Bogoliubov transformations to replace the scattering matrix S.

2.3 Equations of Motion: Real Quadrature Form and Complex Mode Form

Using the quantum Itō rule and the quantum Itō products, and exploiting the canonical commutation relations between the operators in x, the *Heisenberg evolution*

$$x(t) = U(t)^* x U(t) = \begin{bmatrix} U(t)^* q_1 U(t) \\ U(t)^* p_1 U(t) \\ U(t)^* q_2 U(t) \\ U(t)^* p_2 U(t) \\ \vdots \\ U(t)^* q_n U(t) \\ U(t)^* p_n U(t) \end{bmatrix}$$

of the vector x can be obtained [14, 15]. This is given by the QSDE,

$$\begin{aligned} dx(t) &= d(U(t)^* x U(t)) \\ &= A_o x(t)dt + B_o \begin{bmatrix} d\mathcal{A}(t) \\ d\mathcal{A}(t)^\# \end{bmatrix}; x(0) = x, \\ d\mathcal{Y}(t) &= d(U(t)^* \mathcal{A}(t) U(t)) \\ &= C_o x(t)dt + D_o d\mathcal{A}(t), \end{aligned} \tag{2.4}$$

with

$$
\begin{aligned}
A_o &= 2\mathbb{J}_n(R + \Im\{K^*K\}),\\
B_o &= 2\imath\mathbb{J}_n[\, -K^\dagger S \; K^\top S^\# \,],\\
C_o &= K,\\
D_o &= S.
\end{aligned}
$$

Here,

$$
\mathcal{Y}(t) = (\mathcal{Y}_1(t), \ldots, \mathcal{Y}_m(t))^\top = U(t)^* \mathcal{A}(t) U(t)
$$

is a vector of *output fields* that is produced by the interaction of the quantum harmonic oscillators and the incoming quantum fields $\mathcal{A}(t)$. Note that the Heisenberg picture dynamics of $x(t)$ is linear, and $\mathcal{Y}(t)$ has a component which is a linear combination of elements of $x(t)$. The index n in the above will be referred to as the number of *degrees of freedom* or simply the *degree* of the linear quantum stochastic system.

2.3.1 Real Quadrature Form

In certain circumstances, it is convenient to write the dynamics (2.4) in the so-called (real) quadrature form as in [15–17]:

$$
\begin{aligned}
dx(t) &= Ax(t)dt + Bdw(t); \; x(0) = x,\\
dy(t) &= Cx(t)dt + Ddw(t),
\end{aligned} \tag{2.5}
$$

with

$$
\begin{aligned}
w(t) &= 2(\Re\{\mathcal{A}_1(t)\}, \Im\{\mathcal{A}_1(t)\}, \ldots, \Re\{\mathcal{A}_m(t)\}, \Im\{\mathcal{A}_m(t)\})^\top,\\
y(t) &= 2(\Re\{\mathcal{Y}_1(t)\}, \Im\{\mathcal{Y}_1(t)\}, \ldots, \Re\{\mathcal{Y}_m(t)\}, \Im\{\mathcal{Y}_m(t)\})^\top.
\end{aligned}
$$

Here,

$$
A = 2\mathbb{J}_n(R + \Im\{K^*K\}), \tag{2.6}
$$

$$
B = 2\imath\mathbb{J}_n[\, -K^* \; K^\top \,]\mathrm{diag}(S, S^\#)\Gamma_m, \tag{2.7}
$$

$$
C = P_m^\top \begin{bmatrix} K + K^\# \\ -\imath K + \imath K^\# \end{bmatrix}, \tag{2.8}
$$

$$
D = \Gamma_m^{-1}\mathrm{diag}(S, S^\#)\Gamma_m, \tag{2.9}
$$

where P_m denotes a $2m \times 2m$ *permutation matrix* acting as

$$P_m[\, a_1 \; a_2 \; \ldots \; a_{2m-1} \; a_{2m} \,]^\top$$
$$= [\, a_1 \; a_3 \; \ldots \; a_{2m-1} \; a_2 \; a_4 \; \ldots \; a_{2m} \,]^\top,$$

and

$$\Gamma_m = P_m \left(I_m \otimes \frac{1}{2} \begin{bmatrix} 1 & \imath \\ 1 & -\imath \end{bmatrix} \right).$$

Note that the matrices A, B, C, D in the quadrature form are all real and are in a one to one correspondence with the matrices A_o, B_o, C_o, D_o, and the quantum noise vector $w(t)$ satisfies the Itō relationship $dw(t)dw(t)^\top = (I + \imath \mathbb{J}_m)dt$.

Some remarks are in order about the nature of the quadrature quantum noise w. From the self non-demolition property we have that $[w_j(s), w_j(t)] = 0$ (w_j denotes the j-th component of w) for all $s, t \geq 0$ and all j. Indeed,

$$\begin{aligned}
&[w_j(s), w_j(t)] \\
&\quad = [\mathcal{A}_j(t) + \mathcal{A}_j^*(t), \mathcal{A}_j(s) + \mathcal{A}_j^*(s)] \\
&\quad = [\mathcal{A}_j(t), \mathcal{A}_j(s)] + [\mathcal{A}_j(t), \mathcal{A}_j^*(s)] + [\mathcal{A}_j^*(t), \mathcal{A}_j(s)] + [\mathcal{A}_j^*(t), \mathcal{A}_j^*(s)] \\
&\quad = \min(t, s) - \min(t, s) \\
&\quad = 0.
\end{aligned}$$

When the state is the vacuum state, w_j is isomorphic to a standard Wiener process, in the sense that it is a realization of a standard Wiener process as an operator-valued process on a Fock space. This will be demonstrated in Sect. 2.7.2 when Gaussian states of a bosonic field are discussed. However, for any $j = 1, 2, \ldots, m$, $w_{2j-1}(t)$ and $w_{2j}(t)$ do not commute for any t, meaning that they cannot be realized jointly on a common classical probability space. On the other hand, we do have that $[w_{2j-1}(s), w_k(t)] = 0$ for all $k \neq 2j$ and all $s, t \geq 0$, so that the collection of processes $\{w_{2j-1}, w_k, k \neq 2j\}$ is equivalent to a collection of independent standard Wiener processes that can be jointly defined on a common classical probability space. The non-commutativity among components of $w(t)$ is a crucial difference from the usual vector of classical independent Wiener processes.

2.3.2 Complex Mode Form

In other situations, it may be more convenient to represent the dynamics of a linear quantum stochastic system in the (complex) mode form. In this form, instead of writing down the Heisenberg evolution of amplitude-phase quadrature pairs $(q_1, p_1, q_2, p_2, \ldots, q_n, p_n)^\top$, we write down the Heisenberg evolution of the annihilation–creation pairs $(a_1, a_1^*, a_2, a_2^*, \ldots, a_n, a_n^*)^\top$, where $a_j = \frac{1}{2}(q_j + \imath p_j)$ and

$a_j^* = \frac{1}{2}(q_j - \imath p_j)$. In terms of these operators, the quadratic Hamiltonian H and coupling operator L can be expressed as [18]

$$H = a^*\Omega_- a + \frac{1}{2}a^*\Omega_+ a^\# + \frac{1}{2}a^\top \Omega_+^\# a,$$

where $\Omega_+ = [\omega_{jk}^+]$ and $\Omega_- = [\omega_{jk}^-]$ are complex $n \times n$ matrices possessing the symmetries $\Omega_-^* = \Omega_-$ and $\Omega_+^\top = \Omega_+$, and

$$L = C_- a + C_+ a^\#,$$

where $C_\pm \in \mathbb{C}^{m\times n}$. Using these expression for H and L one can write down the QSDE for the Heisenberg evolution of a and $a^\#$. Let

$$\begin{aligned} a(t) &= U(t)^* a U(t) \\ &= (U(t)^* a_1 U(t), U(t)^* a_2 U(t), \ldots, U(t)^* a_n U(t))^\top, \end{aligned}$$

and introduce the *doubled-up matrix*

$$\Delta(E_-, E_+) \triangleq \begin{bmatrix} E_- & E_+ \\ E_+^\# & E_-^\# \end{bmatrix},$$

for any complex matrices $E_\pm$ of the same dimension. We also introduce the notation

$$\mathbb{K}_m \triangleq \begin{bmatrix} I_m & 0 \\ 0 & -I_m \end{bmatrix}, \tag{2.10}$$

and define the $\cdot^\flat$ involution operator that acts on a $2k \times 2l$ matrix X as

$$X^\flat = \mathbb{K}_l X^* \mathbb{K}_k. \tag{2.11}$$

Then the QSDE for $(a(t)^\top, a(t)^*)^\top$ and the output $\mathcal{Y}(t)$ can be expressed in the doubled-up form [18]

$$\begin{aligned} \begin{bmatrix} da(t) \\ da(t)^\# \end{bmatrix} &= \tilde{A} \begin{bmatrix} a(t) \\ a(t)^\# \end{bmatrix} dt + \tilde{B} \begin{bmatrix} d\mathcal{A}(t) \\ d\mathcal{A}(t)^\# \end{bmatrix}, \\ \begin{bmatrix} d\mathcal{Y}(t) \\ d\mathcal{Y}(t)^\# \end{bmatrix} &= \tilde{C} \begin{bmatrix} a(t) \\ a(t)^\# \end{bmatrix} dt + \tilde{D} \begin{bmatrix} d\mathcal{A}(t) \\ d\mathcal{A}(t)^\# \end{bmatrix}, \end{aligned} \tag{2.12}$$

where

$$\begin{aligned} A_\mp &= -\frac{1}{2}(C_-^* C_\mp - C_+^\top C_\pm^\# - \imath \Omega_\mp), \\ \Omega_- &= -\frac{1}{2\imath}(A_- - A_-^*), \end{aligned}$$

$$\Omega_+ = -\frac{1}{2\imath}(A_+ + A_+^\top),$$

$$\tilde{A} = \Delta(A_-, A_+), \tag{2.13}$$

$$\tilde{B} = -\Delta(C_-, C_+)^\flat \Delta(S, 0), \tag{2.14}$$

$$\tilde{C} = \Delta(C_-, C_+), \tag{2.15}$$

$$\tilde{D} = \Delta(S, 0). \tag{2.16}$$

Note that it is necessary to adopt a doubled-up form since the evolutions of $a(t)$ and $a(t)^\#$ are, in general, coupled. The following example illustrates the description of a system in mode form.

Example 2.1 Consider the nondegenerate optical parametric amplifier (NOPA) from Sect. 1.5.2. We consider a lossless version of this device with $\kappa = 0$. Therefore, the device is only coupled to two fields collected in the vector $\mathcal{A}(t)$, and has parameters $\Omega_- = 0_{2\times 2}$, $\Omega_+ = \begin{bmatrix} 0 & \imath\epsilon/2 \\ \imath\epsilon/2 & 0 \end{bmatrix}$, $C_- = \begin{bmatrix} \sqrt{\gamma} & 0 \\ 0 & \sqrt{\gamma} \end{bmatrix}$, $C_+ = 0_{2\times 2}$, and $S = I_2$. The doubled-up form of the evolution of the cavity modes in the vector $a = [\, a_1 \; a_2 \,]^\top$ is

$$\begin{bmatrix} da(t) \\ da(t)^\# \end{bmatrix} = \begin{bmatrix} -\gamma/2 & 0 & 0 & \epsilon/2 \\ 0 & -\gamma/2 & \epsilon/2 & 0 \\ 0 & \epsilon/2 & -\gamma/2 & 0 \\ \epsilon/2 & 0 & 0 & -\gamma/2 \end{bmatrix} \begin{bmatrix} a(t) \\ a(t)^\# \end{bmatrix} dt$$
$$- \begin{bmatrix} \sqrt{\gamma} & 0 & 0 & 0 \\ 0 & 0 & \sqrt{\gamma} & 0 \\ 0 & \sqrt{\gamma} & 0 & 0 \\ 0 & 0 & 0 & \sqrt{\gamma} \end{bmatrix} \begin{bmatrix} d\mathcal{A}(t) \\ d\mathcal{A}(t)^\# \end{bmatrix},$$
$$\begin{bmatrix} d\mathcal{Y}(t) \\ d\mathcal{Y}(t)^\# \end{bmatrix} = \sqrt{\gamma} \begin{bmatrix} a(t) \\ a(t)^\# \end{bmatrix} dt + \begin{bmatrix} d\mathcal{A}(t) \\ d\mathcal{A}(t)^\# \end{bmatrix}.$$

In the physics literature, it is more common for the quantum stochastic dynamics to be written in the heuristic form of a quantum Langevin equation (introduced earlier in Chap. 1) driven by a quantum white noise process $\xi(t)$, where $\mathcal{A}(t) = \int_0^t \xi(s)ds$. For the mode form of the equation of motion, this quantum Langevin equation takes the form

$$\begin{bmatrix} \dot{a}(t) \\ \dot{a}(t)^\# \end{bmatrix} = \tilde{A} \begin{bmatrix} a(t) \\ a(t)^\# \end{bmatrix} + \tilde{B} \begin{bmatrix} \xi(t) \\ \xi(t)^\# \end{bmatrix},$$
$$\begin{bmatrix} \eta(t) \\ \eta(t)^\# \end{bmatrix} = \tilde{C} \begin{bmatrix} a(t) \\ a(t)^\# \end{bmatrix} + \tilde{D} \begin{bmatrix} \xi(t) \\ \xi(t)^\# \end{bmatrix},$$

where η is the output process that propagates from the system after the incoming white noise ξ interacts with the latter. The process η satisfies the same time commutation relations as ξ.

Remark 2.1 Note that for notational expediency, in the monograph we will often not explicitly write the time dependence in the QSDEs describing the evolution of the system.

2.3.3 Transfer Function of Linear Dynamical Quantum Systems

Whether one is working with a linear quantum system in the quadrature form (2.5) or the mode form (2.12), as with classical linear systems, one can define a transfer function. We start with a system $G = (A, B, C, D)$ in the quadrature form (2.5). Following [18], we can define the Laplace transform of an adapted quantum stochastic process such as $w(t)$ as $w[s] = \int_0^\infty e^{-st} dw(t)$. The "particular" solution of (2.5), that does not depend on the initial condition $x(0) = x$, is given by

$$y_p(t) = \int_0^t C e^{A(t-s)} B dw(s) + Dw(t).$$

Now, taking Laplace transforms of both side of the equality gives

$$y_p[s] = (C(sI - A)^{-1}B + D)w[s],$$

and from this we can define the transfer function $\Xi_G(s)$ from $w[s]$ to $y_p[s]$ of G just as for a classical linear system,

$$\Xi_G[s] = C(sI - A)^{-1}B + D.$$

Similarly, if we are working in the mode form (2.12) we define $\mathcal{A}[s] = \int_0^\infty e^{-s\tau} d\mathcal{A}(\tau)$, $\mathcal{Y}[s] = \int_0^\infty e^{-s\tau} d\mathcal{Y}(\tau)$, $\xi[s] = \int_0^\infty e^{-s\tau}\xi(\tau)d\tau$, $\eta[s] = \int_0^\infty e^{-s\tau}\eta(\tau)d\tau$. Since $\mathcal{A}(t) = \int_0^t \xi(\tau)d\tau$ and $\mathcal{Y}(t) = \int_0^t \eta(\tau)d\tau$, note that $\mathcal{A}[s] = \xi[s]$ and $\mathcal{Y}[s] = \eta[s]$. Moreover, $\mathcal{A}^\#[s] = \mathcal{A}[s^*]^\#$ and analogous identities hold for $\mathcal{Y}[s]$, $\xi[s]$ and $\eta[s]$. The transfer function $\tilde{\Xi}_G$ from $(\mathcal{A}[s], \mathcal{A}^*[s])^\top$ to $(\mathcal{Y}[s], \mathcal{Y}^*[s])^\top$ or, equivalently, from $(\xi[s], \xi^*[s])^\top$ to $(\eta[s], \eta^*[s])^\top$, for G in the mode form $G = (\tilde{A}, \tilde{B}, \tilde{C}, \tilde{D})$ is given by

$$\tilde{\Xi}_G[s] = \tilde{C}(sI - \tilde{A})^{-1}\tilde{B} + \tilde{D}.$$

The two transfer functions Ξ_G and $\tilde{\Xi}_G$ for the quadrature and mode forms of G, respectively, are related by the identity

$$\Xi_G[s] = \Gamma_m^{-1}\tilde{\Xi}_G[s]\Gamma_m,$$

where Γ_m is as defined before.

In analogy with the classical case, the transfer function is important in analyzing the input–output behavior of linear quantum systems.

2.4 Inclusion of Idealized Static Transformations on Bosonic Fields: The Bogoliubov Transformation

When $H = 0$ and $L = 0$, a linear unitary transformation of the input field $\mathcal{A}(t)$ is realized,

$$\mathcal{Y}(t) = S\mathcal{A}(t).$$

Such a transformation preserves the differential commutation relations, that is,

$$\left[d\begin{bmatrix}\mathcal{Y}(t)\\ \mathcal{Y}(t)^{\#}\end{bmatrix}, d\begin{bmatrix}\mathcal{Y}(t)\\ \mathcal{Y}(t)^{\#}\end{bmatrix}^{*}\right] = \left[d\begin{bmatrix}\mathcal{A}(t)\\ \mathcal{A}(t)^{\#}\end{bmatrix}, d\begin{bmatrix}\mathcal{A}(t)\\ \mathcal{A}(t)^{\#}\end{bmatrix}^{*}\right] = \mathbb{K}_m dt, \tag{2.17}$$

where for a (column) vector of operators v and w,

$$[v, w^*] = vw^* - (w^{\#}v^{\top})^{\top}.$$

However, a unitary S is not the only matrix that can result in preservation of the differential commutation relations. More generally, one may consider a so-called *Bogoliubov transformation* implemented by a *Bogoliubov matrix* $W = \Delta(W_+, W_-)$ of the form

$$\begin{bmatrix}\mathcal{Y}'(t)\\ \mathcal{Y}'(t)^{\#}\end{bmatrix} = W\begin{bmatrix}\mathcal{A}(t)\\ \mathcal{A}(t)^{\#}\end{bmatrix}, \tag{2.18}$$

such that $\mathcal{Y}'(t)$ satisfies the same differential commutation relation as $\mathcal{A}(t)$. This is guaranteed by the definition of a Bogoliubov matrix, that will now be given.

Definition 2.1 Let $\cdot^{\flat}$ denote the involution defined via (2.11). A complex $2m \times 2m$ matrix W is a Bogoliubov matrix if it is of a doubled-up form $W = \Delta(W_+, W_-)$ for some complex $m \times m$ matrices W_+, W_-, and W is $\flat$-unitary,

$$W^{\flat}W = WW^{\flat} = I_{2m}.$$

Notice that, by definition, $W^{\flat} = W^{-1}$ and W reduces to a unitary matrix when $W_- = 0$. Linear transformations of the form (2.18) preserve the differential commutation relation, so that the output field $\mathcal{Y}'(t)$ is a valid output field. However, when $W_- \neq 0$ it cannot be modeled within the Hudson–Parthasarathy QSDE framework. That is, there does *not* exist a unitary solution $U(t)$ of the QSDE such that $\mathcal{Y}'(t)$ in (2.18) can be written as $\mathcal{Y}'(t) = U(t)^*\mathcal{A}(t)U(t)$.

From a physical point of view, when $W_- \neq 0$ the "pathology" discussed above can be understood from the fact that transformations such as (2.18) represent the action of an idealized infinite-bandwidth squeezing device acting on an incoming boson field $\mathcal{A}(t)$. Squeezing here means that the device reduces the quantum fluctuations on some quadratures of the incoming field $\mathcal{A}(t)$ while increasing them in other quadratures. We will shortly illustrate this point by demonstrating how a Bogoliubov transformation arises as the infinite-bandwidth limit of a degenerate parameteric amplifier. Such an idealized device is not physical since it produces uniform squeezing across the continuum of frequencies contained in $\mathcal{A}(t)$, a process that would require an infinite amount of energy. From a mathematical physics perspective, the linear transformation (2.18) represents a transformation between two fields $\mathcal{A}(t)$ and $\mathcal{Y}'(t)$. The former arises from a Fock space representation of field operators satisfying the canonical commutation relations $[\mathcal{A}(f), \mathcal{A}^*(g)] = \int_0^\infty f(s)^* g(s) ds$ with respect to vacuum state of the field, while the latter arises from a Fock space representation with respect to another state induced by W (e.g., a squeezed field state as postulated in [6, Chap. 6]); see, e.g., [20] for a more in-depth treatment of these representations in the case of squeezed field states. When $W_- \neq 0$ the two representations are not unitarily equivalent, since the transformation involved violates the necessary and sufficient conditions of the Shale's Theorem [1, Theorem 22.11].

Although the transformation (2.18) with W_- represents an idealized non-physical scenario, it is convenient to include in the modeling repertoire as it allows some simplification in describing the system. That is, it allows high-bandwidth dynamic linear devices with fast dynamics to be approximated as static linear transformations without any dynamics. This would be appropriate in instances where the fast device is connected to other linear devices with appreciably lower bandwidth (slower dynamics). We illustrate this in the next example.

Example 2.2 Recall the degenerate parametric amplifier (DPA) introduced in Sect. 1.5.3. Under the assumption of losslessness ($\kappa = 0$), the DPA is modeled as a single oscillator coupled to a single field with $S = I$, $\Omega_- = 0$, $\Omega_+ = \imath\epsilon$, $\epsilon > 0$, $C_- = \sqrt{\gamma}$, and $C_+ = 0$. The matrix $\tilde{A}$ is Hurwitz (all its eigenvalues are in the left half plane) and the system is stable (in the sense that the oscillator's mean photon number is bounded at all times) if we take $\epsilon < \gamma$ as we shall do from now on. The mode form transfer function of the system is

$$\tilde{\Xi}_{\mathrm{DPA}}[s] = \frac{1}{P(s)} \begin{bmatrix} s^2 - (\gamma^2 + \epsilon^2)/4 & -\epsilon\gamma/2 \\ -\epsilon\gamma/2 & s^2 - (\gamma^2 + \epsilon^2)/4 \end{bmatrix},$$

where $P(s) = (s + \gamma/2)^2 - \epsilon^2/4$. The zeros of P are the poles of the transfer function, namely $s = \pm\epsilon/2 - \gamma/2$. In the frequency domain, the output white noise field is

$$\eta[s] = \frac{1}{P[s]} \left(s^2 - (\gamma^2 + \epsilon^2)/4\right) \xi[s] - \frac{1}{2P(s)} \epsilon\gamma\xi^*[s],$$

where $\xi^*[s] = \xi[s^*]^*$. In terms of quadratures $\eta^q = \eta + \eta^*$ and $\eta^p = (\eta - \eta^*)/\imath$, we find that

$$\eta^q[s] = \Xi^q_{\mathrm{DPA}}[s]\eta^q[s], \quad \eta^p[s] = \Xi^p_{\mathrm{DPA}}[s]\eta^p[s],$$

where

$$\Xi^q_{\mathrm{DPA}}[s] = \frac{s - (\gamma + \epsilon)/2}{s + (\gamma - \epsilon)/2} = \frac{1}{\Xi^p_{\mathrm{DPA}}[s]}.$$

In this example we consider the idealized case where $\gamma, \epsilon \to \infty$ (approximated in practice by suitably large values of these parameters) while maintaining the ratio $\frac{\epsilon}{\gamma}$ to be constant; see [6, 10.2.1.g] for a more complete discussion. Rescaling $\gamma = k\gamma_0$ and $\epsilon = k\epsilon_0$ is equivalent to the substitution of γ by γ_0 and ϵ by ϵ_0 and the rescaling of s as $\frac{s}{k}$:

$$\tilde{\Xi}_{\mathrm{DPA}}[s; \gamma = k\gamma_0, \epsilon = k\epsilon_0] = \tilde{\Xi}_{\mathrm{DPA}}[s/k; \gamma_0, \epsilon_0].$$

We now take the limit $k \to \infty$ in which the cavity has an instantaneous response. Thus the cavity's internal dynamics are essentially eliminated. This yields the output $\eta[s]$ given by

$$\eta[s] = -\cosh(r_0)\,\eta[s] - \sinh(r_0)\,\eta^*[s],$$

where

$$r_0 = \ln \frac{\gamma_0 + \epsilon_0}{\gamma_0 - \epsilon_0},$$

as a Bogoliubov transformation of the input. In this limit the DPA can thus be viewed as a static device that outputs a squeezed white noise field from a vacuum white noise source. The transfer function becomes a constant Bogoliubov matrix that is independent of frequency. That is,

$$\begin{aligned}\tilde{\Xi}_{\mathrm{DPA\,static}}[s] &= \lim_{k\to\infty} \tilde{\Xi}_{\mathrm{DPA}}[s/k; \gamma_0, \epsilon_0] \\ &= -\Delta(\cosh r_0, \sinh r_0), \forall s \in \mathbb{C},\end{aligned}$$

and the quadrature transfer functions are the constant functions

$$\Xi^q_{\mathrm{DPA\,static}}[s] = -e^{r_0},\ \Xi^p_{\mathrm{DPA\,static}}[s] = -e^{-r_0}.$$

Therefore, we see that in the high bandwidth regime, the DPA can be approximated by a Bogoliubov transformation implemented by the Bogoliubov matrix

$$W = -\begin{bmatrix} \cosh(r_0) & \sinh(r_0) \\ \sinh(r_0) & \cosh(r_0) \end{bmatrix}.$$

When a linear quantum system (2.12) with $\tilde{D} = I$ is driven by vacuum noise that has passed through a device modelled as a static Bogoliubov transformation $W = \Delta(W_-, W_+)$, as shown in part (a) of Fig. 2.1, then the equation of motion for the system is again of the form (2.12) but now with the matrix $\tilde{D} = \Delta(S, 0)$ being replaced by $\tilde{D} = W$ in (2.13)–(2.16). In particular, we have that

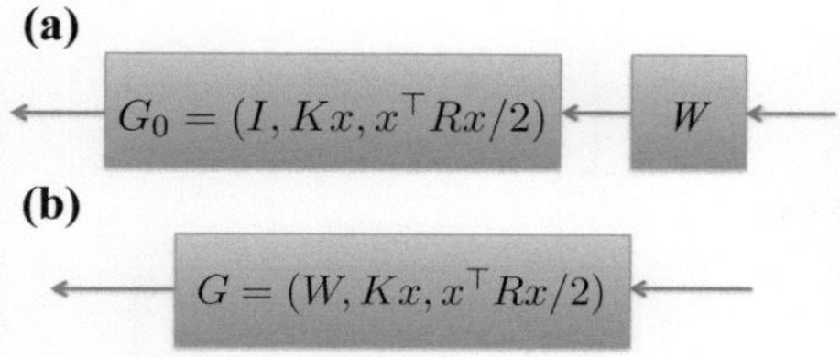

Fig. 2.1 **a** A static device performing a linear transformation W driving a linear quantum system G_0 with scattering matrix $S = I$, and **b** Equivalent representation of the cascaded system in part **a** as a single linear quantum system $G = (W, L, H)$

$$\tilde{A} = -\frac{1}{2}\tilde{C}^\flat \tilde{C} - \imath \tilde{\Omega}, \tag{2.19}$$

$$\tilde{B} = -\tilde{C}^\flat \tilde{D}, \tag{2.20}$$

$$\tilde{C} = \Delta(C_-, C_+), \tag{2.21}$$

$$\tilde{D} = W, \tag{2.22}$$

with

$$\tilde{\Omega} = -\imath \Delta(\imath \Omega_-, \imath \Omega_+).$$

Note the property that $\tilde{\Omega}^\flat = \tilde{\Omega}$. We denote systems of the type depicted in Fig. 2.1 part (a) with the shorthand notation $G = (W, L, H)$, as shown in part (b) of the figure.

2.4.1 *Completely Passive Linear Dynamical Quantum Systems*

This section will introduce a special class of linear quantum systems that shall be referred to as *completely passive* linear quantum stochastic systems, for reasons that will be explained below. This class will appear in several contexts in the monograph and are of interest in applications such linear optical quantum memories. Further discussions relating to this class of systems can be found in, e.g., [21–23] and in Chap. 3 of this monograph.

For $k = 1, \ldots, n$ let $a_k = \frac{q_k + \imath p_k}{2}$ be the annihilation operator for mode k and define $a = (a_1, \ldots, a_n)^\top$. The vector a satisfies the CCR

$$\begin{bmatrix} a \\ a^\# \end{bmatrix} \begin{bmatrix} a^* & a^\top \end{bmatrix} - \left(\begin{bmatrix} a^\# \\ a \end{bmatrix} \begin{bmatrix} a^\top & a^* \end{bmatrix} \right)^\top$$
$$= \mathrm{diag}(I_n, -I_n),$$

and note that,

$$\begin{bmatrix} a \\ a^{\#} \end{bmatrix} = \begin{bmatrix} \Sigma \\ \Sigma^{\#} \end{bmatrix} x,$$

where

$$\Sigma = \begin{bmatrix} \frac{1}{2} & \imath\frac{1}{2} & 0 & 0 & 0 & \dots & 0 \\ 0 & 0 & \frac{1}{2} & \imath\frac{1}{2} & 0 & \dots & 0 \\ \vdots & \ddots & \ddots & \ddots & \ddots & \ddots & \vdots \\ 0 & \dots & \dots & \dots & 0 & \frac{1}{2} & \imath\frac{1}{2} \end{bmatrix}.$$

Moreover, we also have $\begin{bmatrix} \Sigma \\ \Sigma^{\#} \end{bmatrix}^{-1} = 2\left[\, \Sigma^* \ \Sigma^\top \,\right]$ and, from the relation $\begin{bmatrix} \Sigma \\ \Sigma^{\#} \end{bmatrix} 2 \left[\, \Sigma^* \ \Sigma^\top \,\right] = I$, the identities,

$$\Sigma\Sigma^* = I/2 = \Sigma^{\#}\Sigma^\top, \ \Sigma\Sigma^\top = 0 = \Sigma^{\#}\Sigma^*. \tag{2.23}$$

Therefore,

$$x = \begin{bmatrix} \Sigma \\ \Sigma^{\#} \end{bmatrix}^{-1} \begin{bmatrix} a \\ a^{\#} \end{bmatrix} = 2\left[\, \Sigma^* \ \Sigma^\top \,\right] \begin{bmatrix} a \\ a^{\#} \end{bmatrix}.$$

We say that a linear quantum system $G = (S, Kx, (1/2)x^\top Rx)$ with n degrees of freedom is completely passive if $H = (1/2)x^\top Rx = (1/2)a^*\tilde{R}a + c$ and $L = Kx = \tilde{K}a$ for some real constant c, some complex $n \times n$ *Hermitian* matrix $\tilde{R}$, and some complex $m \times n$ matrix $\tilde{K}$. The terminology is motivated by physical considerations. Notice that the Hamiltonian H contains no terms of the form $c_1 a_j^2$, $c_2 a_k^{*2}$, $c_3 a_j a_k$ and $c_4 a_j^* a_k^*$ and, likewise, the coupling operator L contains no terms of the form $c_5 a_k^*$ (where $c_1, \dots, c_5$ denote arbitrary complex constants), with the indices j and k running over $1, \dots, n$. These terms correspond to interactions that require an external source of quanta (e.g., an external pump beam) to implement. Therefore, they cannot be realized using only passive optical components like phase shifters, beam splitters and mirrors, devices that will be discussed in more detail in Chap. 3. This is the physical motivation for referring to these class of systems as completely passive.

In a completely passive system, the Heisenberg picture evolution of the system annihilation operator $a(t)$ is not coupled to $a(t)^{\#}$ or to $\mathcal{A}(t)^{\#}$. In the complex mode form, the evolution of $a(t)$ is given by

$$\begin{aligned} da(t) &= (\imath\tilde{R} - (1/2)\tilde{K}^*\tilde{K})a(t)dt + \tilde{K}^* S d\mathcal{A}(t), \\ d\mathcal{Y}(t) &= \tilde{K}a(t)dt + Sd\mathcal{A}(t). \end{aligned}$$

The transfer function $\Xi[s]$ from $\mathcal{A}[s]$ to $\mathcal{Y}[s]$ given by,

$$\Xi[s] = \left(\tilde{K}\left(sI - (\imath\tilde{R} - (1/2)\tilde{K}^*\tilde{K})\right)^{-1}\tilde{K}^* + I\right) S,$$

has the special property that $\Xi[\imath\omega]$ is unitary for all for all ω for which $\imath\omega$ is not an imaginary eigenvalue of $\imath\tilde{R} - (1/2)\tilde{K}^*\tilde{K}$ [21]. It is easy to see that $\imath\tilde{R} - (1/2)\tilde{K}^*\tilde{K}$ cannot have any eigenvalues in the (open) right half plane, with a positive real part. Moreover, if this matrix has all its eigenvalues in the (open) left half plane (all eigenvalues have negative real part), meaning that the system is stable in a sense that will be discussed in Sect. 2.6, then $\Xi[s]$ is in fact lossless bounded real [22]. This means that $\Xi[s]$ is analytic for all s with $\Re\{s\} > 0$, $\Xi[\imath\omega]$ is unitary for all $\omega \in \mathbb{R}$ and, moreover, $\Xi[s]$ satisfies,

$$\Xi[s]^*\Xi[s] \leq I,$$

for all s with $\Re\{s\} > 0$.

2.5 Physical Realizability Conditions and Parameterizations for Linear Dynamical Quantum Systems

2.5.1 Physical Realizability Conditions for Linear QSDEs

We have now seen that the dynamics of linear quantum stochastic systems in the Heisenberg picture leads to an equation of motion in the form of linear QSDEs, i.e., as given by Eq. (2.5) in the quadrature form and (2.12) in the mode form. These equations can be viewed as a quantum analogue of the equations that describe classical linear stochastic systems. However, unlike classical linear stochastic systems, the matrices A, B, C, D in (2.5) or the matrices $\tilde{A}$, $\tilde{B}$, $\tilde{C}$, $\tilde{D}$ in (2.12) cannot be arbitrary, but are constrained by quantum mechanics so that these equations represent the evolution of a valid (open) quantum system. When the linear QSDE represents a valid quantum system, it is said that the QSDE or the system represented by the QSDE is physically realizable. To avoid unnecessary repetition, a formal definition will be given below for systems in the quadrature form. This definition can be directly adapted to systems in the mode form due to the one-to-one relationship between the two forms.

Definition 2.2 (*Physical realizability*) A system of linear quantum stochastic differential equations of the form (2.5) is said to be physically realizable if there exists a quadratic Hamiltonian $H = (1/2)x^\top Rx$, a coupling operator $L = Kx$, and a Bogoliubov matrix W such that

$$A = 2\mathbb{J}_n(R + \Im\{K^*K\}), \tag{2.24}$$

$$B = 2\imath\mathbb{J}_n[\,-K^* \;\; K^\top\,]W\Gamma_m, \tag{2.25}$$

$$C = P_m^\top \begin{bmatrix} K + K^\# \\ -\imath K + \imath K^\# \end{bmatrix}, \tag{2.26}$$

$$D = \Gamma_m^{-1} W \Gamma_m. \tag{2.27}$$

Essentially, the definition states that if a system of QSDEs (2.5) is physically realizable then there is a linear quantum system $G = (W, L, H)$ for which the Heisenberg evolution of x is given by (2.5). G is, of course, by definition, a physical system. We have the following result [16, 24, 25]

Theorem 2.1 *A system of linear quantum stochastic differential equations of the form (2.5) is physically realizable if and only if*

$$A\mathbb{J}_n + \mathbb{J}_n A^\top + B\mathbb{J}_m B^\top = 0, \tag{2.28}$$
$$\mathbb{J}_n C^\top + B\mathbb{J}_m D^\top = 0, \tag{2.29}$$
$$D\mathbb{J}_m D^\top = \mathbb{J}_m. \tag{2.30}$$

Proof Note that the condition (2.30) is equivalent to the statement that D is a symplectic matrix (hence it is invertible with D^{-1} also symplectic). Without loss of generality, we may reduce the proof to the case with $D = I$ (i.e., $W = I$) for the following reasons. Let $dy_o = D^{-1}Cx(t)dt + dw(t)$, then $y(t) = Dy_\mathrm{o}(t)$. Let $C' = D^{-1}C$ and consider the linear quantum stochastic system,

$$dx(t) = Ax(t)dt + Bdw(t),$$
$$dy_\mathrm{o}(t) = C'x(t)dt + dw(t).$$

We will show that this system is physically realizable if and only if (2.28)–(2.30) is satisfied with C replaced with C' and $D = I$. From this it easily follows that the original system (2.5) is physically realizable if and only if (2.28)–(2.30) holds. This is because y is a symplectic transformation of y_o, so if the system with output y is physically realizable so is the system with output y_o, and vice-versa. Thus for the remainder of the proof we identify C with C' and D with I.

To this end, let us first show that physical realizability of the system $G = (A, B, C, I)$ implies (2.28)–(2.30). Indeed, physical realizability implies that the canonical commutation relation for $x(t)$,

$$x(t)x(t)^\top - (x(t)x(t)^\top)^\top = 2\imath\mathbb{J}_n,$$

and the differential commutation relation for $y(t)$,

$$dy(t)dy(t)^\top - (dy(t)dy(t)^\top)^\top = 2\imath\mathbb{J}_n dt,$$

must hold for all times $t \geq 0$. This follows from the fact that the joint evolution of the system and the bosonic fields is unitary (given by the solution of a Hudson–Parthasarathy QSDE). Since $x(0) = x$ satisfies $xx^\top - (xx^\top)^\top = 2\imath\mathbb{J}_n$, for the commutation relation to hold for all $x(t)$, $t \geq 0$, we must have

$$d(x(t)x(t)^\top - (x(t)x(t)^\top)^\top) = 0.$$

The condition (2.28) follows from this by substituting the expression $dx(t) = Ax(t)dt + Bdw(t)$ and using the quantum Itō product rule, and the Itō table to explicitly compute the differential. By the same procedure and using the expression $dy(t) = Cx(t)dt + dw(t)$, we obtain (2.29). The details of the calculations can be found in [16, 26].

We now consider the converse implication, that (2.28) and (2.29) imply that there exist $R = R^\top \in \mathbb{R}^{2n\times 2n}$ and $K \in \mathbb{C}^{m\times 2n}$ such that (2.24)–(2.27) holds. Indeed, define $R = \frac{1}{4}(-\mathbb{J}_n A + A^\top \mathbb{J}_n)$ and $K = -\frac{\imath}{2}[\,0_{m\times m}\; I_{m\times m}\,]\Gamma_m^{-1} B^\top \mathbb{J}_n$; see [16, 26] for details of the origin of these expressions. Using these definitions and (2.28), we can directly calculate that $\mathbb{J}_n(R + \Im\{K^*K\}) = \mathbb{J}_n R + \mathbb{J}_n \Im\{K^*K\} = A$ and $2\imath\mathbb{J}_n[\,-K^*\; K^\top\,]\Gamma_n = B$. Using the last expression for B and (2.29), we then also get that $C = P_m^\top \begin{bmatrix} K + K^\# \\ -\imath K + \imath K^\# \end{bmatrix}$. This completes the proof. □

It should be noted from the above theorem that, remarkably, physical realizability is equivalent to preservation of canonical commutation relations for $x(t)$ and $y(t)$. The notion of physical realizability and the physical realizability constraints on the systems matrices were first introduced in [16, 26] for the quadrature form.

2.5.2 Parameterization of Linear Dynamical Quantum Systems

From the exposition in the preceding sections, we see that there are two sets of parameters that can be used to parameterize linear quantum stochastic systems with the same number of inputs and outputs. The first is the set of three parameters (W, K, R), with W a $2n \times 2n$ Bogoliubov matrix, K an $m \times 2n$ complex matrix, and R a real symmetric $2n \times 2n$ matrix, that describe the physical parameters of the system including the Hamiltonian H and the linear coupling operator L. The other parameterization is via the system matrices (A, B, C, D) that appear in the QSDE for the Heisenberg evolution (2.5) of the system.

The two parameterizations (W, K, R) and (A, B, C, D) are equivalent, and can be used interchangeably, according to which one may be more convenient for the purpose at hand. This is because of a bijective correspondence between the two parameterizations: to any given (W, K, R) parametrization there corresponds a unique (A, B, C, D) parameterization, and vice-versa. The (A, B, C, D) matrices for a given set of (W, K, R) matrices are given by (2.6)–(2.9) (recall that the class of symplectic matrices have a bijective correspondence with the class of Bogoliubov matrices). On the other hand, the (W, K, R) parameterization can be obtained from a given set of (A, B, C, D) matrices following the proof of Theorem 2.1.

2.5.3 Linear Dynamical Quantum Systems with Less Outputs Than Inputs

In general, not all the outputs of a system may be of interest or can be observed or utilized. In many situations, one may only be interested in certain pairs of the output field quadratures in $y(t)$ [16]. In the most general scenario, one can consider $y(t)$ having an even dimension $2n_y < 2m$ and D is a $2n_y \times 2m$ matrix satisfying $D\mathbb{J}_m D^\top = \mathbb{J}_{n_y}$. That is, we can consider outputs of the form:

$$dy(t) = Cx(t)dt + Ddw(t), \tag{2.31}$$

with $C \in \mathbb{R}^{2n_y \times 2n}$, $D \in \mathbb{R}^{2n_y \times 2m}$ with $n_y < m$. Generalizing the notion discussed in Sect. 2.5.1 (for the case where there are as many outputs as there are inputs), a linear quantum system with output (2.31) is physically realizable if and only if there exist matrices $C' \in \mathbb{R}^{2(m-n_y)\times 2n}$ and $D' \in \mathbb{R}^{2(m-n_y)\times 2m}$ such that the system

$$\begin{aligned} dx(t) &= Ax(t)dt + Bdw(t);\ x(0) = x, \\ dy'(t) &= \begin{bmatrix} C \\ C' \end{bmatrix} x(t)dt + \begin{bmatrix} D \\ D' \end{bmatrix} dw(t), \end{aligned} \tag{2.32}$$

is physically realizable with the same number of inputs and outputs. Therefore the matrices A, B, $[\,C^\top\ (C')^\top\,]^\top$, and $[\,D^\top\ (D')^\top\,]^\top$ satisfy the constraints (2.28)–(2.30) when C and D in (2.29) and (2.30) are replaced by $[\,C^\top\ (C')^\top\,]^\top$ and $[\,D^\top\ (D')^\top\,]^\top$, respectively. We can easily obtain the following necessary and sufficient condition for physical realizability of linear quantum systems with less outputs than inputs see, e.g., [25].

Theorem 2.2 *A linear quantum stochastic system with less outputs than inputs is physically realizable if and only if*

$$A\mathbb{J}_n + \mathbb{J}_n A^\top + B\mathbb{J}_m B^\top = 0, \tag{2.33}$$
$$\mathbb{J}_n C^\top + B\mathbb{J}_m D^\top = 0, \tag{2.34}$$
$$D\mathbb{J}_m D^\top = \mathbb{J}_{n_y}. \tag{2.35}$$

Proof The necessity of (2.33)–(2.35) is immediate from the definition of physically realizable systems with less outputs than inputs given above and the physical realizability conditions for systems with the same number of input and outputs. For the sufficiency, first note that for D satisfying (2.35), it follows from a construction used in the proof of [27, Lemma 6] that a matrix $D' \in \mathbb{R}^{2(m-n_y)\times 2m}$ can be constructed such that the matrix $\tilde{D} = [\,D^\top\ (D')^\top\,]^\top$ is symplectic. Now, define $C' = D'\mathbb{J}_m B^\top \mathbb{J}_n$ and $\tilde{C} = [\,C^\top\ (C')^\top\,]^\top$. Consider now a system $\tilde{G}$ with an equal number of inputs and outputs and system matrices $(A, B, \tilde{C}, \tilde{D})$. From the physical realizability conditions (2.33)–(2.35) and the definition of C' and $\tilde{C}$, it follows that $\tilde{G}$ satisfies (2.28)–(2.30) and is therefore physically realizable. It now follows from definition that the original system with output $y(t)$, of smaller dimension than $w(t)$, is physically realizable. This completes the proof. □

Thus the physical realizability constraints for systems with the same number of outputs as inputs, or less inputs than outputs, have essentially the same form, so the following corollary is immediate.

Corollary 2.1 *An arbitrary linear quantum stochastic system is physically realizable if and only if*

$$A\mathbb{J}_n + \mathbb{J}_n A^\top + B\mathbb{J}_m B^\top = 0, \tag{2.36}$$

$$\mathbb{J}_n C^\top + B\mathbb{J}_m D^\top = 0, \tag{2.37}$$

$$D\mathbb{J}_m D^\top = \mathbb{J}_{n_y}. \tag{2.38}$$

Remark 2.2 All the results relating to linear quantum systems in this chapter depart from the assumption that the position and momentum operators satisfy the default CCR given by (2.3). However, these operators may be defined differently and satisfy a different CCR. For instance, one may wish to take as the position and momentum pair for the j-th oscillator $q_j' = q_j/\sqrt{2}$ and $p_j' = p_j/\sqrt{2}$ satisfying $[q_j', p_j'] = \imath$, and defining $x' = (q_1', p_1', q_2', p_2', \ldots, q_n', p_n')^\top$ as the canonical internal operator that will satisfy the CCR $[x', (x')^\top] = \imath\mathbb{J}_n$. More generally, one may take $x' = \mathsf{T}x$ satisfying $[x', (x')^\top] = 2\imath\mathsf{T}\mathbb{J}_n\mathsf{T}^\top$, with an invertible real $2n \times 2n$ matrix T that would typically be diagonal. All of the structural results for linear quantum systems that have been obtained can be adapted to this different choice of CCR by making the substitutions (given here for the real quadrature form):

$$\mathsf{T}A\mathsf{T}^{-1} \to A,\ \mathsf{T}B \to B,\ C\mathsf{T}^{-1} \to C,$$
$$\mathsf{T}\mathbb{J}_n\mathsf{T}^\top \to \mathbb{J}_n,\ \mathsf{T}^{-\top}R\mathsf{T}^{-1} \to R,\ K\mathsf{T}^{-1} \to K,$$

while the system matrix D and scattering matrix S remain the same.

2.6 Stability of Linear Quantum Systems

As with classical linear systems, one can define a notion of stability for linear quantum systems. We will do this for systems in quadrature form but analogous notions and results also hold for systems in complex mode form. Consider a linear quantum system in quadrature form driven by fields that are all in the vacuum state. Then quantum expectation of $\langle x(t)\rangle$ is given by the ODE

$$\langle \dot{x(t)}\rangle = A\langle x(t)\rangle.$$

The system is said to be:

- *Asymptotically stable* or simply *stable* if $|\langle x(t)\rangle| \to 0$ as $t \to \infty$ for any initial state of the system.
- *Marginally stable* if $|\langle x(t)\rangle|$ does not go to 0 as $t \to \infty$, but remains bounded at all times $t \geq 0$ for any initial state of the system.
- *Unstable* if there exists some initial state of the system such that $|\langle x(t)\rangle| \to \infty$ as $t \to \infty$.

From standard linear state-space theory, we immediately have necessary and sufficient conditions for a system to be (asymptotically) stable, marginally stable and unstable. These are:

- A system is stable if and only A is Hurwitz. That is, all its eigenvalues have a negative real part.
- A system is marginally stable if and only if A has one or more eigenvalues on the imaginary axis with geometric multiplicity equal to its algebraic multiplicity while all other remaining eigenvalues have negative real part.
- A system is unstable if and only if A has at least one eigenvalue with positive real part or one eigenvalue on the imaginary axis with geometric multiplicity less than its algebraic multiplicity.

One can associate a natural energy functional to a linear quantum system. This energy functional $\mathcal{E}(x(t))$ is defined by

$$\mathcal{E}(x(t)) = \sum_{j=1}^{n} \langle q_j(t)^2 + p_j(t)^2 \rangle = \mathrm{Tr}(\langle P(t)\rangle),$$

where $P(t) = (1/2)\left(x(t)x(t)^\top + (x(t)x(t)^\top)^\top\right)$. Note that $P(t) = P(t)^\top$ and $\langle P(t)\rangle \geq 0$. Suppose that the system has some initial energy but not receiving any energy from the external input fields (all are in the vacuum state). Then $P(t)$ satisfies the Lyapunov differential equation:

$$\dot{P}(t) = AP(t) + P(t)A^\top + BB^\top,$$

with initial condition $P(0) = (1/2)\langle xx^\top + (xx^\top)^\top\rangle$. Notice that when the system is stable, A is Hurwitz and $\lim_{t\to\infty} P(t) = P_\infty$ exists, is unique, and given by the unique solution to the algebraic Lyapunov equation,

$$AP_\infty + P_\infty A^\top + BB^\top = 0.$$

Thus, when the system is stable the system's energy in the long time settles to a steady-state value. This is a necessity in applications as unbounded energy growth leads to the system being damaged or destroyed, so ensuring stability is an important goal in the control of linear quantum systems.

2.7 Gaussian States

The oscillators and the bosonic fields to which they can be coupled are defined on an infinite-dimensional Hilbert space. For the oscillator this is the Hilbert space $\otimes_{j=1}^{n} L^2(\mathbb{R}; \mathbb{C}) \equiv L^2(\mathbb{R}^n; \mathbb{C})$, where n is the number of oscillators, while for the bosonic fields it is the Fock space $\mathcal{F}_m = \Gamma_s(L^2(\mathbb{R}_+; \mathbb{C}^m))$, where m is the multiplicity space of the field, i.e., the number of distinct bosonic fields. As such, on these fields there exist observables that have a continuous spectra which are the quantum analogue of real-valued random variables taking values in $\mathbb{R}$. Moreover, one can define states on the oscillator or field that are the quantum analogue of the Gaussian states in classical probability theory.

Gaussian states are defined differently for single-mode oscillators (a collection of oscillators with a single well-defined frequency) and bosonic fields, since the latter supports an infinite number of modes (in frequency). In this section, we provide a definition of Gaussian quantum states, beginning with the case of a collection of single-mode oscillators and then proceeding to define Gaussian states for bosonic fields. The treatment here closely follows [18, 19].

Gaussian states have a special status with respect to linear quantum systems. Due to their linear dynamics, linear quantum systems have the property that they preserve Gaussian states. This means that if the oscillators are initialized in a Gaussian state and input fields are in a Gaussian state, then the joint state of the oscillators and output fields will be in Gaussian state at all times in the system's evolution. This will be discussed in more detail in Sect. 4.3.2 when quantum Kalman filters for linear quantum systems are introduced.

2.7.1 Gaussian State of a Collection of Single-Mode Oscillators

Consider a collection of n independent harmonic oscillators on the collective Hilbert space $L^2(\mathbb{R}^n; \mathbb{C}) = L^2(\mathbb{R}; \mathbb{C})^{\otimes n}$, with annihilation operators $a_1, a_2, \ldots, a_n$. As before, we let $a = [\, a_1 \; a_2 \; \ldots \; a_n \,]^\top$, and introduce $\mathcal{S}_n$ to denote the set of all annihilation operators for n independent oscillators, so $a \in \mathcal{S}(n)$. A quantum state on $L^2(\mathbb{R}^n; \mathbb{C})$ is said to be Gaussian if

$$\left\langle \exp\left(\imath \left\{ [\, u^* \; u^\top \,] \begin{bmatrix} a \\ a^\# \end{bmatrix} \right\} \right) \right\rangle = \exp\left\{ -{}^1\!/\!{}_2 [\, u^* \; u^\top \,] F \begin{bmatrix} u \\ u^\# \end{bmatrix} + \imath [\, u^* \; u^\top \,] \begin{bmatrix} \langle a \rangle \\ \langle a^\# \rangle \end{bmatrix} \right\},$$

for all $u \in \mathbb{C}^n$ and for a positive semidefinite Hermitian matrix $F \geq 0$ given by

$$F = \left\langle \begin{bmatrix} a - \langle a \rangle \\ (a - \langle a \rangle)^{\#} \end{bmatrix} \begin{bmatrix} (a - \langle a \rangle)^* & (a - \langle a \rangle)^\top \end{bmatrix} \right\rangle = \begin{bmatrix} I + N^\top & M \\ M^* & N \end{bmatrix} \tag{2.39}$$

with

$$N_{jk} = \langle (a_j - \langle a_j \rangle)^* (a_k - \langle a_k \rangle) \rangle, \quad M_{jk} = \big\langle (a_j - \langle a_j \rangle)(a_k - \langle a_k \rangle) \big\rangle. \tag{2.40}$$

We have the properties that $N = N^*$ and $M = M^\top$, and $F \geq 0$ implies that $N \geq 0$. F is referred to as the "covariance matrix" of the Gaussian state. Note that the definition of a Gaussian state given above is based on a quantum generalization of the characteristic function of a classical Gaussian probability distribution, which uniquely identifies the distribution.

Let us now consider zero-mean states satisfying $\langle a_j \rangle = 0$ for all j. The joint vacuum state of the n oscillators is a special state corresponding to $N = 0$, $M = 0$, with $F = F_{\text{vac}}$ given by

$$F_{\text{vac}} = \begin{bmatrix} I & 0 \\ 0 & 0 \end{bmatrix}. \tag{2.41}$$

Here the subscript vac stands for "vacuum."

For a fixed $N \geq 0$, M is constrained by the condition that $F \geq 0$. When $n = 1$, N and M are scalars and the condition $F \geq 0$ is equivalent to $N \geq 0$ with $|M|^2 \leq N(N+1)$. More generally, we can perform the diagonalization $V^* N V = \text{diag}(N_1, \ldots, N_n)$ for some unitary V, in which case new modes $a' = Va$ can be defined. Then N_j can be interpreted as the average number of quanta in the mode a'_j. Note that, in general, one cannot simultaneously diagonalize N and M.

2.7.1.1 Generalized Araki–Woods Representation

Given a zero-mean Gaussian state characterized by the covariance matrix F in (2.39), modes having that state can be constructed via canonical transformations of vacuum modes. That is, we will show that for any Gaussian state for which a has covariance F given by (2.39), there exists a $2n \times 4n$ matrix $\tilde{S}_0$ such that

$$\begin{bmatrix} a \\ a^{\#} \end{bmatrix} = \tilde{S}_0 \begin{bmatrix} a_0 \\ a_0^{\#} \end{bmatrix} \tag{2.42}$$

where $a_0 = \begin{bmatrix} a_1 \\ a_2 \end{bmatrix} \in \mathcal{S}(n+n)$ has vacuum statistics, and

$$\tilde{S}_0 \tilde{S}_0^\flat = I. \tag{2.43}$$

Indeed, we will construct $\tilde{S}_0 = \Delta(E_-^0, E_+^0)$ for some $n \times 2n$ matrices E_-^0, E_+^0. This can be viewed as a generalization of a construction by Araki and Woods [28] for non-squeezed thermal states, see [20, 29].

2.7.1.2 Construction of Araki–Woods Vacuum Representation

Step 1: Diagonalize N. Determine a unitary matrix $V \in \mathbb{C}^{n\times n}$ such that $V^*NV = \text{diag}\,(N_1, \ldots, N_n)$, with the eigenvalues ordered such that $N_1 \geq \cdots \geq N_n \geq 0$. We can restrict our attention to the case with N diagonalized in this manner.
Step 2: Ignore zero eigenvalues. Take the first n_+ eigen values to be strictly positive, with the remaining $n_0 = n - n_+$ being zero. With respect to the eigen decomposition $\mathbb{C}^n = \mathbb{C}^{n_+} \oplus \mathbb{C}^{n_0}$, decompose F as

$$F = \begin{bmatrix} I + N_{++} & 0 & M_{++} & M_{+0} \\ 0 & I & M_{0+} & M_{00} \\ M_{++}^\top & M_{+0}^\top & N_{++} & 0 \\ M_{0+}^\top & M_{00}^\top & 0 & 0 \end{bmatrix}.$$

However, if a positive semidefinite matrix has a zero on a diagonal then every entry on the corresponding row and column must vanish[3] so that in fact,

$$F \equiv \begin{bmatrix} I + N_{++} & 0 & M_{++} & 0 \\ 0 & I & 0 & 0 \\ M_{++}^\top & 0 & N_{++} & 0 \\ 0 & 0 & 0 & 0 \end{bmatrix}.$$

We can thus restrict our attention to the case where N is diagonal and positive definite (thus invertible).
Step 3: Explicit Construction. We begin by noting the constraint $I + N \geq MN^{-1}M^*$, which follows from

$$0 \leq \begin{bmatrix} I & -MN^{-1} \\ 0 & 0 \end{bmatrix} F \begin{bmatrix} I & -MN^{-1} \\ 0 & 0 \end{bmatrix}^* = \begin{bmatrix} I + N - MN^{-1}M^* & 0 \\ 0 & 0 \end{bmatrix}.$$

Now, introduce the matrices X, Y, Z defined by:

$$X = \sqrt{I + N - MN^{-1}M^*},$$
$$Y = \sqrt{N} = \text{diag}\left(\sqrt{N_1}, \ldots, \sqrt{N_n}\right),$$

[3]To see why this is the case, consider a complex $k \times k$ matrix $E \geq 0$ with $E_{11} = 0$. Taking $u(t) = (tx_1, x_2, \ldots, x_k)^\top$ we have that $0 \leq u(t)^* Eu(t) = 2t\text{Re}\{\sum_{j>1} x_1^* E_{1j} x_j + \sum_{j,k>1} x_j^* E_{jk} x_k\}$. However, for this inequality to hold for all real t it must be that $\Re\{\sum_{j>1} x_1^* E_{1j} x_j\} = 0$. Now, replacing t with $\imath t$ shows that $\Im\{\sum_{j>1} x_1^* E_{1j} x_j\}$ also vanishes. As this is true for any x_j, it must be the case that $E_{1j} = E_{j1}^* = 0$ for all $j > 1$.

$$Z = MY^{-1}.$$

Note that $Y = Y^\top$ and from $Z = MY^{-1}$ we have that $YZ^\top = M^\top = M = ZY = ZY^\top$. The matrices satisfy

$$XX^* - YY^* + ZZ^* = I \text{ and } YZ^\top = ZY^\top. \tag{2.44}$$

Now take b_1 and b_2 to be any two distinct modes in $\mathcal{S}(n)$ (hence they are commuting). Fix the state to be the joint vacuum state for these two modes. Then we have the representation

$$a = Xb_1 + Yb_2^\# + Zb_2. \tag{2.45}$$

Indeed, it can be straightforwardly verified that

$$\langle a^\# a^\top \rangle = Y^2 = N,$$
$$\langle aa^\top \rangle = ZY^\top = ZY = M.$$

We have thus constructed the representation (2.42) with $\tilde{S}_0 = \Delta(E_-^0, E_+^0)$ and

$$E_-^0 = \left[X \; Z \right], \quad E_+^0 = \left[0 \; Y \right].$$

Property (2.43) follows from (2.44).

2.7.2 *Gaussian States of the Field and Their Fock Space Representation*

Consider m continuous-mode bosonic fields indexed by $j = 1, 2, \ldots, m$ with annihilation field operators $b_j(t)$ satisfying the field commutation relations $[b_j(t), b_k(t')^*] = \delta_{jk}\delta(t - t')$ and $[b_j(t), b_k(t')] = 0$. Let us introduce the shorthand notation,

$$\breve{b} = \begin{bmatrix} b \\ b^\# \end{bmatrix}.$$

A Gaussian state of a continuous-mode bosonic field is state with characteristic function of the form

$$\left\langle \exp\left(\imath \int_0^\infty \breve{u}(t)^* \breve{b}(t)\, dt\right)\right\rangle$$
$$= \exp\left(-1/2 \int_0^\infty \breve{u}(t)^* F \breve{u}(t)\, dt + \imath \int_0^\infty \breve{u}(t)^* \langle \breve{b}(t) \rangle dt\right),$$

for any $u \in L^2(\mathbb{R}_+; \mathbb{C}^m)$. Here, as in the single-mode case, F is a $2m \times 2m$ Hermitian matrix that is again of the form (2.39) with the entries of N and M specified by the correlation functions

$$\left\langle \left(b_j(t) - \langle b_j(t) \rangle \right)^* \left(b_k(t') - \langle b_k(t') \rangle \right) \right\rangle = N_{jk}\, \delta(t - t'),$$
$$\left\langle \left(b_j(t) - \langle b_j(t) \rangle \right) \left(b_k(t') - \langle b_k(t') \rangle \right) \right\rangle = M_{jk}\, \delta(t - t').$$

That is,

$$\left\langle \left(\breve{b}(t) - \langle \breve{b}(t) \rangle \right) \left(\breve{b}(t') - \langle \breve{b}(t') \rangle \right)^* \right\rangle \equiv F\delta\left(t - t'\right). \tag{2.46}$$

Following the treatment in [18], we now consider a zero-mean Gaussian state $\omega_{N,M}(\cdot) = \langle \cdot \rangle$ of n continuous-mode bosonic fields with $\langle \breve{b}(t) \rangle = 0$ for $t \geq 0$. An important special case is a joint vacuum Gaussian state, when $N = 0$ and $M = 0$. This is the state when the fields are all empty (contain no photons). The vacuum state for the field is completely specified by the characteristic function,

$$\left\langle \exp\left(\imath \int_0^\infty \breve{u}\,(t)^*\, \breve{b}\,(t)\, dt \right) \right\rangle_{\text{vac}} = \exp\left(-1/2 \int_0^\infty \breve{u}\,(t)^*\, F_{\text{vac}} \breve{u}\,(t)\, dt \right),$$

for any $u \in L^2(\mathbb{R}_+; \mathbb{C}^m)$. For concreteness, let us use this definition to show that any component $w_j(t)$ of the quadrature noise $w(t)$ is a standard Wiener process under the vacuum state, as claimed in Sect. 2.3.2. First, from its definition we immediately have that (i) $w_j(0)^2 = 0$ and $\langle w_j(t) \rangle = 0$. Now, take $u_\lambda = \lambda e_j 1_{[t_1,t_2]}$ with $\lambda \in \mathbb{R}$ and $0 \leq t_1 \leq t_2$ (recall that e_j is a unit column vector with all entries 0 except for a 1 as the j-th component). From the definition of a vacuum state given above,

$$\begin{aligned}
&\left\langle \exp\left(\imath \lambda (w_j(t_2) - w_j(t_1)) \right) \right\rangle_{\text{vac}} \\
&= \left\langle \exp\left(\imath \int_{t_1}^{t_2} dw_j(t) \right) \right\rangle_{\text{vac}} \\
&= \left\langle \exp\left(\imath \int_0^\infty u_\lambda(t)^\top (b(t) + b(t)^\#) dt \right) \right\rangle_{\text{vac}} \\
&= \left\langle \exp\left(-1/2 \int_0^\infty \|u_\lambda(t)\|^2 dt \right) \right\rangle_{\text{vac}} \\
&= \exp\left(-1/2 \lambda^2 (t_2 - t_1) \right), \ \forall \lambda \in \mathbb{R}.
\end{aligned}$$

Therefore, (ii) the increment $w_j(t_2) - w_j(t_1)$ has the characteristic function of a zero-mean Gaussian random variable with variance $t_2 - t_1$. Moreover, by adaptedness we have that (iii) $\langle (w(t_4) - w(t_3))(w(t_2) - w(t_1)) \rangle = \langle (w(t_4) - w(t_3)) \rangle \langle (w(t_2) - w(t_1)) \rangle = 0$ for all $t_1 < t_2 < t_3 < t_4$, so that increments at disjoint time intervals are uncorrelated. Since $[w_j(t), w_i(s)] = 0$ for all $s, t \geq 0$, from (i)–(iii) we conclude that $w_j(t)$ has precisely the statistics of a standard Wiener process and is therefore a Fock space realization of the latter.

For field operators $b_1(t), b_2(t), \ldots, b_m(t)$ in a zero-mean jointly Gaussian state, we define $\mathcal{B}_j(f) = \int_0^\infty f(s)^* b_j(s)ds$ for any $f \in L^2(\mathbb{R}_+; \mathbb{C})$ and its adjoint process $\mathcal{B}_j^*(f) = \int_0^\infty f(s)b_j(s)^* ds$ as they are well-defined and more regular mathematical objects, and can be manipulated using the Hudson–Partharathy quantum stochastic calculus. We introduce these notations to distinguish them from the earlier notations $\mathcal{A}_j(f)$ and $\mathcal{A}_j^*(f)$ that were assigned for the special case of a vacuum field state, if field j is in the vacuum state then one can simply set $\mathcal{B}_j(f)$ to be $\mathcal{A}_j(f)$. The operators $\mathcal{B}_j(f)$ and $\mathcal{B}_j^*(f)$ satisfy the same commutation relations as $\mathcal{A}(f)$ and $\mathcal{A}^*(f)$, $[\mathcal{B}_j(f), \mathcal{B}_k^*(g)] = \delta_{jk} \int_0^\infty f(s)^* g(s)ds$, and concrete realizations of $\mathcal{B}(f) = (\mathcal{B}_1(f), \mathcal{B}_2(f), \ldots, \mathcal{B}_m(f))^\top$ and $\mathcal{B}(f)^\# = (\mathcal{B}_1^*(f), \mathcal{B}_2^*(f), \ldots, \mathcal{B}_m^*(f))^\top$ on a suitable Hilbert space are dependent on the state of the field. However, for arbitrary Gaussian states one can relate the associated realizations of $\mathcal{B}(f)$ and $\mathcal{B}^*(f)$ to the vacuum state representation of these operators, via the generalized Araki–Woods representation from the previous section. In particular, any operator $\mathcal{B}(f)$ associated with a zero-mean Gaussian state $\omega_{N,M}$ has a Fock space representation of the form

$$\mathcal{B}(f) = X\mathcal{A}_1(f) + Y\mathcal{A}_2(f)^\# + Z\mathcal{A}_2(f), \tag{2.47}$$

with complex $m \times m$ matrices X, Y, and Z as constructed in the previous section, determined by the values of the parameters N and M of $\omega_{N,M}$, and where $\mathcal{A}_1(f)$ and $\mathcal{A}_2(f)$ are two independent vacuum annihilation processes that can each be realized on a distinct copy of the Fock space $\mathcal{F}_m$. Note that by construction X, Y, Z guarantee that the commutation relations $[\mathcal{B}_j(f), \mathcal{B}_k^*(g)] = \delta_{jk} \int_0^t f(s)^* g(s)ds$ hold for any $f, g \in L^2(\mathbb{R}_+; \mathbb{C})$.

The Itō table for a jointly Gaussian state of the fields can be directly constructed by exploiting the generalized Araki–Woods representation (2.47) and the vacuum Itō table. Recall that $\mathcal{A}_1(f)$ and $\mathcal{A}_2(f)$ in (2.47) are vacuum representations on distinct copies of the Fock space $\mathcal{F}_m$. The extended Itō table for the integrated operators $\mathcal{B}_j(t) = \mathcal{B}_j(1_{[0,t]})$ and $\mathcal{B}_j^*(t) = \mathcal{B}_j^*(1_{[0,t]})$ when the field is in an arbitrary Gaussian state is then

$$\begin{array}{c|ll} \times & d\mathcal{B}_k^* & d\mathcal{B}_k \\ \hline d\mathcal{B}_j & (\delta_{jk} + N_{kj})dt & M_{jk}dt \\ d\mathcal{B}_j^* & M_{kj}^* dt & N_{jk}dt. \end{array} \tag{2.48}$$

Note that in general Gaussian states, the counting process Λ need not be defined. We can also define a QSDE of the Hudson–Parthasarathy type but in which the vacuum field operators $\mathcal{A}$ and $\mathcal{A}^\#$ are replaced by field operators $\mathcal{B}$ and $\mathcal{B}^\#$ corresponding to a non-vacuum zero-mean jointly Gaussian state of the field. This yields the QSDE (without the counting process Λ),

$$dU(t) = \big(-(\imath H + 1/2(L^*(I+N^\top)L + L^\top N L^\# - L^* M L^\# - L^\top M^\# L))dt + d\mathcal{B}(t)^* L - L^* d\mathcal{B}(t)\big)U(t) \tag{2.49}$$

with initial condition $U(0) = I$. As with the vacuum case, after interaction with the system, $\mathcal{B}$ is transformed to $\mathcal{B}^{\text{out}}$ according to $\mathcal{B}^{\text{out}}(t) = U(t)^*\mathcal{B}(t)U(t)$. Using (2.47), the QSDE can be expressed in terms of the vacuum operator $\mathcal{A}(t) = [\, \mathcal{A}_1(t)^\top \; \mathcal{A}_2(t)^\top \,]^\top$,

$$dU(t) = (-(\imath H + L_{N,M}^* L_{N,M}/2)dt + d\mathcal{A}(t)^* L_{N,M} - L_{N,M}^* d\mathcal{A}(t))U(t), \tag{2.50}$$

with

$$L_{N,M} = \begin{bmatrix} X^* L \\ Z^* L - Y^\top L^\# \end{bmatrix}.$$

Example 2.3 An example of a non-vacuum zero-mean Gaussian state of a single bosonic field is a *squeezed vacuum* field state where the parameters $M \neq 0$ and $N \neq 0$ satisfy the identity $|M|^2 = N(N+1)$. Such a field can be approximately produced in the laboratory using the DPA that was treated in Example 2.2. In the limit discussed in that example, where the DPA becomes idealized with an infinite bandwidth, its output is an ideal squeezed Gaussian field state satisfying the quantum Itō rule with

$$N = \sinh^2 r_0 = \frac{4\kappa_0\epsilon_0}{(\kappa_0^2 - \epsilon_0^2)^2},$$

$$M = \cosh r_0 \sinh r_0 = \frac{2\kappa_0\epsilon_0(\kappa_0^2 - \epsilon_0^2)}{(\kappa_0^2 - \epsilon_0^2)^2}.$$

It can be easily verified that the parameters M and N above satisfy $|M|^2 = N(N+1)$. In practice, treating the output of the DPA as a squeezed bosonic field will be valid as long as it is driving a quantum system with sufficiently slower dynamics (sufficiently lower bandwidth) than the DPA itself.

2.7.3 Coherent States

The Gaussian states that we have discussed so far have a common feature, they all have zero mean. We shall now discuss an important class of nonzero-mean Gaussian states that model the output beam from a laser and can facilitate quantum feedback control. These are the coherent states, and we will now give describe the coherent states of single-mode oscillators and bosonic fields.

2.7.4 Coherent States of a Single-Mode Oscillator

A coherent state of a single-mode oscillator is a pure state of the oscillator that is indexed by a complex number α and denoted by $|\alpha\rangle$. The coherent state $|\alpha\rangle$ is a normalized eigenvector of the oscillator's annihilation operator corresponding to the eigenvalue α, $a|\alpha\rangle = \alpha|\alpha\rangle$. In a coherent state, the annihilation operator has mean $\langle a\rangle = \langle\alpha|a|\alpha\rangle = \alpha$ and mean photon number of $\langle\alpha|a^*a|\alpha\rangle = |\alpha|^2$, while the complex "covariance" matrix F defined in (2.39) corresponds to that of a vacuum state, with $N = M = 0$. Therefore, the vacuum state of an oscillator is a particular coherent state with $\alpha = 0$.

A coherent state with $\alpha \neq 0$ can be generated form the vacuum state $|0\rangle$ by applying a unitary displacement operator $D(\alpha) = \exp(\alpha a^* - \alpha^* a)$, $|\alpha\rangle = D(\alpha)|0\rangle$. We have that $D(\alpha)^* a D(\alpha) = \alpha + a$, and we notice the duality between the Schrödinger picture state transformation $|0\rangle \stackrel{D(\alpha)}{\mapsto} |\alpha\rangle$ and Heisenberg picture operator transformation $b \stackrel{D(\alpha)^*(\cdot)D(\alpha)}{\mapsto} b + \alpha I$, $\langle\alpha|a|\alpha\rangle = \langle 0|a + \alpha|0\rangle$ for all $\alpha \in \mathbb{C}$. Therefore, α is also referred to as the amplitude of the coherent state.

2.7.5 Coherent States of a Bosonic Field

Coherent states of a bosonic field are directly related to the exponential vectors and can be interpreted in analogous way to the coherent states of single-mode oscillator. The coherent state of a single bosonic field is a pure state of the field that is indexed by a function $f \in L^2(\mathbb{R}_+; \mathbb{C})$ and denoted by $|f\rangle$. It is simply a normalized version of the exponential vector $e(f)$,

$$|f\rangle = \frac{e(f)}{\|e(f)\|} = e(f)\exp\left(-\|f\|^2/2\right).$$

It follows that that $A(g)|f\rangle = \langle g, f\rangle|f\rangle$, thus $|f\rangle$ is an eigenvector of $A(g)$ for any $g \in L^2(\mathbb{R}_+; \mathbb{C})$ corresponding to the eigenvalue $\langle g, f\rangle$. More formally, in terms of the annihilation field operator $b(t)$ (satisfying $[b(t), b(t')^*] = \delta(t-t')$ and $[b(t), b(t')] = 0$), we have that $b(t)|f\rangle = f(t)|f\rangle$ for all $t \geq 0$, mirroring the property of coherent states of a single-mode oscillator.

The analogue of the displacement operator $D(\alpha)$ for bosonic fields is the Weyl operator $W(g)$ defined through its action on the exponential vectors,

$$W(g)e(f) = \exp(-\imath\langle g, f\rangle - \|f\|^2/2 - \|g\|^2/2)e(f + g).$$

From its definition, $W(f)$ is a unitary operator on $\Gamma_s(L^2(\mathbb{R}_+, \mathbb{C}))$ and we have that $W(f)|\Omega\rangle = |f\rangle$, where $|\Omega\rangle = e(0)$ is the vacuum state of the bosonic filed. Therefore, any coherent state can be generated from $e(0)$ via $W(f)$. In the bosonic case, we also have the duality between the Schroedinger state transformation $|\Omega\rangle \overset{W(g)}{\mapsto} |f\rangle$, with the Heisenberg picture transformation

$$\mathcal{A}(g) \overset{W(g)^*(\cdot)W(g)}{\mapsto} W(g)^*\mathcal{A}(f)W(g), \langle f|\mathcal{A}(g)|f\rangle = \langle 0|W(g)^*\mathcal{A}(f)W(g)|0\rangle.$$

Note that $W(f) = W(f_{t]}) \otimes W(f_{[t})$. The Weyl operator $W(f_{t]})$ with time index t is an adapted process and satisfies the QSDE,

$$dW(f_{t]}) = \left(-{}^1\!/_2|f(t)|^2dt + f(t)d\mathcal{A}(t)^* - f(t)^*d\mathcal{A}(t)\right) W(f_{t]}), \quad W(f_{0]}) = I.$$

Thus we have that $W(f) = \lim_{t\to\infty} W(f_{t]})$. The time-indexed Weyl operator is realized in the laboratory by an electro-optic modulator (EOM) to produce the state $W(f_{t]})|\Omega\rangle$. In the Heisenberg picture, the EOM implements the displacement $b(t) \mapsto b(t) + f(t)I$, where $b(t)$ is a vacuum bosonic annihilation operator. Coherent states of a single-mode oscillator can be created in the steady-state by driving a stable linear quantum system with a coherent bosonic field. In the laboratory, this would be implemented by driving the inputs of the system with EOMs.

The treatment above can be easily extended to multiple quantum fields that are each in a coherent state. If there are m fields each in a coherent state with amplitudes $f_1, f_2, \ldots, f_m$ then we replace the Hilbert space $L^2(\mathbb{R}_+; \mathbb{C})$ with $L^2(\mathbb{R}_+; \mathbb{C}^m)$, replace f with the vector $f = (f_1, f_2, \ldots, f_m)^\top$, and generalize the Weyl operator $W(f_{t]})$ according to the solution of the QSDE,

$$dW(f_{t]}) = \left(-{}^1\!/_2\|f(t)\|^2dt + d\mathcal{A}(t)^*f(t) - f(t)^*d\mathcal{A}(t)\right) W(f_{t]}), \quad W(f_{0]}) = I.$$

Here $\mathcal{A}(t)$ is now a vector of m annihilation processes, one for each field as before.

Since coherent fields can be generated from the vacuum, a system driven by coherent fields can be analyzed by referring it back to the vacuum. Suppose the system $G = (S, L, H)$ is driven by coherent fields with amplitude vector f. Then we can take all fields to be in the vacuum state and consider the joint unitary evolution of the system and vacuum fields given by $U_f(t) = U(t)W(f_{t]})$, where $U(t)$ is the unitary evolution of G given by the Hudson–Parthasarathy QSDE. By applying the quantum Itō stochastic calculus, it is easy to show that $U_f(t)$ satisfies the QSDE,

$$\begin{aligned} dU_f(t) = \Big(-\big(\imath(H + \Im\{L^*Sf(t)\}) + {}^1\!/_2(L + f(t)I)^*(L + f(t)I)\big)dt \\ + d\mathcal{A}(t)^*(L + f(t)I) - (L + f(t)I)^*Sd\mathcal{A}(t) \\ + \mathrm{Tr}\big((S - I)d\Lambda(t)^\top\big)\Big) U_f(t), \end{aligned}$$

with initial condition $U_f(0) = I$. The vacuum driven system is thus $G_{\mathrm{vac}} = (S, L + fI, H + \Im\{L^*Sf\})$. Let $j_t(\cdot) = U(t)^* \cdot U(t)$ as in Sect. 2.1.4 and also let $j_t^f(\cdot) = U_f(t)^* \cdot U_f(t)$. The equivalence between G and the vacuum referred system G_{vac} is in the sense that

$$\langle f|j_t(X)|f\rangle = \langle \Omega|j_t^f(X)|\Omega\rangle, \forall X \in B(\mathfrak{h}) \text{ and } \forall t \geq 0.$$

That is, for all $t \geq 0$, the quantum expectation of $j_t(X)$ under the coherent field $|f\rangle$ coincides with the quantum expectation of $j_t^f(X)$ under the vacuum state, for any bounded operator X on $\mathfrak{h}$.

Finally, we remark that the QSDEs for the Weyl operator and G_{vac} can have time-dependent parameters, $S(t)$, $L(t)$ and $H(t)$. For instance, in the case of G_{vac}, we have $L(t) = L + f(t)I$, and $H(t) = H + \Im\{L^* S f(t)\}$, while S is not time-dependent. This is a slightly more general model than the one presented in Sect. 2.1.4. However, this extension is valid and defines a unitary solution of the QSDE when $S(t)$, $L(t)$, and $H(t)$ are adapted processes that are bounded for each t, $H(t)$ self-adjoint and $S(t)$ unitary for all $t \geq 0$ [30].

References

1. K.R. Parthasarathy, *An Introduction to Quantum Stochastic Calculus* (Birkhauser, Berlin, 1992)
2. P.A. Meyer, *Quantum Probability for Probabilists*, 2nd edn. (Springer, Berlin, 1995)
3. L. Bouten, R. van Handel, M.R. James, An introduction to quantum filtering. SIAM J. Control Optim. **46**, 2199–2241 (2007)
4. R.L. Hudson, K.R. Parthasarathy, Quantum Ito's formula and stochastic evolution. Commun. Math. Phys. **93**, 301–323 (1984)
5. C. Gardiner, M. Collett, Input and output in damped quantum systems: quantum stochastic differential equations and the master equation. Phys. Rev. A **31**, 3761–3774 (1985)
6. C.W. Gardiner, P. Zoller, *Quantum Noise: A Handbook of Markovian and Non-Markovian Quantum Stochastic Methods with Applications to Quantum Optics*, 3rd edn. (Springer, Berlin, 2004)
7. A. Silberfarb, I. Deutsch, Continuous measurement with travelling-wave probes. Phys. Rev. A **68**, 013817 (2003)
8. F. Fagnola, On quantum stochastic differential equations with unbounded coefficients. Probab. Theory Relat. Fields **86**, 501–516 (1990)
9. J. Gough, M.R. James, The series product and its application to quantum feedforward and feedback networks. IEEE Trans. Autom. Control **54**(11), 2530–2544 (2009)
10. M. Gregoratti, The Hamiltonian operator associated to some quantum stochastic differential equations. Commun. Math. Phys. **254**, 489–512 (2001)
11. M. Gregoratti, Erratum: The Hamiltonian operator associated with some quantum stochastic evolutions. Commun. Math. Phys. **264**, 563–564 (2006)
12. A.M. Chebotarev, Quantum stochastic differential equation is unitarily equivalent to a symmetric boundary problem for the Schrödinger equation. Math. Notes **61**(4), 510–518 (1997)
13. E. Wong, B. Hajek, *Stochastic Processes in Engineering Systems* (Springer, New York, 1985)
14. S.C. Edwards, V.P. Belavkin, Optimal quantum filtering and quantum feedback control (2005), arXiv preprint, http://arxiv.org/pdf/quant-ph/0506018
15. V.P. Belavkin, S.C. Edwards, Quantum filtering and optimal control, in *Quantum Stochastics and Information: Statistics, Filtering and Control (University of Nottingham, UK, 15–22 July 2006)*, ed. by V.P. Belavkin, M. Guta (World Scientific, Singapore, 2008), pp. 143–205
16. M.R. James, H.I. Nurdin, I.R. Petersen, H^∞ control of linear quantum stochastic systems. IEEE Trans. Autom. Control **53**(8), 1787–1803. Reprinted, with permission, © 2008 IEEE (2008)
17. H.M. Wiseman, G.J. Milburn, *Quantum Measurement and Control* (Cambridge University Press, Cambridge, 2010)

18. J.E. Gough, M.R. James, H.I. Nurdin, Squeezing components in linear quantum feedback networks. Phys. Rev. A **81**, 023804. Reprinted, with permission, © (2010) by the American Physical Society (2010)
19. H.I. Nurdin, Quantum filtering for multiple input multiple output systems driven by arbitrary zero-mean jointly Gaussian input fields. Russ. J. Math. Phys. **21**(3), 386–398. © 2014 Springer. Reprinted with permission of Springer (2014)
20. H. Hellmich, R. Honegger, C. Köstler, B. Kümmerer, A. Rieckers, Couplings to classical and non-classical squeezed white noise as stationary Markov processes, vol. 38 (Publ. RIMS, Kyoto Univ, 2002), pp. 1–31
21. J. Gough, R. Gohm, M. Yanagisawa, Linear quantum feedback networks. Phys. Rev. A **78**, 061204 (2008)
22. A.I. Maalouf, I.R. Petersen, Bounded real properties for a class of annihilation-operator linear quantum systems. IEEE Trans. Autom. Control **56**(4), 786–801 (2011)
23. J. Gough, G. Zhang, On realization theory of quantum linear systems. Automatica **59**, 139–151 (2015)
24. H.I. Nurdin, M.R. James, A.C. Doherty, Network synthesis of linear dynamical quantum stochastic systems. SIAM J. Control Optim. **48**(4), 2686–2718 (2009)
25. H.I. Nurdin, Structures and transformations for model reduction of linear quantum stochastic systems. IEEE Trans. Autom. Control **59**(9), 2413–2425. Reprinted, with permission, © 2014 IEEE (2014)
26. H.I. Nurdin, Topics in classical and quantum linear stochastic systems, Ph.D. dissertation, The Australian National University (2007)
27. H.I. Nurdin, Network synthesis of mixed quantum-classical linear stochastic systems, in *Proceedings of the 2011 Australian Control Conference (AUCC)*. Engineers Australia (2011), pp. 68–75
28. H. Araki, E.J. Woods, Representations of the canonical commutation relations describing a nonrelativistic infinite free Bose gas. J. Math. Phys. **4**, 637–662 (1963)
29. J. Gough, Quantum white noise and the master equation for Gaussian reference states. Russ. J. Math. Phys. **10**(2), 142–148 (2003)
30. L. Bouten, R. van Handel, On the separation principle of quantum control, in *Quantum Stochastics and Information: Statistics, Filtering and Control (University of Nottingham, UK, 15–22 July 2006)*, ed. by V.P. Belavkin, M. Guta (World Scientific, Singapore, 2008), pp. 206–238

Chapter 3
Realization Theory for Linear Dynamical Quantum Systems

Abstract This chapter presents a realization/network synthesis theory for linear quantum systems. This theory addresses the question of how one can go from an abstract description of a linear quantum system to a concrete realization of the system using quantum optical devices. Two distinct types of realization problems are introduced and treated: the strict (or hard) realization problem and the transfer function (or soft) realization problem. The system to be realized is decomposed into simpler subsystems, and how these subsystems can be realized, at least approximately, in the quantum optical setting is developed. In particular, it is shown that simpler realizations can be obtained for completely passive linear quantum systems.

In the previous chapter, the class of linear quantum systems was introduced, and the notion of physical realizability and parameterizations for this class has been elaborated upon. A natural question that follows from the theory of Chap. 2, and which is the theme of this chapter, is, given an abstract mathematical description of a linear quantum system in terms of linear QSDEs, what physical system would it correspond to? In other words, what physical system would realize an abstractly defined linear quantum system? To motivate this question further, we begin with a brief overview of aspects of linear electrical network synthesis that are relevant to our purposes in this chapter.

Well known in electrical engineering, and in systems and control theory in particular, is that a classical (continuous time, causal, linear time invariant) electrical network described by a set of (coupled) linear ordinary differential equations can be represented in different ways, for example with a frequency domain or transfer function representation, a modern state-space representation and, more recently, using Willems' behavioral representation [1]. In the case where the system is initially at rest (all initial conditions are zero), one can switch between the transfer

---

The Introduction, Sect. 3.1 and the associated appendices contain materials adapted from [3]
Section 3.2 and the associated appendices contain some materials reprinted, with permission, from [13] © 2010 IEEE.
Section 3.3 contain materials reprinted, with permission, from [16] © 2010 IEEE.

H.I. Nurdin and N. Yamamoto, *Linear Dynamical Quantum Systems*,
Communications and Control Engineering, DOI 10.1007/978-3-319-55201-9_3

function and state-space representations for any given network. However, although one can associate a unique transfer function representation to a state-space representation, the converse is not true; for a given transfer function, there are infinitely many state-space representations. The state-space representation can be made to be unique (up to a similarity transformation of the state-space matrices) by requiring that the representation be of minimal order (i.e., the representation is both controllable and observable). The *synthesis* or *realization* problem in linear electrical networks theory deals with the inverse scenario where one is presented with a transfer function or state-space description of a linear system and would like to synthesize or build such a system from various linear electrical components such as resistors, capacitors, inductors, and op-amps. A particularly advantageous feature of the state-space representation, since it is given by a set of first-order ordinary differential equations, is that how a system can be *systematically* synthesized can be inferred directly from the representation. For instance, consider the following state-space system,

$$\begin{aligned} \frac{dx(t)}{dt} &= \begin{bmatrix} 2 & 5 \\ -2 & -4 \end{bmatrix} x(t) + \begin{bmatrix} 1 \\ 0.1 \end{bmatrix} u(t) \\ y(t) &= \begin{bmatrix} 0 & 1 \end{bmatrix} x(t) + u(t), \end{aligned} \tag{3.1}$$

where $x(t)$ is the state, $u(t)$ is the input signal, and $y(t)$ is the output signal. In an electrical circuit, $u(t)$ could be the voltage at certain input ports of the circuit and $y(t)$ could be the voltage at another set of ports of the circuit, different from the input ports. This system can be implemented according to the schematic shown in Fig. 3.1. This schematic can then be used to implement the system at the hardware level, Fig. 3.2 [2, Chap. 13]. However, linear electrical network synthesis is a mature subject that deals with much more than just how one can obtain *some* realization of a particular system. For instance, it also addresses the fundamental issues such as how a passive network, a network that does not require an external source of energy, can also be synthesized using only passive electrical components. The primary objective in this chapter is to develop an analogously systematic method for synthesizing arbitrarily complex linear quantum systems described by a linear QSDE. A general

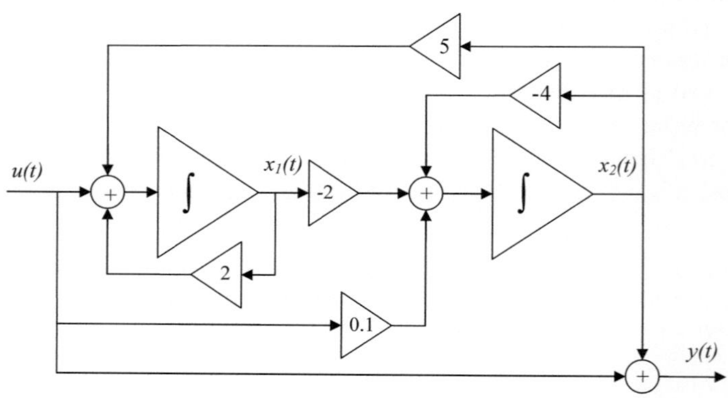

Fig. 3.1 Schematic for the implementation of the classical system (3.1). Figure adapted from [3]

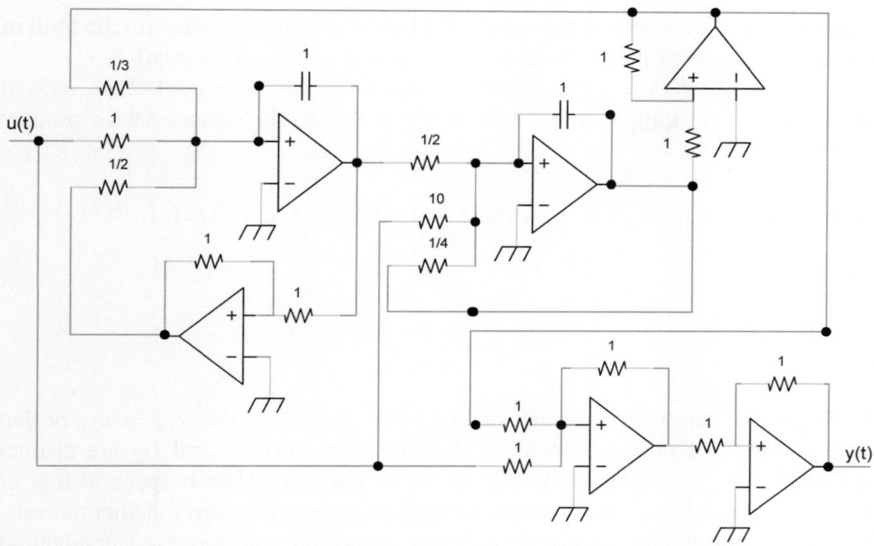

Fig. 3.2 Hardware implementation of the schematic diagram shown in Fig. 3.1. Figure adapted from [3] © 2009 Society for Industrial and Applied Mathematics. Reprinted with permission. All rights reserved

theory for *static* linear quantum networks that do not contain dynamical components was developed in [4], whereas the synthesis theory for linear dynamical quantum systems was initiated in [3]. In the context of linear quantum systems, there are two distinct types of realization problems that can be considered:

1. *Strict* or *hard* realization. This is the problem of realizing the parameters (S, K, R) of a given linear quantum system G.
2. Transfer function or *soft* realization. This is the problem of realizing the transfer function $\Xi_G[\cdot]$ of a quantum system G.

Notice that a solution to the strict realization problem also solves the transfer function realization problem, but the converse is false. In the soft problem, there are additional degrees of freedom that can be exploited to achieve the realization. Namely, there is freedom to transform to a different collection of internal degrees of freedom that may simplify the realization problem. This will be made clear in Sect. 3.3. We begin by considering the strict realization problem for linear quantum systems, following the development in [3].

3.1 Architecture for Strict Realization

3.1.1 The Concatenation and Series Product and Reducible Quantum Networks

We first recall the formalisms of concatenation product, series product, and reducible networks (with respect to the series product) that were introduced in [5] for the manipulation of networks of general open Markov quantum systems (in the limit of zero time delays for fields to propagate between network components).

Let $G_1 = (S_1, K_1 x_{1,0}, (1/2)x_{1,0}^\top R_1 x_{1,0})$ and $G_2 = (S_2, K_2 x_{2,0}, (1/2)x_{2,0}^\top R_2 x_{2,0})$ be two linear quantum systems, where $x_{k,0} = x_k(0)$. The concatenation product $G_1 \boxplus G_2$ is defined as

$$G_1 \boxplus G_2 = \left(S_{1\boxplus 2}, (K_1 x_{1,0}, K_2 x_{2,0})^\top, (1/2)x_{1,0}^\top R_1 x_{1,0} + (1/2)x_{2,0}^\top R_2 x_{2,0}\right),$$

where

$$S_{1\boxplus 2} = \begin{bmatrix} S_1 & 0 \\ 0 & S_2 \end{bmatrix}.$$

We emphasize here that the definition allows the possibility that $x_{1,0} = x_{2,0}$ or that some components of $x_{1,0}$ coincide with those of $x_{2,0}$. If G_1 and G_2 are distinct oscillators (i.e., the components of $x_{1,0}$ act on a distinct Hilbert space to that of the components of $x_{2,0}$), the concatenation product simply groups together the variables of two non-interacting linear quantum systems to form a larger linear quantum system.

One can also pass the output of a system G_1 as the input to system G_2, as long as both systems have the same number of input and output channels. This kind of operation, which is the cascading or loading of G_2 onto G_1, is represented by the series product $G_2 \triangleleft G_1$ defined by

$$G_2 \triangleleft G_1 = \Big(S_2 S_1, K_2 x_{2,0} + S_2 K_1 x_{1,0}, (1/2)x_{1,0}^\top R_1 x_{1,0} + (1/2)x_{2,0}^\top R_2 x_{2,0} + \frac{1}{2\imath} x_{2,0}^\top (K_2^* S_2 K_1 - K_2^\top S_2^\# K_1^\#) x_{1,0}\Big).$$

The resulting system $G_2 \triangleleft G_1$ is again a linear quantum system with scattering matrix, coupling operator, and Hamiltonian as given by the above formula.

With concatenation and series products having been defined, we now arrive at the notion of a reducible network with respect to the series product, which we shall henceforth refer to more simply as just a reducible network. Suppose that a network consists of l linear quantum systems $G_k = (S_k, L_k, H_k)$, with $L_k = K_k x_{k,0}$ and $H_k = (1/2)x_{k,0}^\top R_k x_{k,0}$, $k = 1, \ldots, l$. A reducible quantum network $\mathcal{N}$ is simply a network that can be constructed from $G_1, G_2, \ldots, G_l$ by specifying a direct interaction Hamiltonian $H^d = \sum_j \sum_{k=j+1} x_{j,0}^\top R_{jk} x_{k,0}$ ($R_{jk} \in \mathbb{R}^{2\times 2}$) and a list $\mathcal{S} = \{G_k \triangleleft G_j\}$ of series connections among G_j and G_k, $j \neq k$, with the condition that a list of connections containing algebraic loops, such as $\{G_2 \triangleleft G_1, G_3 \triangleleft G_2, G_1 \triangleleft G_3\}$, are disallowed. Here, algebraic loops refer to loops in which the field entering a system G_j as a particular input channel can return via some path as a component of the same input to G_j. The network $\mathcal{N}$ is again a linear quantum system and is denoted by $\mathcal{N} = \{\{G_k\}_{k=1,\ldots,l}, H^d, \mathcal{S}\}$. If $\mathcal{N}_0 = \{\{G_k\}_{k=1,\ldots,l}, 0, \mathcal{S}\} = (S_0, L_0, H_0)$ is a reducible network, then $\mathcal{N}$ is simply $\mathcal{N}_0$ equipped with the direct interaction

Hamiltonian H^d, $\mathcal{N} = \mathcal{N}_0 \boxplus (0, 0, H) = (S_0, L_0, H_0 + H^d)$. Algebraic loops are subtle and are beyond the scope of the theory in [5]. They can, however, be treated in the more general framework of quantum feedback networks developed in [6]. This section exploits the tools in [5], so the discussion is restricted to reducible networks. Fortunately, this is in fact sufficient to develop a network synthesis theory of linear quantum systems as will be shown in Sect. 3.1.2. A network synthesis theory can also be developed following the more general theory of [6] and will be pursued in Sect. 3.2.

Two important decompositions of a linear quantum system based on the series product that will be exploited in the ensuing development are

$$(S, L, H) = (I, L, H) \triangleleft (S, 0, 0), \tag{3.2}$$
$$(S, L, H) = (S, 0, 0) \triangleleft (I, S^* L, H), \tag{3.3}$$

where $(S, 0, 0)$ represents a static passive linear network implementing the unitary matrix S.

3.1.2 Main Synthesis Theorem

Let us temporarily restrict our attention to the case of two distinct linear quantum systems coupled to m independent bosonic fields, with m output channels: an n_1 degree-of-freedom linear quantum system $G_1 = (S_1, L_1, H_1)$ with canonical operators $x_1 = (q_{1,1}, p_{1,1}, \ldots, q_{1,n_1}, p_{1,n_1})^\top$, Hamiltonian operator $H_1 = (1/2)x^\top R_1 x_1$, coupling operator $L_1 = K_1 x_1$, and scattering matrix S_1, and, similarly, an n_2 degree-of-freedom linear quantum system $G_2 = (S_2, L_2, H_2)$ with canonical operators $x_2 = (q_{2,1}, p_{2,1}, \ldots, q_{2,n_2}, p_{2,n_2})^\top$, Hamiltonian operator $H_2 = (1/2)x_2^\top R_2 x_2$, coupling operator $L_2 = K_2 x_2$, and unitary scattering matrix S_2.

Consider now a reducible quantum network $\mathcal{N}_{12}$ given by $\mathcal{N}_{12} = \{\{G_1, G_2\}, H_{12}^d, G_2 \triangleleft G_1\}$, as shown in Fig. 3.3, where H_{12}^d is a direct interaction Hamiltonian term between G_1 and G_2 given by

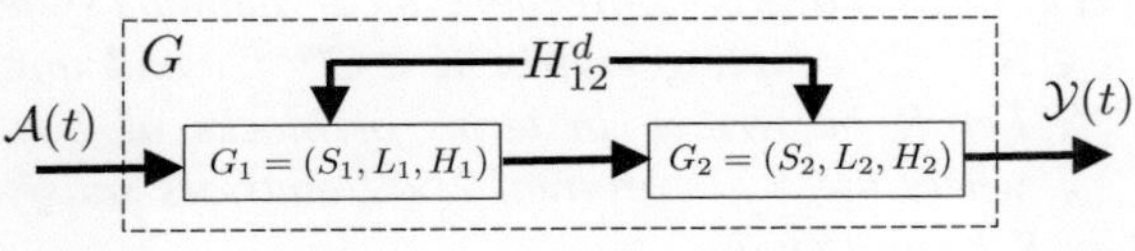

Fig. 3.3 Cascade connection of G_1 and G_2 with indirect interaction H_{12}^d. Figure adapted from [3]

$$\begin{aligned} H_{12}^d &= (1/2)x_1^\top R_{12}x_2 + (1/2)x_2^\top R_{12}^\top x_1 - \frac{1}{2\imath}(L_2^* S_2 L_1 - L_1^* S_2^* L_2), \\ &= x_2^\top R_{12}^\top x_1 - \frac{1}{2\imath}(L_2^* S_2 L_1 - L_2^\top S_2^\# L_1^\#), \\ &= x_2^\top \left(R_{12}^\top - \frac{1}{2\imath}(K_2^* S_2 K_1 - K_2^\top S_2^\# K_1^\#) \right) x_1, \end{aligned}$$

with $R_{12} \in \mathbb{R}^{2\times 2}$ and the second equality following from the fact that elements of L_1 commute with those L_2. Also, note that the matrix $\frac{1}{2\imath}(K_2^* S_2 K_1 - K_2^\top S_2^\# K_1^\#)$ is real.

Some straightforward algebraic manipulations (see [5] for details) then show that

$$\mathcal{N}_{12} = (S_2 S_1, S_2 L_1 + L_2, H_1 + H_2 + H_{12}^f + H_{12}^d),$$

where $H_{12}^f = \frac{1}{2\imath}(L_2^* S_2 L_1 - L_1^* S_2^* L_2)$. Now, let us inspect the Hamiltonian term of $\mathcal{N}_{12}$. After substituting in the definition of H_1, H_2, H_{12}^d, and H_{12}^f, we have that

$$H_1 + H_2 + H_{12}^f + H_{12}^d = (1/2)[\, x_1^\top \; x_2^\top \,] \begin{bmatrix} R_1 & R_{12} \\ R_{12}^\top & R_2 \end{bmatrix} \begin{bmatrix} x_1 \\ x_2 \end{bmatrix}.$$

Letting $x = (x_1^\top, x_2^\top)^\top$, $S_{2\leftarrow 1} = S_2 S_1$, and defining

$$R = \begin{bmatrix} R_1 & R_{12} \\ R_{12}^\top & R_2 \end{bmatrix}, \tag{3.4}$$

$$K = [\, S_2 K_1 \; K_2 \,], \tag{3.5}$$

we see that

$$\mathcal{N}_{12} = \left(S_{2\leftarrow 1}, Kx, \frac{1}{2}x^\top R x \right). \tag{3.6}$$

Therefore, $\mathcal{N}_{12} = (S_{2\leftarrow 1}, L_{2\leftarrow 1}, H_{2\leftarrow 1})$, with $S_{2\leftarrow 1} = S_2 S_1$, $L_{2\leftarrow 1} = Kx$, and $H_{2\leftarrow 1} = (1/2)x^\top R x$. By repeated application of the above construction, we can prove the following synthesis theorem.

Theorem 3.1 *Let G be an n degree-of-freedom linear quantum system with Hamiltonian matrix $R \in \mathbb{R}^{2n\times 2n}$, coupling matrix $K \in \mathbb{C}^{m\times 2n}$, and unitary scattering matrix $S \in \mathbb{C}^{m\times m}$. Let R be written in terms of blocks of 2×2 matrices as $R = [R_{jk}]_{j,k=1,\ldots,n}$, where the R_{jk}'s are real 2×2 matrices satisfying $R_{kj} = R_{jk}^\top$ for all j, k, and let K be written as*

$$K = [\, K_1 \; K_2 \; \ldots \; K_n \,],$$

where, for each j, $K_j \in \mathbb{C}^{m\times 2}$. Let $G_j = (S_j, \tilde{K}_j x_j, (1/2)x_j^\top R_{jj} x_j)$ for $j = 1, \ldots, n$ be independent linear quantum systems with canonical operators $x_j = (q_j, p_j)^\top$, m output fields, Hamiltonian matrix R_{jj}, coupling matrix $\tilde{K}_j$, and scattering matrix

S_j. Also, define $S_{k\leftarrow j}$ for $j \leq k+1$ as $S_{k\leftarrow j} = \overleftarrow{\prod}_{l=j}^{k} S_l = S_k \cdots S_{j+1} S_j$ for $j < k$, $S_{k\leftarrow k} = S_k$, and $S_{k\leftarrow k+1} = I_{m\times m}$, and let H^d be a direct interaction Hamiltonian given by

$$H^d = \sum_{j=1}^{n-1} \sum_{k=j+1}^{n} x_k^\top \left(R_{jk}^\top - \frac{1}{2\imath} (\tilde{K}_k^* S_{k\leftarrow j+1} \tilde{K}_j - \tilde{K}_k^\top S_{k\leftarrow j+1}^\# \tilde{K}_j^\#) \right) x_j. \tag{3.7}$$

If $S_1, \ldots, S_n$ satisfies $S_n S_{n-1} \cdots S_1 = S$ and $\tilde{K}_k$ satisfies $\tilde{K}_k = S_{n\leftarrow k+1}^* K_k$ for $k = 1, \ldots, n$, then the reducible network of harmonic oscillators $\mathcal{N}$ given by $\mathcal{N} = \{\{G_1, \ldots, G_n\}, H^d, \{G_2 \lhd G_1, G_3 \lhd G_2, \ldots, G_n \lhd G_{n-1}\}\}$ is equivalent to G. That is, G can be synthesized as a series connection $G_n \lhd \ldots \lhd G_2 \lhd G_1$ of n one degree-of-freedom linear quantum systems, along with a suitable bilinear direct interaction Hamiltonian involving the canonical operators of these systems. In particular, if $S = I_{m\times m}$ (no scattering), then S_k can be chosen to be $S_k = I_{m\times m}$ and $\tilde{K}_k$ can be chosen to be $\tilde{K}_k = K_k$ for $k = 1, \ldots, n$.

Proof Let $H_j = (1/2) x_j^\top R_{jj} x_j$, $L_j = \tilde{K}_j x_j$, and

$$H_k^f = \sum_{j=2}^{k} \left(L_j^* \sum_{l=1}^{j-1} S_{j\leftarrow l+1} L_l - \sum_{l=1}^{j-1} L_l^* S_{j\leftarrow l+1}^* L_j \right), \quad k \geq 2.$$

We start with the series connection $G_{12} = G_2 \lhd G_1$. It is given by

$$G_{12} = (S_2 S_1, S_2 L_1 + L_2, H_1 + H_2 + H_2^f).$$

Repeating this calculation recursively for $G_{123} = G_3 \lhd G_{12}, G_{1234} = G_4 \lhd G_{123}, \ldots, G_{12\ldots n} = G_n \lhd G_{12\ldots(n-1)}$, we finally obtain that

$$G_{12\ldots n} = \left(S_{n\leftarrow 1}, \sum_{k=1}^{n} S_{n\leftarrow k+1} L_k, \sum_{k=1}^{n} H_k + H_n^f \right).$$

Now, noting that H_n^f may be rewritten as

$$\begin{aligned} H_n^f &= \frac{1}{2\imath} \sum_{j=1}^{n-1} \sum_{k=j+1}^{n} (L_k^* S_{k\leftarrow j+1} L_j - L_j^* S_{k\leftarrow j+1}^* L_k) \\ &= \frac{1}{2\imath} \sum_{j=1}^{n-1} \sum_{k=j+1}^{n} (L_k^* S_{k\leftarrow j+1} L_j - L_k^\top S_{k\leftarrow j+1}^\# L_j^\#) \\ &= \frac{1}{2\imath} \sum_{j=1}^{n-1} \sum_{k=j+1}^{n} x_k^\top (\tilde{K}_k^* S_{k\leftarrow j+1} \tilde{K}_j - \tilde{K}_k^\top S_{k\leftarrow j+1}^\# \tilde{K}_j^\#) x_j, \end{aligned}$$

where the second equality holds since L_j commutes with L_k whenever $j \neq k$, we have that

$$\sum_{k=1}^{n} H_k + H_n^f + H^d = (1/2) \sum_{j=1}^{n} \sum_{k=1}^{n} x_j R_{jk} x_k = (1/2) x^\top R x,$$
$$x = (x_1^\top, x_2^\top, \ldots, x_n^\top)^\top.$$

Therefore, if $S_1, \ldots, S_n$ and $\tilde{K}_1, \ldots, \tilde{K}_n$ satisfy the conditions stated in the theorem, we find that $\mathcal{N} = \{\{G_1, \ldots, G_n\}, H_d, \{G_2 \triangleleft G_1, G_3 \triangleleft G_2, \ldots, G_n \triangleleft G_{n-1}\}$ is given by

$$\mathcal{N} = \left(S, Kx, (1/2)x^\top Rx\right).$$

That is, $\mathcal{N}$ is a linear quantum system with Hamiltonian matrix R, coupling matrix K, and scattering matrix S, and is therefore equivalent to G. This completes the proof of the synthesis theorem. □

Thus, according to the theorem, synthesis of an arbitrary n degree-of-freedom linear dynamical quantum system is in principle possible if the following two requirements can be fulfilled:

1. Arbitrary one degree-of-freedom linear quantum systems $G = (I, L, H)$ can be synthesized. It follows from this that general one degree-of-freedom linear quantum systems $G' = (S, L, H)$ can be synthesized as $G' = (I, L, H) \triangleleft (S, 0, 0)$.
2. The bilinear interaction Hamiltonian H^d as given by (3.7) can be synthesized.

3.1.3 Systematic Synthesis of Linear Quantum Systems

This section details the construction of arbitrary one degree-of-freedom linear quantum systems and implementation of bilinear direct interactions among the canonical operators of these systems, at least approximately. The focus is on realization in the quantum optical domain, using various linear and nonlinear quantum optical components, though realizations in other physical platforms are also possible so long as they support linear dynamics in the Heisenberg picture.

We begin with a description of some key quantum optical components that will be required for the synthesis. This is followed by a discussion of general synthesis of one degree-of-freedom linear quantum systems and finally by a discussion of the implementation of bilinear direct interaction Hamiltonians between distinct one degree-of-freedom linear quantum systems.

3.1.3.1 Essential Quantum Optical Components

Optical Cavities

An optical cavity is depicted in Fig. 3.4. It is essentially a system of fully reflecting or partially transmitting mirrors in which light can be trapped, either bounced repeatedly from the mirrors to form a standing wave or circulating inside the cavity (as in a ring cavity). If partially transmitting mirrors are present, then light can escape from the cavity, producing an output field and also acting as source of loss.

An optical cavity has a Hamiltonian of the form $H_{cav} = \omega_{cav} a^* a$, where ω_{cav} is the resonance frequency of the cavity and $a = \frac{q+\imath p}{2}$ is the (non-self-adjoint) cavity annihilation operator (or cavity mode) satisfying $[a, a^*] = 1$. Here, $q = a + a^*$ is the position operator of the cavity mode (also called the *amplitude quadrature* of the mode), and $p = -\imath a + \imath a^*$ is the momentum operator of the cavity mode (also called the *phase quadrature* of the mode). If M is a partially transmitting mirror, then photons escaping the cavity through this mirror can be modeled as the interaction of M with a vacuum bosonic noise field $\mathcal{A}(t)$ incident at this mirror via the idealized Hamiltonian $H_{\rm int}$ of the form (1.24), taking $j = 1$ and $L = \sqrt{\kappa} a$, where κ is a positive constant called the mirror decay rate or coupling coefficient. When there are several leaky mirrors, then the collective escape of photons through these mirrors can be modeled by a sum of such interaction Hamiltonians, one for each mirror and with each mirror interacting with its own distinct incident vacuum bosonic field. The total cavity Hamiltonian is then just the sum of H_{cav} and all the interaction Hamiltonians. Note that, in general, the field incident at a transmitting mirror need not be a vacuum field, but can be other types of fields, such as a coherent laser beam. Nonetheless, the interaction of the cavity mode with such fields via the mirrors will still be governed by (1.24) with a coupling operator of the form $L = \sqrt{\kappa} a$.

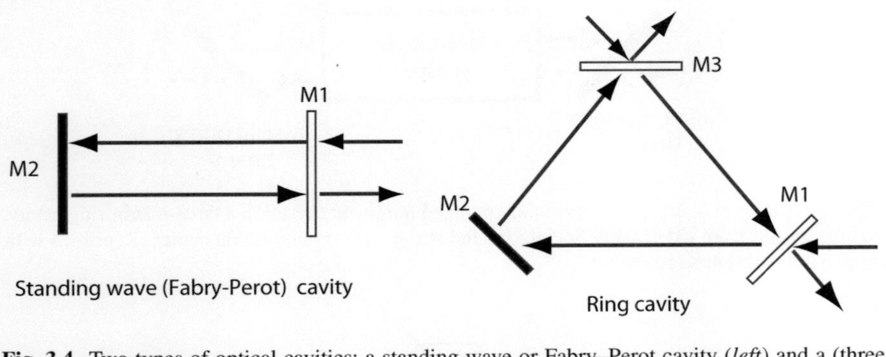

Fig. 3.4 Two types of optical cavities: a standing wave or Fabry–Perot cavity (*left*) and a (three mirror) ring cavity (*right*). *Arrows* indicate the direction of propagation of light in the cavity. *Black rectangles* denote fully reflecting mirrors, while *white rectangles* denote partially transmitting mirrors. Figure adapted from [3] © 2009 Society for Industrial and Applied Mathematics. Reprinted with permission. All rights reserved

Degenerate Parametric Amplifier

An optical cavity can be equipped with a $\chi^{(2)}$ nonlinear optical crystal and a classical pump beam in the configuration of a degenerate parametric amplifier (DPA), following the treatment in [7, Sect. 10.2] and as introduced earlier in Sect. 1.5.3. However, in this chapter, DPA will be used to refer to the internal dynamics of the device in Sect. 1.5.3 rather than its input-output dynamics. The pump beam serves as a source of quanta that facilitates an interaction in the nonlinear crystal in which photons of the pump beam are annihilated to create photons of the cavity mode. In an optical cavity, such as the ring cavity shown in Fig. 3.5, a crystal can be positioned in one arm of the cavity (e.g., in the arm between mirrors M1 and M2) and shone with a strong coherent pump beam of (angular) frequency $\omega_p = 2\omega_r$, where ω_r is some reference frequency. The mirrors at the end of the arms should be chosen such that they fully transmit light beams of frequency ω_p. A schematic representation of a DPA (a nonlinear crystal with a classical pump) is shown in Fig. 3.6.

Remark 3.1 In the remaining figures, black rectangles will be used to denote mirrors that are fully reflecting at the cavity frequency and fully transmitting at the pump

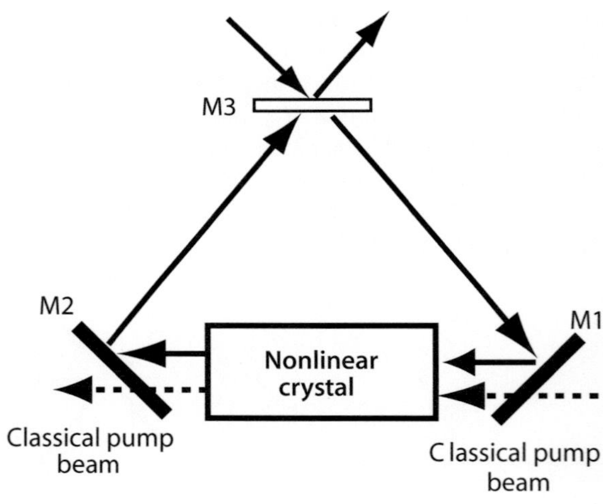

Fig. 3.5 A DPA consisting of a classically pumped nonlinear crystal in a three-mirror ring cavity. Figure adapted from [3] © 2009 Society for Industrial and Applied Mathematics. Reprinted with permission. All rights reserved

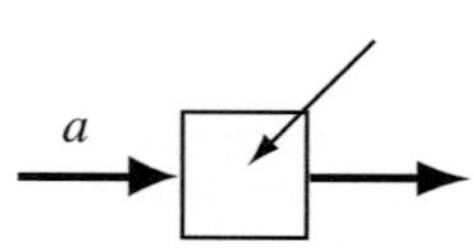

Fig. 3.6 Schematic representation of a DPA. The *white rectangle* symbolizes the nonlinear crystal, while the *diagonal arrow* into the *rectangle* denotes the pump beam. Figure adapted from [3] © 2009 Society for Industrial and Applied Mathematics. Reprinted with permission. All rights reserved

frequency (whenever a pump beam is employed), while white rectangles denote partially transmitting mirrors at the cavity frequency.

Let $a = \frac{q+\imath p}{2}$ be the cavity mode, and let the cavity frequency be $\omega_{cav} = \omega_r + \Delta$, where Δ is the detuning from the reference frequency ω_r. Under the assumption that the pump beam is intense and undepleted in the interaction, it may be treated as being classical, in which case the crystal-pump-cavity interaction can be modeled by the (time-varying) Hamiltonian $H(t) = \omega_{cav} a^* a + \frac{\imath}{2}(\epsilon e^{-i\omega_p t}(a^*)^2 - \epsilon^* e^{\imath\omega_p t} a^2)$ [7, Eq. 10.2.1], where ϵ is a complex number representing the effective pump amplitude (i.e., the amplitude of the beam scaled by a constant of the nonlinear crystal). By moving to a rotating frame with respect to $\omega_r = \omega_p/2$ (i.e., applying the transformation $a \mapsto a e^{-\imath\omega_p t/2}$; see [7, Sect. 10.2.1] for a derivation of the equations of motion of the DPA in the rotating frame), H can be rewritten as

$$H = \Delta a^* a + \frac{\imath}{2}(\epsilon (a^*)^2 - \epsilon^* a^2)$$

and expressed compactly as $H = (1/2) x_0^\top R x_0 + c$ (recall $x_0 = (q, p)^\top$), where

$$R = 1/2 \begin{bmatrix} \Delta + \frac{\imath}{2}(\epsilon - \epsilon^*) & \frac{1}{2}(\epsilon + \epsilon^*) \\ \frac{1}{2}(\epsilon + \epsilon^*) & \Delta - \frac{\imath}{2}(\epsilon - \epsilon^*) \end{bmatrix}, \tag{3.8}$$

and c is a real number. Since c merely contributes a phase factor that has no effect on the overall dynamics of the system operators, it can simply be ignored. That is, in the rotating frame, the DPA can be treated as a harmonic oscillator with fixed quadratic Hamiltonian $(1/2) x_0^\top R x_0$.

Two-Mode Squeezing

Beams propagating inside two cavities can be made to interact by having their paths intersect inside a $\chi^{(2)}$ nonlinear optical crystal and pumping the crystal with a pump beam as a source of quanta. For instance, in a $\chi^{(2)}$ optical crystal in which two cavities interact with an undepleted classical pump beam, as depicted in Fig. 3.7, the interaction can be modeled by the Hamiltonian

$$H(t) = \frac{\imath}{2}(\epsilon e^{-\imath\omega_p t} a_1^* a_2^* - \epsilon^* e^{\imath\omega_p t} a_1 a_2),$$

where ϵ is a complex number representing the effective amplitude of the pump beam and ω_p is the pump beam frequency. Moving to a rotating frame at half the pump frequency by applying the transformation $a_1 \mapsto a_1 e^{\imath(\omega_p/2)t}$ and $a_2 \mapsto a_2 e^{\imath(\omega_p/2)t}$, $H(t)$ can be expressed in this new frame in the time-invariant form $H = \frac{\imath}{2}(\epsilon a_1^* a_2^* - \epsilon^* a_1 a_2)$. This kind of Hamiltonian is called a *two-mode squeezing Hamiltonian*, as it affects simultaneous squeezing in the quadratures of two independent

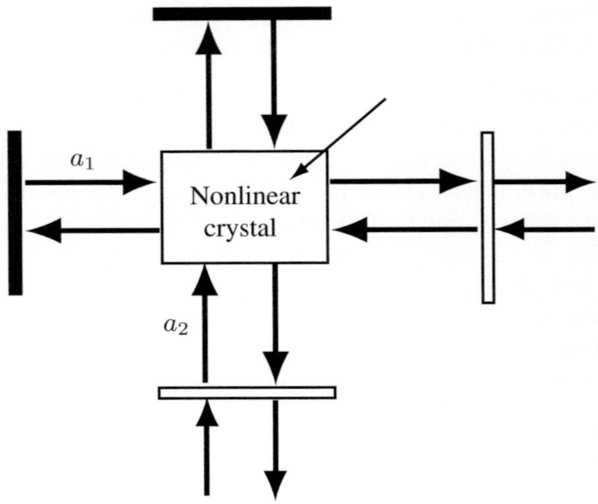

Fig. 3.7 Two cavity modes interacting in a single classically pumped nonlinear crystal. The *diagonal arrow* into the crystal denotes the pump beam. Figure adapted from [3]

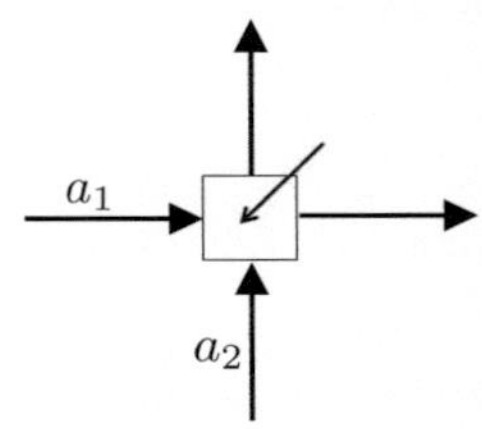

Fig. 3.8 Schematic representation of a two-mode squeezer. Figure adapted from [3]

modes that are linear combinations of a_1 and a_2. It will play an important role later on in this chapter. For further details on the physical theory of two-mode squeezing, see [8, 9] (Fig. 3.8).

Remark 3.2 It is implicitly assumed that the equations of motion for a collection of linear quantum systems are given with respect to a common rotating frame of frequency ω_r, including the transformation of all bosonic fields $\mathcal{A}_i(t)$ according to $\mathcal{A}_i(t) \mapsto \mathcal{A}_i(t)e^{\imath\omega_r t}$, and that all classical pumps employed are of frequency $\omega_p = 2\omega_r$. This is a natural setting in quantum optics where a rotating frame is essential for obtaining linear time-invariant QSDE models for active devices that require an external source of quanta. In a control setting, this means both the quantum plant and the controller equations have been expressed in the same rotating frame.

Static Linear Optical Devices and Networks

Static linear optical devices implement a constant frequency-independent linear transformation of fields. Such transformations are represented by a complex square matrix and can be applied to a set of independent incoming single-mode fields $a = (a_1, a_2, \ldots a_m)^\top$ to produce an equal number $a' = (a'_1, a'_2, \ldots a'_m)^\top$ of independent outgoing fields. The incoming and outgoing fields satisfy the CCR for single-mode oscillators. However, the incoming fields may also be vacuum bosonic fields $\mathcal{A}(t) = (\mathcal{A}_1(t), \mathcal{A}_2(t), \ldots, \mathcal{A}_m(t))^\top$ with outgoing bosonic fields (that need no longer be in the vacuum state) $\mathcal{Y}(t) = (\mathcal{Y}_1(t), \mathcal{Y}_2(t), \ldots, \mathcal{Y}_m(t))^\top$. For bosonic fields, the commutation relations are $[d\mathcal{A}_j(t), d\mathcal{A}_k(t)] = 0$ and $[d\mathcal{A}_j(t), d\mathcal{A}_k(t)^*] = \delta_{jk}dt$. To avoid cumbersome and unnecessary repetitions, in the following we restrict the discussion to the operation of a static linear optical device on single-mode fields. The operation is completely analogous for bosonic incoming and outgoing fields, requiring only the obvious substitutions $a \to \mathcal{A}(t)$, $a' \to \mathcal{Y}(t)$, $[a_j, a_k] = 0 \to [d\mathcal{A}_j(t), d\mathcal{A}_k(t)] = 0$, and $[a_j, a_k^*] = \delta_{jk} \to [d\mathcal{A}_j(t), d\mathcal{A}_k(t)^*] = \delta_{jk}dt$, etc.

The action of a static linear optical device can mathematically be expressed as

$$\begin{bmatrix} a' \\ a'^{\#} \end{bmatrix} = Q \begin{bmatrix} a \\ a^{\#} \end{bmatrix}; \; Q = \begin{bmatrix} Q_1 & Q_2 \\ Q_2^{\#} & Q_1^{\#} \end{bmatrix},$$

where $Q_1, Q_2 \in \mathbb{C}^{m \times m}$, and Q is a Bogoliubov matrix as elaborated in Chap. 2. As a consequence, the output fields a' satisfy the same commutation relations as a.

In the special case where $Q_2 = 0$, the device does not mix creation and annihilation operators of the fields, and it follows that Q_1 is necessarily a complex *unitary* matrix. Such devices are said to be *static passive* linear optical devices by the fact that they do not require an external source of quanta for their operation. It is well known that any static passive network can be constructed using only beam splitters and mirrors (e.g., see references 2–4 in [10]). In all other cases, the devices are *static active*. Specific passive and static devices that will be of interest will be discussed in the sections to follow.

Phase Shifter

A phase shifter is a device that imparts a phase shift on the incoming field. If the input field is a, then a phase shifter outputs the field $a' = e^{\imath\theta}a$ for some real number θ, called the *phase shift*. This device is schematically represented by the symbol shown in Fig. 3.9. A phase shifter with a single input field is a static passive device with transformation matrix Q_{PS} given by

$$Q_{PS} = \begin{bmatrix} e^{\imath\theta} & 0 \\ 0 & e^{-\imath\theta} \end{bmatrix}.$$

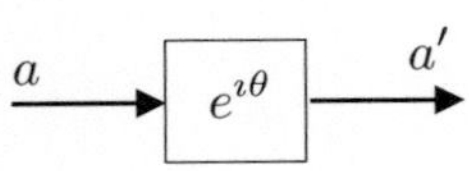

Fig. 3.9 Phase shifter with a phase shift of θ radians. Figure adapted from [3] © 2009 Society for Industrial and Applied Mathematics. Reprinted with permission. All rights reserved

Beam Splitter

A beam splitter is a static passive device that takes two input fields a_1 and a_2 and linearly combines them to produce two output fields a_1' and a_2' such that the number of photons is conserved: $a_1^* a_1 + a_2^* a_2 = (a_1')^* a_1' + (a_2')^* a_2'$. The transformation performed by a beam splitter is given by

$$Q_{BS} = \begin{bmatrix} B & 0 \\ 0 & B^\# \end{bmatrix},$$

where B is a unitary matrix given by

$$B = e^{\iota \Xi/2} \begin{bmatrix} e^{\iota\Psi/2} & 0 \\ 0 & e^{-\iota\Psi/2} \end{bmatrix} \begin{bmatrix} \cos(\Theta)/2 & \sin(\Theta)/2 \\ -\sin(\Theta)/2 & \cos(\Theta)/2 \end{bmatrix} \begin{bmatrix} e^{\iota\Phi/2} & 0 \\ 0 & e^{-\iota\Phi/2} \end{bmatrix}.$$

Here, Ξ, Θ, Φ, Ψ are real numbers. Θ is called the *mixing angle* of the beam splitter and the most important parameter. Φ and Ψ introduce a phase difference in the two incoming and outgoing modes, respectively, while Ξ introduces an overall phase shift on both modes.

The operation of a beam splitter with $\Xi = \Psi = \Phi = 0$ can be modeled by an effective Hamiltonian $H_{BS}^0 = \iota\Theta(a_1^* a_2 - a_1 a_2^*)$ (see [4, Sect. 4.1] for details), in the sense that

$$Q_{BS} \begin{bmatrix} a \\ a^\# \end{bmatrix} = \exp(\iota H_{BS}^0) \begin{bmatrix} a \\ a^\# \end{bmatrix} \exp(-\iota H_{BS}^0),$$

where $a = (a_1, a_2)^\top$. By considering phase-shifted inputs $a_1 \to a_1 e^{\iota\frac{\theta+\Phi}{2}}$ and $a_2 \to a_2 e^{\iota\frac{\theta-\Phi}{2}}$ (θ being an arbitrary real number), it follows that a beam splitter with $\Xi = 0$ and $\Psi = -\Phi$ will have the effective Hamiltonian $H_{BS} = \iota\Theta(e^{-\iota\Phi} a_1^* a_2 - e^{\iota\Phi} a_1 a_2^*) = \alpha a_1^* a_2 + \alpha^* a_1 a_2$, with $\alpha = \iota\Theta e^{-\iota\Phi}$. This is the most general type of beam splitter that will be employed in the ensuing realization theory. A beam splitter with a Hamiltonian of the form H_{BS} is represented schematically by the symbol in Fig. 3.10.

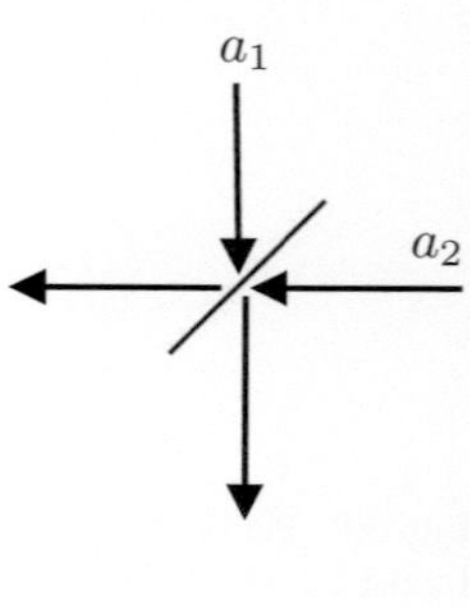

Fig. 3.10 Schematic representation of a beam splitter. Figure adapted from [3] © 2009 Society for Industrial and Applied Mathematics. Reprinted with permission. All rights reserved

Squeezer

Squeezing of a field is an operation in which the variance of one quadrature of a mode a, either q or p, is squeezed or attenuated at the expense of increasing the variance of the other quadrature. This type of operation is performed with a device called a *squeezer*. An ideal squeezer realizes the transformation

$$Q_{squeezer} = \begin{bmatrix} \cosh(s) & e^{\imath\theta}\sinh(s) \\ e^{-\imath\theta}\sinh(s) & \cosh(s) \end{bmatrix},$$

where s and θ are real parameters. The parameter s is referred to as the squeezing parameter, while θ is called the phase angle. For $s < 0$, the squeezer squeezes the amplitude quadrature of $e^{-\imath\theta/2}a$ (a phase-shifted version of a), while if $s > 0$, it squeezes the phase quadrature and then shifts the phase of the squeezed field by $\theta/2$. It is easy to see that $Q_{squeezer}^{-1}$ is given by

$$Q_{squeezer}^{-1} = \begin{bmatrix} \cosh(s) & -e^{\imath\theta}\sinh(s) \\ -e^{-\imath\theta}\sinh(s) & \cosh(s) \end{bmatrix}.$$

Single-mode squeezing can be performed on the cavity mode inside a DPA, while a squeezer for a bosonic field is an idealized device of infinite bandwidth that can be realized approximately by a high bandwidth DPA, see the discussion in Example 2.3. A squeezer with parameters s, θ is schematically represented by the symbol shown in Fig. 3.11.

Static Optical Linear Networks

It is well known that an arbitrary static linear optical network has a decomposition as a cascade of simpler static networks. In particular, any Bogoliubov matrix Q can be constructively decomposed as [10]:

$$\begin{aligned} Q &= \exp\begin{bmatrix} A_1 & 0 \\ 0 & A_1^\# \end{bmatrix} \exp\begin{bmatrix} 0 & D \\ D & 0 \end{bmatrix} \exp\begin{bmatrix} A_3 & 0 \\ 0 & A_3^\# \end{bmatrix} \\ &= \begin{bmatrix} \exp A_1 & 0 \\ 0 & \exp A_1^\# \end{bmatrix} \begin{bmatrix} \cosh D & \sinh D \\ \sinh D & \cosh D \end{bmatrix} \begin{bmatrix} \exp A_3 & 0 \\ 0 & \exp A_3^\# \end{bmatrix}, \end{aligned}$$

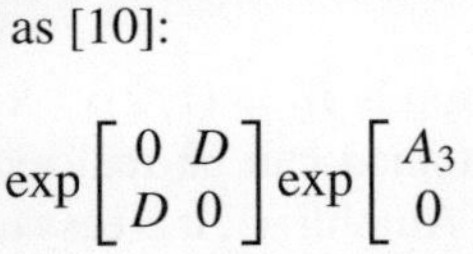

Fig. 3.11 Schematic representation of a squeezer. Figure adapted from [3] © 2009 Society for Industrial and Applied Mathematics. Reprinted with permission. All rights reserved

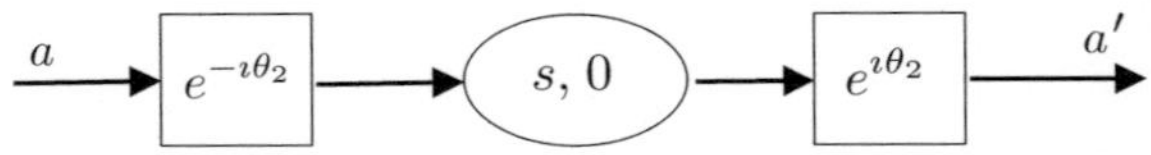

Fig. 3.12 Implementation of a squeezer with arbitrary phase angle employing a squeezer with a zero phase angle and two phase shifters. Figure adapted from [3] © 2009 Society for Industrial and Applied Mathematics. Reprinted with permission.

where A_1 and A_3 are skew symmetric complex matrices and D is a real diagonal matrix. The first and third factors in the decomposition correspond to passive static networks implementable by beam splitters and mirrors, while the second factor corresponds to an independent collection of squeezers (with trivial phase angles) each acting on a distinct field. Thus, fields going through a static linear optical network undergo a sequence of three operations: They are initially mixed by a passive network, then they undergo squeezing, and finally, they undergo a final passive transformation. In the special case where the entire network is passive, the D matrix is zero and the second factor becomes an identity matrix. As a simple illustration, a squeezer with arbitrary phase angle θ can be built by sandwiching a squeezer with phase angle 0 between a $-\theta/2$ phase shifter at its input and a $\theta/2$ phase shifter at its output, respectively. This is shown in Fig. 3.12.

3.1.3.2 Synthesis of One Degree-of-Freedom Linear Quantum Systems

We will now discuss the systematic realization of one degree-of-freedom linear quantum systems with a trivial scattering matrix, $S = I$. Such systems are completely described by a real symmetric Hamiltonian matrix $R = R^\top \in \mathbb{R}^{2\times 2}$ and complex coupling matrix $K \in \mathbb{C}^{m\times 2}$. Therefore, to realize these systems, one needs to be able to realize both R and K. Below, we will describe the quantum optical realization of one degree-of-freedom linear quantum systems that is based around a ring cavity structure, such as shown in Fig. 3.4, utilizing fully reflecting and partially transmitting mirrors, and placing beam splitters and nonlinear optical crystals at appropriate locations between the mirrors.

The Hamiltonian of the system is $H = (1/2)x^\top Rx$. In a one degree-of-freedom setup, such a quadratic Hamiltonian can be realized with a DPA as discussed in section "Degenerate Parametric Amplifier". It is easy to see from (3.8) that by suitably choosing the complex effective pump amplitude parameter ϵ and the cavity detuning parameter Δ of the DPA, one can realize any real symmetric matrix R. In fact, for any particular R, the choice of parameters ϵ and Δ for its realization is *unique*. For instance, to realize

$$R = \begin{bmatrix} 1 & -2 \\ -2 & 0.5 \end{bmatrix},$$

one solves the set of simultaneous linear equations

$$\Delta - \Im\{\epsilon\} = 2, \quad \Re\{\epsilon\} = -4, \quad \Delta + \Im\{\epsilon\} = 1$$

for $\Delta, \Im\{\epsilon\}, \Re\{\epsilon\}$ to yield the unique solution $\Delta = 3/2$ and $\epsilon = -4 - \imath/2$.

We now turn our attention to the realization of the coupling operator $L = Kx_0$. We begin by writing $K = [\, K_1^\top \; \ldots \; K_m^\top \,]^\top$, where $K_l \in \mathbb{C}^{1\times 2}$ for each $l = 1, \ldots, m$. Each K_l represents the coupling of the oscillator to the bosonic field $\mathcal{A}_l(t)$, and so it suffices to study how to realize the coupling to just one field. To this end, let us now consider only a single bosonic field $\mathcal{A}(t)$ that is coupled to the oscillator via the coupling operator $L = Kx_0$ for some $K \in \mathbb{C}^{1\times 2}$. It will be more convenient to express $L = \alpha q + \beta p$ as $L = \tilde{\alpha} a + \tilde{\beta} a^*$, with $\tilde{\alpha} = \frac{\alpha - \imath\beta}{2}$ and $\tilde{\beta} = \frac{\alpha + \imath\beta}{2}$, and a and a^* being the oscillator's annihilation and creation operators, respectively.

Partly taking inspiration from a scheme of Wiseman and Milburn for quantum non-demolition measurement of the position operator of a cavity [11], we consider the scheme shown in Fig. 3.13. The scheme relies upon using additional mirrors to realize an auxiliary cavity mode b with a resonance frequency coinciding with the reference frequency $\omega_r = \omega_p/2$. In a rotating frame at frequency ω_r, the auxiliary mode b interacts with the mode a via a two-mode squeezer with pump beam frequency ω_p and a beam splitter that collectively realize an interaction Hamiltonian H_{ab} of the form

$$H_{ab} = \frac{\imath}{2}(\epsilon_1 a^* b^* - \epsilon_1^* ab) + \frac{\imath}{2}(\epsilon_2 a^* b - \epsilon_2^* ab^*), \tag{3.9}$$

where ϵ_1 is the effective pump amplitude of the two-mode squeezer and ϵ_2 is given by $\epsilon_2 = 2\Theta e^{-\imath\Phi}$, where Θ is the mixing angle of the beam splitter and Φ is the relative phase introduced between the input fields by the beam splitter. If we take

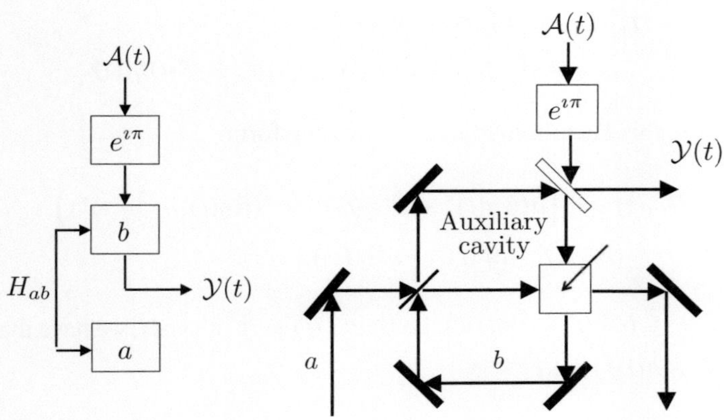

Fig. 3.13 Scheme for (approximate) implementation of a coupling $L = \tilde{\alpha} a + \tilde{\beta} a^*$ to cavity mode a using an auxiliary cavity b (whose dynamics is adiabatically eliminated), a two-mode squeezer, and a beam splitter with the appropriate parameters. The *left figure* is a block diagram showing the fast mode b interacting with the slow mode a via the direct interaction Hamiltonian H_{ab}, implemented by the two-mode squeezer and the beam splitter, and also interacting with a 180° phase-shifted input field $\mathcal{A}(t)$ to produce the output field $\mathcal{Y}(t)$. The *right figure* details the physical implementation of the block diagram. Figure adapted from [3] © 2009 Society for Industrial and Applied Mathematics. Reprinted with permission. All rights reserved

the coupling coefficient γ_2 of the partially transmitting mirror M on b to be large, such that b loses photons at a much faster rate than a and settles to the vacuum state very quickly, one can take the limit $\gamma_2 \to \infty$ to adiabatically eliminate b and obtain a reduced dynamics that only involves a. As shown in Appendix "Appendix A: Adiabatic Elimination of Coupled Cavity Modes" using the singular perturbation theory for QSDEs developed in [12], after eliminating b the resulting coupling operator to a is given by

$$L = \frac{1}{\sqrt{\gamma_2}}(-\epsilon_2^* a + \epsilon_1 a^*).$$

Thus, if we now choose $\epsilon_1, \epsilon_2, \gamma_2$ with γ_2 large and such that

$$\tilde{\alpha} = -\frac{\epsilon_2^*}{\sqrt{\gamma_2}}, \ \tilde{\beta} = \frac{\epsilon_1}{\sqrt{\gamma_2}}, \tag{3.10}$$

we can approximately realize any coefficients $\tilde{\alpha}$ and $\tilde{\beta}$ in a linear coupling operator $L = \tilde{\alpha} a + \tilde{\beta} a^*$. Note that the π radian phase shifter in front of $\mathcal{A}(t)$ in Fig. 3.13 is required to compensate for the scattering term in the unitary model that arises after the adiabatic elimination procedure (detailed in Appendix "Appendix A: Adiabatic Elimination of Coupled Cavity Modes").

For the special case where $\tilde{\alpha}, \tilde{\beta}$ satisfy $|\tilde{\alpha}| > |\tilde{\beta}| \geq 0$, there is an alternative realization of the linear coupling that is based on pre- and postprocessing with squeezed bosonic fields. Let $\gamma = |\tilde{\alpha}|^2 - |\tilde{\beta}|^2 > 0$, and consider the interaction Hamiltonian

$$\begin{aligned} H_{\text{int}}(t) &= \imath(L\eta(t)^* - L^*\eta(t)) \\ &= \imath((\tilde{\alpha} a + \tilde{\beta} a^*)\eta(t)^* - (\tilde{\alpha} a^* + \tilde{\beta}^* a)\eta(t)). \end{aligned}$$

We can rewrite this Hamiltonian in the following form:

$$\begin{aligned} H_{\text{int}}(t) &= \imath\left(a(\tilde{\alpha}\eta(t)^* - \tilde{\beta}^*\eta(t)) - a^*(\tilde{\alpha}\eta(t) - \tilde{\beta}\eta(t)^*)\right) \\ &= \imath\sqrt{\gamma}(a\eta'(t)^* - a^*\eta'(t)), \end{aligned}$$

where $\eta'(t) = \frac{1}{\sqrt{\gamma}}(\tilde{\alpha}\eta(t) - \tilde{\beta}\eta(t)^*)$. Letting $\mathcal{Z}(t) = \int_0^t \eta'(s)ds$, we have that $\mathcal{Z}(t) = \frac{1}{\sqrt{\gamma}}(\tilde{\alpha}\mathcal{A}(t) - \tilde{\beta}\mathcal{A}(t)^*)$, and

$$\begin{bmatrix} \mathcal{Z}(t) \\ \mathcal{Z}(t)^* \end{bmatrix} = Q \begin{bmatrix} \mathcal{A}(t) \\ \mathcal{A}(t)^* \end{bmatrix}, \quad Q = \begin{bmatrix} \frac{\tilde{\alpha}}{\sqrt{\gamma}} & -\frac{\tilde{\beta}}{\sqrt{\gamma}} \\ -\frac{\tilde{\beta}^*}{\sqrt{\gamma}} & \frac{\tilde{\alpha}}{\sqrt{\gamma}} \end{bmatrix}.$$

The key idea is reinterpreting the Hamiltonian in the form $H_{\text{int}}(t) = \imath\sqrt{\gamma}(a\eta'(t)^* - a^*\eta'(t))$, which describes the interaction of the oscillator with the new field $\mathcal{Z}(t)$. Since $|\tilde{\alpha}|^2 - |\tilde{\beta}|^2 = \gamma > 0$, we have that $|\alpha/\sqrt{\gamma}|^2 - |\beta/\sqrt{\gamma}|^2 = 1$; therefore, Q is Bogoliubov. In fact, $\mathcal{Z}(t)$ can be obtained from $\mathcal{A}(t)$ by passing the latter

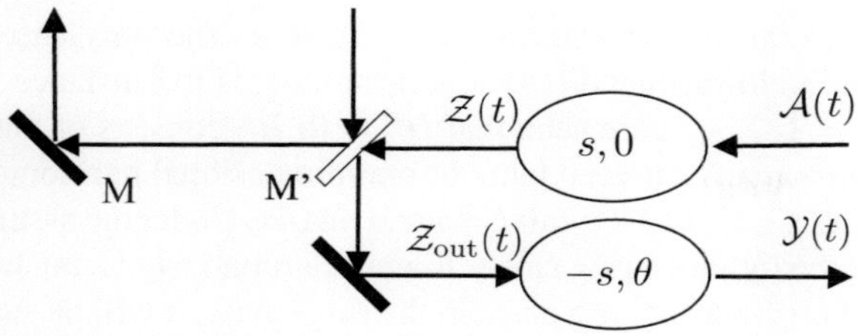

Fig. 3.14 Scheme for implementation of a coupling $L = \tilde{\alpha}a + \tilde{\beta}a^*$ with $\tilde{\alpha} > 0$ and $\tilde{\alpha} > |\tilde{\beta}|$. Here, $s = -\text{arccosh}(\tilde{\alpha}/\sqrt{\gamma})$, $\theta = \arg(\tilde{\beta})$ and the mirror M' has coupling coefficient $\gamma = \sqrt{\tilde{\alpha}^2 - |\tilde{\beta}|^2}$. Figure adapted from [3]

through a squeezer with the appropriate parameters (section "Squeezer"). Thus, $\mathcal{Z}(t)$ is a squeezed version of $\mathcal{A}(t)$. The required squeezer parameters are $s = -\text{arccosh}(\tilde{\alpha}/\sqrt{\gamma})$ and $\theta = \arg\tilde{\beta}$. The field $\mathcal{Z}(t)$ satisfies $[d\mathcal{Z}(t), d\mathcal{Z}(t)^*] = dt$ and the Itō rules,

$$\begin{bmatrix} d\mathcal{Z}(t) \\ d\mathcal{Z}(t)^* \end{bmatrix} \begin{bmatrix} d\mathcal{Z}(t) \; d\mathcal{Z}(t)^* \end{bmatrix} = Q \begin{bmatrix} 0 & 1 \\ 0 & 0 \end{bmatrix} Q^\top dt.$$

The interaction Hamiltonian H_{int} can be implemented in one arm of a ring cavity with a fully reflecting mirror M and a partially transmitting mirror M' with coupling coefficient γ, with $\mathcal{Z}(t)$ incident on M'. After the interaction, an output field $\mathcal{Z}_{out}(t)$ is produced by M' given by

$$\begin{aligned} \mathcal{Z}_{out}(t) &= U(t)^*\mathcal{Z}(t)U(t) \\ &= \frac{\tilde{\alpha}}{\sqrt{\gamma}} U(t)^*\mathcal{A}(t)U(t) - \frac{\tilde{\beta}}{\sqrt{\gamma}} U(t)^*\mathcal{A}(t)^*U(t). \end{aligned}$$

However, the actual output of interest is the field $\mathcal{Y}(t) = U(t)^*\mathcal{A}(t)U(t)$ that is produced when the oscillator interacts directly with $\mathcal{A}(t)$. Since Q and its inverse Q^{-1} are both Bogoliubov, $\mathcal{Y}(t)$ can be recovered from $\mathcal{Z}_{\text{out}}(t)$ by exploiting the relation,

$$\begin{bmatrix} \mathcal{Y}(t) \\ \mathcal{Y}(t)^* \end{bmatrix} = Q^{-1} \begin{bmatrix} \mathcal{Z}_{\text{out}}(t) \\ \mathcal{Z}_{\text{out}}(t)^* \end{bmatrix},$$

which follows directly from the fact that $(\mathcal{Z}(t), \mathcal{Z}(t)^*)^\top = Q(\mathcal{A}(t), \mathcal{A}(t)^*)^\top$. The complete implementation of this linear coupling is shown in Fig. 3.14.

3.1.3.3 Engineering the Interactions Between One-Dimensional Open Quantum Harmonic Oscillators

Following Theorem 3.1, the second ingredient for synthesizing a general linear quantum system is being able to realize a direct interaction Hamiltonian H^d given by

(3.7) between several quantum harmonic oscillators. The only case where this is not required, and field-mediated interactions suffice, is the instance when R_{jk} and L_j and S_j, $j, k = 1, \ldots, n$, are such that $H^d = 0$. H^d consists of the sum of direct interaction Hamiltonians between pairs of one-dimensional harmonic oscillators of the form $H_{kl} = x_k^\top C_{kl} x_l$ $(k \neq l)$ with C_{kl} a real matrix. Under the assumption that the time it takes for the light in a ring cavity to make a round trip is much faster than the timescales of all processes taking place in the ring cavity, it will be sufficient to only consider how to realize H_{kl} for any pair of one-dimensional harmonic oscillators and realizing all of them simultaneously in a network. Now, let $a_j = (p_j + \imath q_j)/2$ and $a_j^* = (p_j - \imath q_j)/2$ for $j = k, l$, and express H_{kl} as

$$H_{kl} = \epsilon_1 a_k^* a_l + \epsilon_1^* a_k a_l^* + \epsilon_2 a_k^* a_l^* + \epsilon_2^* a_k a_l$$

for some complex numbers ϵ_1 and ϵ_2. The first component $H_{kl}^1 = \epsilon_1 a_k^* a_l + \epsilon_1^* a_k a_l^*$ can be realized using a beam splitter with mixing angle $\Theta = |\epsilon_1|$, $\Phi = -\arg(\epsilon_1) + \frac{\pi}{2}$, $\Psi = -\Phi$, and $\Xi = 0$ (see section "Beam Splitter"). The second component $H_{kl}^2 = \epsilon_2 a_k^* a_l^* + \epsilon_2^* a_k a_l$ can be realized by interacting a_k and a_l in a two-mode squeezing process as described in section "Two-Mode Squeezing", inside a $\chi^{(2)}$ nonlinear crystal driven by a classical pump beam of frequency $2\omega_r$ and effective pump amplitude $-2\imath\epsilon_2$. The overall Hamiltonian H_{kl} is then realized by positioning the arms of the two ring cavities (with canonical operators x_k and x_l) such that their circulating light beams intersect at two locations. At one location, a beam splitter is placed to realize H_{kl}^1 and a pumped nonlinear crystal is placed at the second to realize H_{kl}^2, respectively. An example of this scheme is depicted in Fig. 3.15.

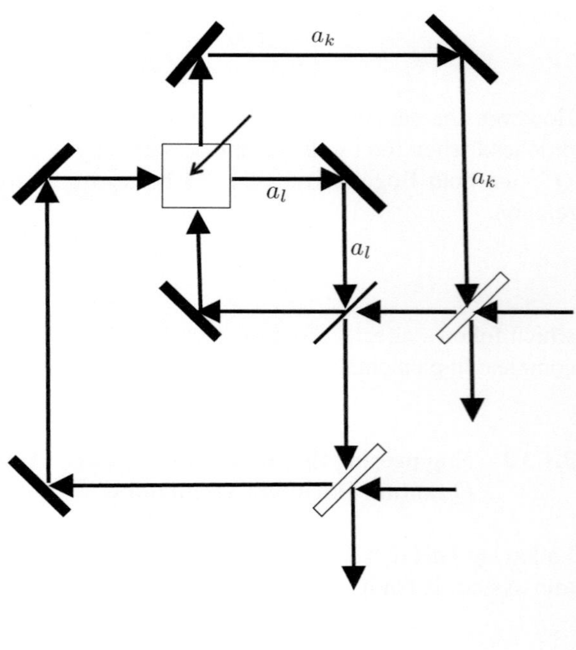

Fig. 3.15 Example implementation of the total direct interaction $H_{kl} = H_{kl}^1 + H_{kl}^2$ between the modes a_k and a_l of two ring cavities. Figure adapted from [3] © 2009 Society for Industrial and Applied Mathematics. Reprinted with permission. All rights reserved

3.1.4 Illustrative Synthesis Example

We now illustrate the network synthesis theory by applying the complete synthesis procedure to two degree-of-freedom linear quantum system. The system under consideration is coupled to a single external bosonic noise field $\mathcal{A}(t)$ and given by $G = (I_{4\times 4}, Kx, x^\top \mathrm{diag}(R_1, R_2)x)$, with $x = (q_1, p_1, q_2, p_2)^\top$, $K = [3/2\ \imath/2\ 1\ \imath]$, $R_1 \begin{bmatrix} 2 & 0.5 \\ 0.5 & 3 \end{bmatrix}$, and $R_2 = \begin{bmatrix} 1 & 0 \\ 0 & 1 \end{bmatrix}$.

Let $G_1 = (I_{2\times 2}, K_1x_1, (1/2)x_1^\top R_1x_1)$ and $G_2 = (I_{2\times 2}, K_2x_2, (1/2)x_2^\top R_2x_2)$ be two independent one degree-of-freedom linear quantum systems with $x_1 = (q_1, p_1)^\top$, $x_2 = (q_2, p_2)^\top$, $K_1 = [3/2\ \imath/2]$, and $K_2 = [1\ \imath]$. From Theorem 3.1, we have that G can be realized as the reducible network $G = \{\{G_1, G_2\}, H_{12}^d, G_2 \triangleleft G_1\}$, with the direct interaction Hamiltonian H_{12}^d between G_1 and G_2 given by (see (3.7))

$$\begin{aligned} H_{12}^d &= -\frac{1}{2\imath}x_2^\top (K_2^* K_1 - K_2^\top K_1^\#)x_1 \\ &= \frac{1}{2}x_2^\top \begin{bmatrix} 0 & -1 \\ 3 & 0 \end{bmatrix} x_1. \end{aligned}$$

This network is depicted in Fig. 3.3.

We shall now demonstrate how G_1, G_2, and H_{12}^d can be individually realized to synthesize the overall system G.

3.1.4.1 Synthesis of G_1 and G_2

From Sect. 3.1.3.2, we have that $R_1 = \begin{bmatrix} 2 & 0.5 \\ 0.5 & 3 \end{bmatrix}$ can be realized as a DPA with parameters $\Delta = 5$ and $\epsilon = 1 + \imath$, while $L_1 = K_1x_1$ can be (approximately) realized by the first scheme proposed in Sect. 3.1.3.2 and shown in Fig. 3.13. To achieve the desired coupling parameter, we can set the coupling coefficient of the mirror M to be $\gamma_2 = 100$, $\epsilon = 10$, and the beam splitter to have a mixing angle of -10, with all other parameters equal to 0 (recall that we require γ_2 to be sufficiently larger than the other parameters so that the auxiliary mode can be adiabatically eliminated). Overall, G_1 can be implemented around the ring cavity structure depicted in Fig. 3.16. G_2 can be realized similarly. The Hamiltonian $H_2 = (1/2)x_2^\top R_2x_2$ can be realized in the same manner as H_1 with the choice $\Delta = 2$ and $\epsilon = 0$. Since $\epsilon = 0$, to realize R_2 it suffices to have a cavity that is detuned from ω_r, the reference frequency in Remark 3.2, by an amount $\Delta = 2$. The coupling operator $L_2 = q_2 + \imath p_2 = 2a_2$, where a_2 is the annihilation operator/cavity mode of G_2, can be realized simply with a partially transmitting mirror with coupling coefficient $\kappa = 4$, through which an impinging field $\mathcal{A}_2(t)$ interacts with the cavity mode a_2 to produce an outgoing field $\mathcal{Y}_2(t)$. The implementation of G_2 is shown in Fig. 3.17.

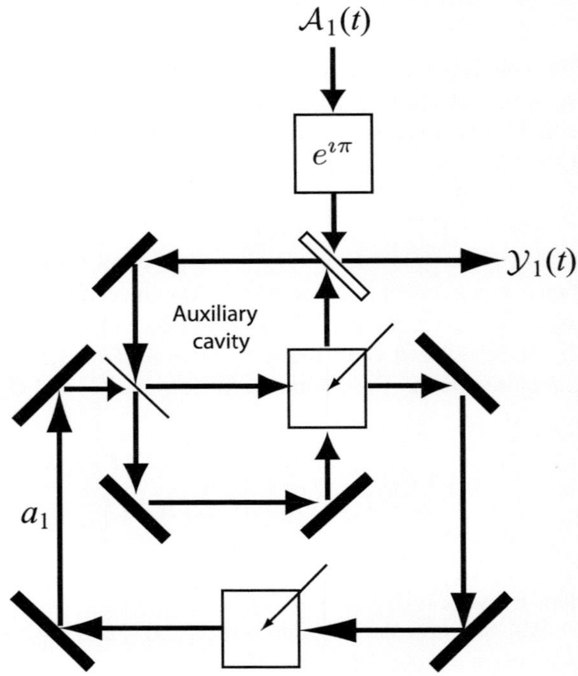

Fig. 3.16 Realization of G_1. Figure adapted from [3] © 2009 Society for Industrial and Applied Mathematics. Reprinted with permission.

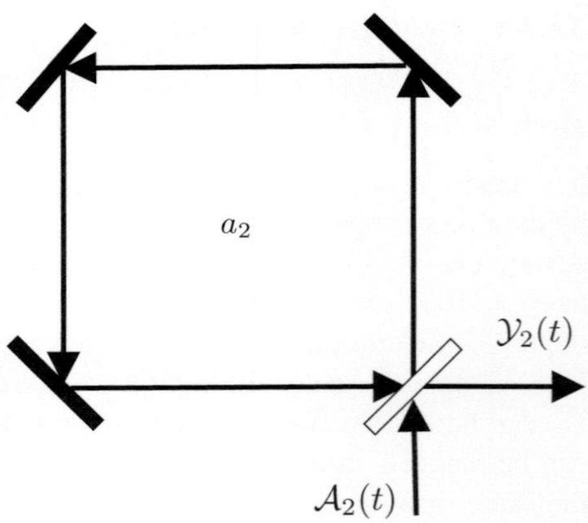

Fig. 3.17 Realization of G_2. Figure adapted from [3] © 2009 Society for Industrial and Applied Mathematics. Reprinted with permission.

3.1.4.2 Synthesis of H_{12}^d

Finally, we come to consider the realization of the direct interaction Hamiltonian H_{12}^d given by $H_{12}^d = \frac{1}{2}x_2^\top \begin{bmatrix} 0 & -1 \\ 3 & 0 \end{bmatrix} x_1$. We can rewrite H_{12}^d in terms of the cavity modes a_1 and a_2 as $H_{12}^d = -\imath(a_1^* a_2 - a_1 a_2^*) + 2\imath(a_1^* a_2^* - a_1 a_2)$. Let $H_{12,1}^d = -\imath(a_1^* a_2 - a_1 a_2^*)$ and $H_{12,2}^d = 2\imath(a_1^* a_2^* - a_1 a_2)$ so that $H_{12}^d = H_{12,1}^d + H_{12,2}^d$. The component $H_{12,1}^d = -\imath(a_1^* a_2 - a_1 a_2^*)$ can be realized using a beam splitter with mixing angle $\Theta = -1$ and all other parameters equal to 0 (section "Beam Splitter"). The second component $H_{12,2}^d = 2\imath(a_1^* a_2^* - a_1 a_2)$ corresponds to two-mode squeezing between a_k and a_l in a $\chi^{(2)}$ nonlinear crystal using a classical pump beam of frequency $\omega_p = 2\omega_r$ and effective amplitude $\epsilon = 4$.

3.1.4.3 Complete Realization of $G = \{\{G_1, G_2\}, H_{12}^d, G_2 \triangleleft G_1\}$

The system G can now be realized by (i) positioning the arms of the two (ring) cavities of G_1 and G_2 such that their internal light beams intersect at two points

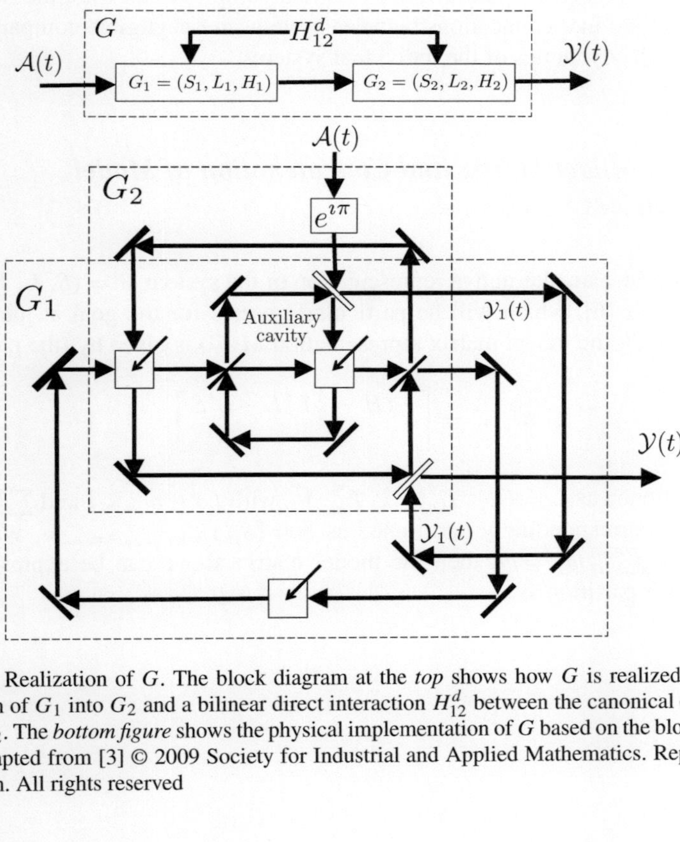

Fig. 3.18 Realization of G. The block diagram at the *top* shows how G is realized by a series connection of G_1 into G_2 and a bilinear direct interaction H_{12}^d between the canonical operators of G_1 and G_2. The *bottom figure* shows the physical implementation of G based on the block diagram. Figure adapted from [3] © 2009 Society for Industrial and Applied Mathematics. Reprinted with permission. All rights reserved

which are occupied by a beam splitter and two-mode squeezer realizing $H^d_{12,1}$ and $H^d_{12,2}$, respectively, and (ii) passing the output $\mathcal{Y}_1(t)$ of G_1 as input to G_2. The realization is shown in Fig. 3.18.

3.2 Architecture for Strict Realization Using Quantum Feedback Networks

We have shown that any linear quantum system can, in principle, be synthesized by a cascade of simple one degree-of-freedom linear quantum subsystems together with a direct interaction Hamiltonian between the canonical operators of the system. However, we have also seen that direct bilinear interaction Hamiltonians between subsystems are challenging to realize experimentally, possibly requiring some complex spatial arrangement and orientation of the subsystems. Thus, it becomes important to investigate alternative synthesis methods, at least approximate ones. Here, we describe one such method by exploiting the quantum feedback network formalism in [6], following the original proposal in [13]. In this scheme, the direct interaction Hamiltonians are approximately realized by suitable field interconnections among the systems. The approximation relies on the assumption that the time delays for establishing field interconnections between systems are negligible compared to the timescale of the dynamics of the individual systems.

3.2.1 *The Model Matrix and Concatenation of Model Matrices*

We will introduce an alternative representation of the system $G = (S, L, H)$ called a *model matrix* [6], which will be particularly useful for the goal at hand. For a given system G, the model matrix representation $M(G)$ is given by (the partitioned matrix):

$$M(G) = \begin{bmatrix} -\imath H - \frac{1}{2}L^*L & -L^*S \\ L & S \end{bmatrix}. \tag{3.11}$$

If L is partitioned as $L = (L_1^\top, L_2^\top, \ldots, L_{n_{\text{out}}}^\top)^\top$, with $L_j \in m_j \times 1$ and $\sum_{j=1}^{n_{\text{out}}} m_j = m$, and S is correspondingly partitioned as $S = [S_{jk}]_{j=1,\ldots,n_{\text{out}},k=1,\ldots,n_{\text{in}}}$ with $S_{jk} \in \mathbb{C}^{m_j \times m'_k}$ and $\sum_{k=1}^{n_{\text{in}}} m'_k = m$, then the model matrix above can be expressed with respect to this partition as:

$$M(G) = \begin{bmatrix} -\imath H - \frac{1}{2}\sum_{j=1}^{n_{\text{out}}} L_j^* L_j & -\sum_{j=1}^{n_{\text{out}}} L_j^* S_{j1} & \dots & -\sum_{j=1}^{n_{\text{out}}} L_j^* S_{jn_{\text{in}}} \\ L_1 & S_{11} & \dots & S_{1n_{\text{in}}} \\ \vdots & \vdots & \ddots & \vdots \\ L_{n_{\text{out}}} & S_{n_{\text{out}}1} & \dots & S_{n_{\text{out}}n_{\text{in}}} \end{bmatrix}. \quad (3.12)$$

Given a partitioning of $M(G)$, such as the one above, we can assign a *unique label* to each row and column of the partition. For instance, for the partitioning (3.12) we may attach the labels $s_0, s_1, \dots, s_{n_{\text{out}}}$ for the first, second, ..., $n_{\text{out}} + 1$-th row of $M(G)$, respectively, and $r_0, r_1, \dots, r_{n_{\text{in}}}$ for the first, second, ..., $n_{\text{in}} + 1$-th column of $M(G)$, respectively. With respect to this labeling scheme, block elements of $M(G)$ can be identified as,

$$M_{s_0 r_0}(G) = -\imath H - \frac{1}{2}\sum_{j=1}^{n_{\text{out}}} L_j^* L_j;\ M_{s_0 r_k}(G) = -\sum_{j=1}^{n_{\text{out}}} L_j^* S_{jk},\ k > 0;$$
$$M_{s_j r_0}(G) = L_j,\ j > 0;\ M_{s_j r_k} = S_{jk},\ j, k > 0.$$

Additionally, the original physical system G can be readily identified from the entries of its model matrix $M(G)$. For the sake of brevity, we will often omit G and write $M(G)$ simply as M and denote its entries as $M_{\alpha\beta}$, with α ranging over row labels and β ranging over the column labels. Also, the parameters (S, L, H) of G can be equivalently viewed as the parameters of the model matrix M.

When there are several model matrices, they can be concatenated together to form the model matrix of a larger system. This concatenation is the analogue, in terms of model matrices, of the concatenation product for open Markov quantum systems and is again denoted by the symbol $\boxplus$. If $G_1 = (S_1, L_1, H_1)$ and $G_2 = (S_2, L_2, H_2)$, then the concatenation $M(G_1) \boxplus M(G_2)$ is defined as:

$$M(G_1) \boxplus M(G_2) = \begin{bmatrix} -\imath H_1 - \imath H_2 - (1/2)L_1^* L_1 - (1/2)L_2^* L_2 & -L_1^* S_1 & -L_2^* S_2 \\ L_1 & S_1 & 0 \\ L_2 & 0 & S_2 \end{bmatrix}$$
$$= M(G_1 \boxplus G_2).$$

3.2.2 Edges, Elimination of Edges, and Reduced Markov Models

Following [6], one can associate a row partition labeled s_k with $k > 0$ with an output port s_k (having multiplicity n_{s_k}), while a column partition r_j with $j > 0$ can be associated with an input port r_j (having multiplicity n_{r_j}). If a system has a decomposition as the concatenation of several subsystems, an output port s_k from one subsystem can be connected to an input port r_j of another subsystem (possibly the same subsystem to which the output port belongs). This connection forms what we shall call an *internal edge* and denote by (s_k, r_j). Of course, for the connection to make sense, the ports s_k and r_j are required to have the same multiplicity. Such an edge then represents a *channel* from port s_k to port r_j. All ports that are part of an internal edge or channel are referred to collectively as *internal ports*, and fields entering or leaving such ports are called *internal fields*. All other input and output ports that do not belong to an internal edge are viewed as having semi-infinite edges (that do not terminate at either an input port or an output port) and are referred to as *external ports*. The associated semi-infinite edges are referred to as *external edges*. Fields entering or leaving external ports are called *external fields*.

Consider an internal edge (s_k, r_j). In such an edge, there is an associated finite time delay required for the signal from port s_k to travel to port r_j. Due to these time delays, concatenated systems possessing internal edges are longer Markov, despite each element in the concatenation being Markov. Fortunately, as shown in [6], in the limit that all time delays on internal edges go to zero, the non-Markov system converges to a *reduced* Markov model. Thus, when the time delays are negligible compared to the timescale of the dynamics of the systems connected by the internal edges, the reduced Markov model is an effective approximation of the non-Markov model for the network. We recall the following result:

Theorem 3.2 ([6, Theorem 3.1 and Lemma 16]) *Let $\tau_{(s_k,r_j)}$, $j, k > 0$, be the time delay for an internal edge (s_k, r_j) and assume that $I - S_{kj}$ is invertible. Then in the limit that $\tau_{(s_k,r_j)} \downarrow 0$, $M(G)$ with the edge (s_k, r_j) connected reduces to a simplified model matrix $M_{\rm red}$ with input ports labeled $r_0, r_1, \ldots, r_{j-1}, r_{j+1}, \ldots, r_{n_{\rm out}}$ and output ports labeled $s_0, s_1, \ldots, s_{k-1}, s_{k+1}, \ldots, s_{n_{\rm in}}$ (i.e., the connected ports r_j and s_k are removed from the labeling and the associated row and column removed from $M(G)$). The block entries of $M_{\rm red}$ are given by:*

$$(M_{\rm red})_{\alpha\beta} = M_{\alpha\beta} + M_{\alpha r_j}(1 - S_{kj})^{-1} M_{s_k \beta},$$

with $\alpha \in \{s_0, s_1, \ldots, s_{n_{\rm out}}\}\backslash\{s_k\}$ and $\beta \in \{r_0, r_1, \ldots, r_{n_{\rm in}}\}\backslash\{r_j\}$. $M_{\rm red}$ is the model matrix of a linear quantum stochastic system $G_{\rm red}$ with parameters:

$$\begin{aligned}(S_{\rm red})_{pq} &= S_{pq} + S_{pj}(I - S_{kj})^{-1} S_{kq},\\ (L_{\rm red})_p &= L_p + S_{pj}(I - S_{kj})^{-1} L_k,\\ H_{\rm red} &= H + \sum_{p=1}^{n_{\rm out}} \Im\{L_p^* S_{pj}(I - S_{kj})^{-1} L_p\},\end{aligned}$$

for all $p \in \{1, 2, \ldots, n_{\rm out}\}\backslash\{k\}$ and $q \in \{1, 2, \ldots, n_{\rm in}\}\backslash\{j\}$.

Internal edges may be sequentially eliminated, by allowing the time delays along these edges to go to zero one at a time, leading to a corresponding sequence of reduced model matrices. The order in which the edge eliminations are performed is irrelevant so that the final reduced model matrix is in fact unique [6, Lemma 17]. Suppose that M can be partitioned as:

$$\begin{bmatrix} -\imath H - (1/2)L_{\mathrm{i}}^* L_{\mathrm{i}} - (1/2)L_{\mathrm{e}}^* L_{\mathrm{e}} & -L_{\mathrm{i}}^* S_{\mathrm{ii}} - L_{\mathrm{i}}^* S_{\mathrm{ei}} & -L_{\mathrm{e}}^* S_{\mathrm{ie}} - L_{\mathrm{i}}^* S_{\mathrm{ii}} \\ L_{\mathrm{i}} & S_{\mathrm{ii}} & S_{\mathrm{ie}} \\ L_{\mathrm{e}} & S_{\mathrm{ei}} & S_{\mathrm{ee}} \end{bmatrix}, \tag{3.13}$$

where the subscript i refers to "internal" and e to "external", meaning that parameters with subscript i or ii pertain to internal ports, those with subscript e or ee pertain to external ports, while S_{ie} and S_{ei} pertain to scattering of internal fields to external fields and vice versa, respectively. A so-called *adjacency matrix* can be introduced to conveniently encode the interconnection among internal input and output ports. Let n_{i} denote the total multiplicity of internal input and output ports. We may also view a port with multiplicity k as k distinct ports of multiplicity 1, as necessary. Suppose that we number these multiplicity 1 ports consecutively starting from 1. Then, an adjacency matrix η is an $n_{\mathrm{i}} \times n_{\mathrm{i}}$ square matrix whose entries are either 1 or 0 with $\eta(j,k) = 1$ ($j, k \in \{1, 2, \ldots, n_{\mathrm{i}}\}$) only if the jth output port and the kth input port form a channel or internal edge. Clearly, there can be at most one element of any row or column of η that can take on the value 1. Internal edges can be simultaneously eliminated as follows:

Theorem 3.3 ([6, Sect. 5]) *Suppose that M has a partitioning based on internal and external components as in (3.13) and that connections between internal ports have been encoded in an adjacency matrix η. If $(\eta - S_{\mathrm{ii}})^{-1}$ exists, then the reduced model matrix M_{red} after simultaneous elimination of all internal edges has the parameters:*

$$\begin{aligned} S_{\mathrm{red}} &= S_{\mathrm{ee}} + S_{\mathrm{ei}}(\eta - S_{\mathrm{ii}})^{-1} S_{\mathrm{ie}}, \\ L_{\mathrm{red}} &= L_{\mathrm{e}} + S_{\mathrm{ei}}(\eta - S_{\mathrm{ii}})^{-1} L_{\mathrm{i}}, \\ H_{\mathrm{red}} &= H + \sum_{j=\mathrm{i,e}} \Im\{L_j^* S_{j\mathrm{i}}(\eta - S_{\mathrm{ii}})^{-1} L_{\mathrm{i}}\}. \end{aligned}$$

3.2.3 Main Synthesis Results

For $j = 1, \ldots, n$, let $G_{jk} = (S_{jk}, L_{jk}, 0)$ for $k = 1, \ldots, n, k \neq j$, and $G_{jj} = (S_{jj}, L_j, H_j)$, with $S_{jk} \in \mathbb{C}^{c_{jk} \times c_{jk}}$, $c_{jk} = c_{kj}$ and $c_{jj} = m$, $L_{jk} = K_{jk} x_j$, and $H_j = (1/2) x_j^\top R_j x_j$ with $R_j = R_j^\top \in \mathbb{R}^{2\times 2}$. Here $x_j = (q_j, p_j)^\top$ is as defined before. Let $G_j = \boxplus_{k=1}^n G_{jk}$ for $j = 1, \ldots, n$, and note that $G_j = (S_j, L_j, H_j)$ with $S_j = \mathrm{diag}(S_{j1}, S_{j2}, \ldots, S_{jn})$, $L_j = (L_{j1}^\top, L_{j2}^\top, \ldots, L_{jn}^\top)^\top$, and H_j as already defined.

Consider the model matrix M for the concatenated system $G = \boxplus_{j=1}^n G_j$ (see Fig. 3.19). Partition M correspondingly according to the given concatenation decomposition of G. We label the first $n+1$ rows of M as $s_{00}, s_{11}, \ldots, s_{1n}$, the next n rows

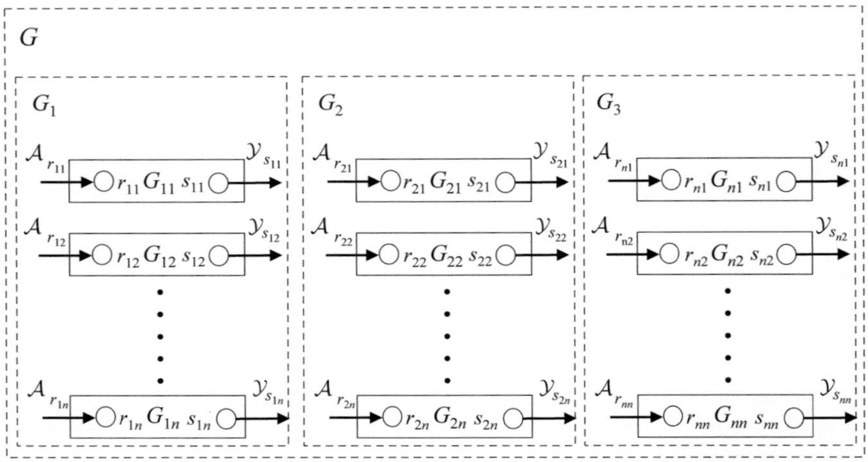

Fig. 3.19 The concatenation decomposition of G as $G = \boxplus_{j=1}^{n} \boxplus_{k=1}^{n} G_{jk}$. Figure adapted from [13] © 2010 IEEE

as $s_{21}, \ldots, s_{2n}$, and so on until the last n rows are labeled $s_{n1}, s_{n2}, \ldots, s_{nn}$. In an analogous fashion, we label the first $n+1$ columns of M as $r_{00}, r_{11}, \ldots, r_{1n}$, the next n columns as $r_{21}, \ldots, r_{2n}$, and so on until the last n columns $r_{n1}, r_{n2}, \ldots, r_{nn}$. For notational clarity, we will sometimes write a bracket around one of both of the subscripts of r or s (e.g., as in $s_{(n-1)k}$ or $r_{(k+1)(k+1)}$).

Theorem 3.4 *Let the output port s_{jk} be connected to the input port r_{kj} to form an internal edge/channel $e_{jk} = (s_{jk}, r_{kj})$ for all $j, k = 1, \ldots, n$, $j \neq k$. Assuming that $\begin{bmatrix} -S_{jk} & I \\ I & -S_{kj} \end{bmatrix}$ is invertible $\forall j, k = 1, \ldots, n$, $j \neq k$, then the reduced model matrix M_{red} obtained by allowing the delays in all internal edges $\{e_{jk},\ j \neq k\}$ go to zero has parameters given by:*

$$
\begin{aligned}
S_{\text{red}} &= \text{diag}(S_{11}, S_{22}, \ldots, S_{nn}), \\
L_{\text{red}} &= (L_{11}^{\top}, L_{22}^{\top}, \ldots, L_{nn}^{\top})^{\top}, \\
H_{\text{red}} &= \sum_{k=1}^{n} H_k + \sum_{j=1}^{n-1} \sum_{k=j+1}^{n} \Im\left\{[\, L_{jk}^{*} \; L_{kj}^{*} \,] \begin{bmatrix} I & -S_{jk} \\ -S_{kj} & I \end{bmatrix}^{-1} \begin{bmatrix} L_{jk} \\ L_{kj} \end{bmatrix}\right\}.
\end{aligned}
$$

The proof of the theorem is given in the Appendix, "Appendix B: Proof of Theorem 3.4". We also exploit the following lemma:

Lemma 3.1 *For any real 2×2 matrix R and unitary complex numbers S_{12} and S_{21} satisfying $S_{12}S_{21} \neq 1$, there exist 1×2 complex matrices K_1 and K_2 such that $R - \Im\{\frac{S_{12}}{1-S_{12}S_{21}} K_1^{*} K_2 + \frac{S_{21}}{1-S_{12}S_{21}} K_1^{\top} K_2^{\#}\} = 0$. In fact, a pair K_1 and K_2 satisfying this is given by $K_1 = [\, \kappa \; \imath\kappa \,]$ with κ an arbitrary nonzero real number* *and* $K_2 = 2\imath[\,1\ 0\,][-K_1^*\Delta^*\ K_1^\top\Delta\,]^{-1}R$, *where* $\Delta = 2\frac{S_{21}-S_{12}^*}{|1-S_{12}S_{21}|^2}$. *Or, alternatively,* $K_2 = [\,\kappa\ \imath\kappa\,]$ *and* $K_1 = 2\imath[\,1\ 0\,][\,K_2^*\Delta\ -K_2^\top\Delta^*\,]^{-1}R^\top$.

See "Appendix C: Proof of Lemma 3.1" in the appendix for a proof of the lemma. As a corollary to the above theorem and lemma, we have the following result:

Corollary 3.1 *Let* $c_{jk} = c_{kj} = 1$ *whenever* $j \neq k$, *and* $c_{jj} = m$ *for all* $j, k = 1, \ldots, n$. *Also, let* $S_{jj} = I_m$, $S_{kj} = e^{\imath\theta_{kj}}$ *and* $S_{jk} = e^{\imath\theta_{jk}}$ *with* $\theta_{kj}, \theta_{jk} \in [0, 2\pi)$ *satisfying* $\theta_{kj} + \theta_{jk} \neq 0$, $K_{jj} = K_j$ *and* $R_j = R_{jj} - 2\mathrm{sym}\left(\sum_{k=1,k\neq j}^{n} \Im\left\{\frac{1}{1-S_{jk}S_{kj}} K_{jk}^* K_{jk}\right\}\right)$ *(*K_j *and* R_{jj} *given), where* $\mathrm{sym}(A) = (1/2)(A + A^\top)$, *and the pair* (K_{jk}, K_{kj}) *(*$j \neq k$*) be given by:*

$$\begin{aligned} K_{jk} &= [\,\kappa_{jk}\ \imath\kappa_{jk}\,], \\ K_{kj} &= 2\imath[\,1\ 0\,][-K_{jk}^*\Delta_{jk}^*\ K_{jk}^\top\Delta_{jk}\,]^{-1}(R_{jk} - \Im\{K_j^\top K_k^\#\}), \end{aligned} \tag{3.14}$$

or

$$\begin{aligned} K_{kj} &= [\,\kappa_{jk}\ \imath\kappa_{jk}\,], \\ K_{jk} &= 2\imath[\,1\ 0\,][\,K_{kj}^*\Delta_{jk}\ -K_{kj}^\top\Delta_{jk}^*\,]^{-1}(R_{jk} - \Im\{K_j^\top K_k^\#\})^\top, \end{aligned} \tag{3.15}$$

where $R_{jk} = R_{kj}^\top \in \mathbb{R}^{2\times 2}$, $\Delta_{jk} = 2\frac{S_{kj}-S_{jk}^*}{|1-S_{kj}S_{jk}|^2}$ *and* κ_{jk} *is an arbitrary nonzero real constant for all* j, k. *Then the reduced Markov model* $G_{\mathrm{red}} = (S_{\mathrm{red}}, L_{\mathrm{red}}, H_{\mathrm{red}})$ *has the decomposition* $G_{\mathrm{red}} = \boxplus_{k=0}^{n} G_{\mathrm{red,k}}$ *with* $G_{\mathrm{red},0} = (0, 0, H_{\mathrm{red}})$ *and* $G_{\mathrm{red,k}} = (S_{kk}, L_{kk}, 0)$ *for* $k = 1, \ldots, n$. *Moreover, the network* $G_{\mathrm{net}} = (S_{\mathrm{net}}, L_{\mathrm{net}}, H_{\mathrm{net}})$ *formed by forming the series product of* $G_{\mathrm{red,n}} \triangleleft \ldots \triangleleft G_{\mathrm{red},2} \triangleleft G_{\mathrm{red},1}$ *within the concatenated system* G_{red} *and defined by* $G_{\mathrm{net}} = G_{\mathrm{red},0} \boxplus (G_{\mathrm{red,n}} \triangleleft \ldots \triangleleft G_{\mathrm{red},2} \triangleleft G_{\mathrm{red},1})$ *is a linear quantum stochastic system with parameters given by:*

$$\begin{aligned} S_{\mathrm{net}} &= I_m, \\ L_{\mathrm{red}} &= Kx,\ K = [\,K_1\ K_2\ \ldots\ K_n\,], \\ H_{\mathrm{red}} &= (1/2)x^\top Rx,\ R = [R_{jk}]_{j,k=1,\ldots,n}. \end{aligned}$$

In other words, G_{net} *realizes a linear quantum stochastic system with the above parameters.*

Remark 3.3 The series connection $G_{\mathrm{red,n}} \triangleleft \ldots \triangleleft G_{\mathrm{red},2} \triangleleft G_{\mathrm{red},1}$ is equivalent to forming and subsequently eliminating the internal edges $\{(s_{kk}, r_{(k+1)(k+1)});\ k = 1, \ldots, n-1\}$.

The proof of this corollary is given in Appendix "Appendix D: Proof of Corollary 3.1". It shows that any linear quantum system can be realized by the quantum network G_{net} constructed following the prescription of the corollary. Finally, any system $(S, Kx, (1/2)x^\top Rx)$ can be constructed as $(S, Kx, (1/2)x^\top Rx) = (I, Kx, (1/2)x^\top Rx) \triangleleft (S, 0, 0)$. The latter is the cascade of a static network realizing the unitary scattering matrix S and the system $(I, Kx, (1/2)x^\top Rx)$.

Example 3.1 Consider a two degree-of-freedom linear quantum system $G_{\mathrm{sys}} = (I, Kx, (1/2)x^\top Rx)$ $(x = (q_1, p_1, q_2, p_2)^\top)$ with

$$K = [\, K_1 \; K_2 \,] = [\, 3/2 \;\; \imath/2 \;\; 1 \;\; \imath \,] \; (K_j \in \mathbb{C}^{1\times 2},\; j = 1, 2);$$

$$R = \begin{bmatrix} R_{11} & R_{12} \\ R_{12}^\top & R_{22} \end{bmatrix} = \begin{bmatrix} 2 & 0.5 & 1 & 1 \\ 0.5 & 3 & -1 & -1 \\ 1 & -1 & 1 & 0 \\ 1 & -1 & 0 & 1 \end{bmatrix} \; (R_{jk} \in \mathbb{R}^{2\times 2},\; j, k = 1, 2).$$

Let $x_1 = (q_1, p_1)^\top$ and $x_2 = (q_2, p_2)^\top$. Define

$$G_1 = \left(\mathrm{diag}(I_m, S_{12}), \begin{bmatrix} K_{11} \\ K_{12} \end{bmatrix} x_1, (1/2)x_1^\top R_1 x_1 \right),$$

and

$$G_2 = \left(\mathrm{diag}(S_{21}, I_m), \begin{bmatrix} K_{21} \\ K_{22} \end{bmatrix} x_2, (1/2)x_2^\top R_2 x_2 \right),$$

with parameters that will be determined in the following. Set $\theta_{12} = 0$ and $\theta_{21} = \pi/2$, so that $\theta_{12} + \theta_{21} \neq 0$ as prescribed in Corollary 3.1. Then, set $S_{12} = e^{\imath\theta_{12}} = 1$, $S_{21} = e^{\imath\theta_{21}} = \imath$, $K_{11} = K_1 = [\, 3/2 \;\; 1/2 \,]$, and $K_{22} = K_2 = [\, 1 \;\; \imath \,]$. Calculate $\Delta_{12} = 2\frac{S_{21} - S_{12}^*}{|1 - S_{12}S_{21}|^2} = -1 + \imath$ and set $\kappa_{12} = 1$. By Corollary 3.1, we then set $K_{12} = [\, \kappa_{12} \;\; \imath\kappa_{12} \,]$ $= [\, 1 \;\; \imath \,]$ and compute $K_{21} = [\, 1.25 - 0.25\imath \;\; 1.75 + 0.75\imath \,]$,

$$R_1 = \begin{bmatrix} 1 & 0.5 \\ 0.5 & 2 \end{bmatrix},$$

and

$$R_2 = -\begin{bmatrix} 0.625 & 2 \\ 2 & 2.625 \end{bmatrix}.$$

Therefore, all the parameters of G_1 and G_2 have now been determined. Label the ports of G_1 and G_2 according to the convention adopted in this section. Concatenate G_1 and G_2 and form the internal edges (s_{12}, r_{21}) and (s_{21}, r_{12}). Then, eliminate these edges to form G_{red}. Finally, eliminating the edge (s_{11}, r_{22}) from G_{red} (see Remark 3.3) yields G_{net} as an approximate realization of G_{sys}. The realization is depicted in Fig. 3.20. G_1 and G_2 can be physically realized in the quantum optic domain following the constructions in Sect. 3.1.3. A quantum optical circuit as a physical realization of G_{sys} is shown in Fig. 3.21.

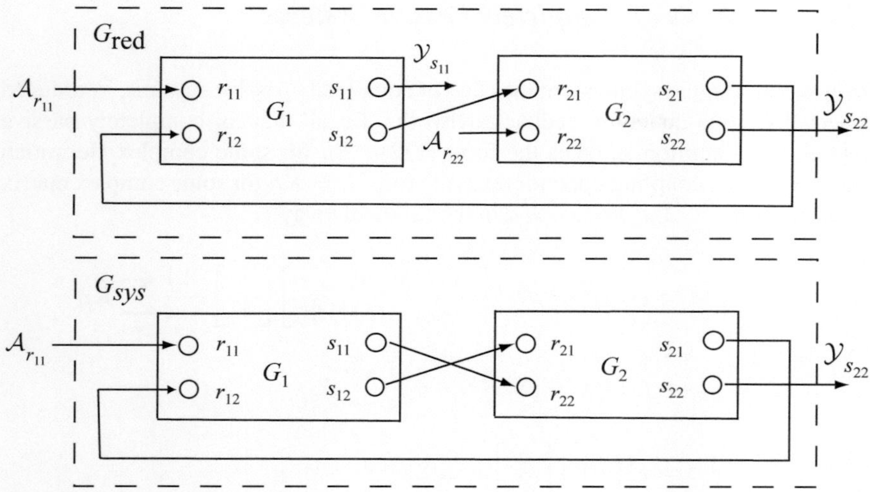

Fig. 3.20 Realization of G_{sys} via a quantum feedback network. In the *top figure*, the internal edges (s_{12}, r_{21}) and (s_{21}, r_{12}) are formed and eliminated to obtain a reduced Markov model G_{red}. Then in the *bottom figure*, a series (cascade) connection is formed by the eliminating the internal edge (s_{11}, r_{22}) to realize G_{sys}. Figure adapted from [13] © 2010 IEEE

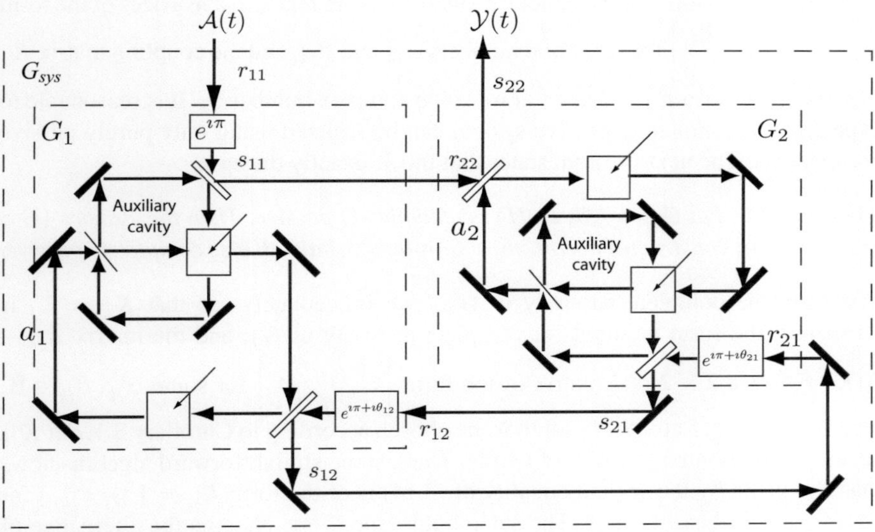

Fig. 3.21 A quantum optical circuit that realizes G_{sys} according to the quantum feedback network in Fig. 3.20. Here $a_j = \frac{q_j + \imath p_j}{2}$ is cavity mode of the optical cavity around which the physical realization of the oscillator G_j is based, $j = 1, 2$. The *dashed box* labeled G_j is the part of the circuit realizing G_j, and the input and output ports of G_j are indicated by their respective labels. Figure adapted from [24]

3.2.4 Synthesis of Completely Passive Systems

We now consider the synthesis of the class of completely passive systems, introduced in Sect. 2.4.1, via quantum feedback networks. Recall that for completely passive systems, the Hamiltonian takes the form $(1/2)a^*\tilde{R}a$ for some complex Hermitian matrix $\tilde{R}$ and the coupling operator takes the form $L = \tilde{K}a$ for some complex matrix $\tilde{K}$. Let us write $(1/2)a^*\tilde{R}a$ and $\tilde{K}a$ in the following way:

$$(1/2)a^*\tilde{R}a = (1/2)\begin{bmatrix} a^* & a^\top \end{bmatrix}\begin{bmatrix} (1/2)\tilde{R} & 0_{n\times n} \\ 0_{n\times n} & (1/2)\tilde{R}^\# \end{bmatrix}\begin{bmatrix} a \\ a^\# \end{bmatrix} - \frac{1}{4}\sum_{j=1}^{n}\tilde{R}_{jj}$$

$$= (1/2)x^\top\Re\{\Sigma^*\tilde{R}\Sigma\}x - \frac{1}{4}\sum_{j=1}^{n}\tilde{R}_{jj}$$

$$\tilde{K}a = \tilde{K}\begin{bmatrix} I & 0 \end{bmatrix}\begin{bmatrix} a \\ a^\# \end{bmatrix} = \tilde{K}\Sigma x.$$

Therefore, $H = (1/2)x^\top Rx$ with $R = \Re\{\Sigma^*\tilde{R}\Sigma\}$, and $L = Kx$ with $K = \tilde{K}\Sigma$.

From the above, it easily follows that for any completely passive system, the block diagonal elements R_{jj} of $R = [R_{ij}]_{i,j=1,\ldots,n}$ must be *diagonal* of the form $\lambda_j I_2$ for some $\lambda_j \in \mathbb{R}$, the off-diagonal block elements R_{jk} are real 2×2 matrices of the form $R_{jk} = \begin{bmatrix} \alpha_{jk} & \beta_{jk} \\ -\beta_{jk} & \alpha_{jk} \end{bmatrix}$ for some real numbers α_{jk} and β_{jk}, and the coupling matrix K_j to x_j is of the form $K_j = [\,\gamma_j \;\; \imath\gamma_j\,]$ for some complex number γ_j. It is reasonable to expect that a completely passive system can be realized using only purely passive components. The next theorem states that this is exactly the case.

Theorem 3.5 *Let $G_{\rm sys} = (S, L, H)$ be completely passive. Then the systems $\{G_j;\ j = 1,\ldots,n\}$ constructed according to Corollary 3.1 are all also completely passive.*

Proof By the complete passivity of $G_{\rm sys}$, we immediately see that $K_{jj} = K_j$ is already of the form required for complete passivity of G_j, and the matrix $R_{jk} - \Im\{K_j^\top K_k^\#\}$ is a 2×2 real matrix of the form $\begin{bmatrix} \alpha'_{jk} & \beta'_{jk} \\ -\beta'_{jk} & \alpha'_{jk} \end{bmatrix}$ for some $\alpha'_{jk}, \beta'_{jk} \in \mathbb{R}$, whenever $j \neq k$. Let κ_{jk}, S_{jk}, and S_{kj} be chosen according to Corollary 3.1. Set K_{jk} according to the first equality of (3.14). Then, some straightforward algebra shows that K_{kj} given by the second equality of (3.14) is of the form $K_{kj} = [\,\gamma'_{kj} \;\; \imath\gamma'_{kj}\,]$ for some $\gamma'_{kj} \in \mathbb{C}$, just like K_{jk}. The same holds true if one chooses the alternative of setting K_{kj} according to the first equality of (3.15) and computing K_{jk} according to the second equality of (3.15). Finally, given this special form of K_{jk}, it is easily inspected that $\mathrm{sym}\big(\Im\{\frac{1}{1-S_{jk}S_{kj}}K_{jk}^*K_{jk}\}\big)$ is a diagonal matrix of the form $\lambda_{jk}I_2$ for some $\lambda_{jk} \in \mathbb{R}$ for all $j, k, j \neq k$ (recall that $\mathrm{sym}(A) = (1/2)(A + A^\top)$), and since R_{jj} is also diagonal of this form (again from the passivity of $G_{\rm sys}$), we have that $R_j = R_{jj} - 2\sum_{k=1,k\neq j}^{n}\mathrm{sym}\big(\Im\{\frac{1}{1-S_{jk}S_{kj}}K_{jk}^*K_{jk}\}\big)$ is again of the same form, for all j. Therefore, it now follows that each subsystem G_j of Sect. 3.2.3 with parameters determined according to Corollary 3.1 is completely passive. □

We conclude the discussion with an example synthesis of a completely passive system.

Example 3.2 Consider a completely passive two degree-of-freedom linear quantum system $G_{\rm sys} = (I, Kx, (1/2)x^\top Rx)$ $(x = (q_1, p_1, q_2, p_2)^\top)$ with

$$K = [\, K_1\ K_2 \,] = [\, -3+\imath\ \ -3\imath - 1\ \ 1\ \ \imath \,]\ (K_j \in \mathbb{C}^{1\times 2},\ j = 1, 2);$$

$$R = \begin{bmatrix} R_{11} & R_{12} \\ R_{12}^\top & R_{22} \end{bmatrix} = \begin{bmatrix} 2 & 0 & 1 & 4 \\ 0 & 2 & -4 & 1 \\ 1 & -4 & 1 & 0 \\ 4 & 1 & 0 & 1 \end{bmatrix} \quad (R_{jk} \in \mathbb{R}^{2\times 2},\ j, k = 1, 2).$$

Setting $\kappa_{12} = 1$, $S_{12} = 1$, $S_{21} = \imath$, and $K_{12} = [\, \kappa_{12}\ \imath\kappa_{12} \,] = [\, 1\ \imath \,]$, from Corollary 3.1 we obtain $K_{11} = K_1$, $K_{22} = K_2$, $K_{21} = [\, 0.5 - 0.5\imath\ \ 0.5 + 0.5\imath \,]$, $R_1 = 0_{2\times 2}$, and $R_2 = 0.5 I_2$. A passive optical circuit that realizes $G_{\rm sys}$ is illustrated in Fig. 3.22.

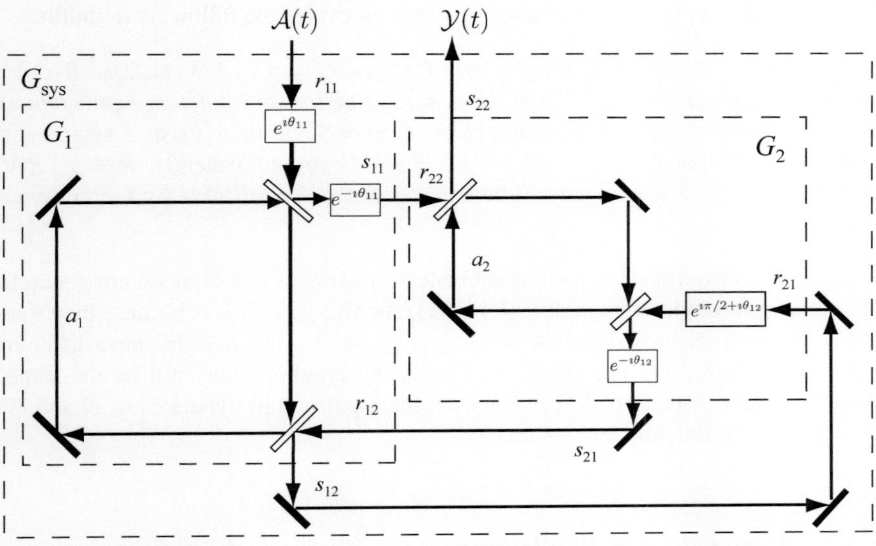

Fig. 3.22 A quantum optical circuit that realizes the passive system $G_{\rm sys}$ of this example. The circuit consists of only passive optical components: mirrors and phase shifters. Here, $\theta_{11} = \pi - \arctan(1/3)$ and $\theta_{21} = -\frac{\pi}{4}$, and the values of all other parameters of devices in this circuit can be determined according to Sect. 3.1. Figure adapted from [24]

3.3 Transfer Function Realization

In Sect. 3.1, we have considered a *strict* type of realization problem for linear quantum systems. It is concerned with synthesizing a linear quantum system described by a given triplet of parameters, $\{S, L = Kx, H = (1/2)x^\top Rx\}$. Strict realizability is relevant to the situation where the dynamics of $x(t)$ is tied to an implementation of a quantum information processing algorithm and has to be realized as given. However, in some linear quantum control problems, including robust disturbance attenuation [14] and LQG synthesis [15] (to be discussed in Chap. 5), the internal dynamics are unimportant. In such problems, what is important is the transfer function of the system as determined by the system matrices (A, B, C, D). Thus, there is freedom to modify the internal dynamics without modifying the transfer function since the latter is invariant under the similarity transformation $(A, B, C, D) \mapsto (VAV^{-1}, VB, CV^{-1}, D)$ for any invertible matrix V. However, for linear quantum systems, the transformation matrix V is restricted to be a real *symplectic* matrix. That is, V satisfies the condition $V\mathbb{J}_nV^\top = \mathbb{J}_n$. This is to guarantee that the transformed variable $Z(t) = VX(t)$ satisfies the CCR for quantum oscillators: $Z(t)Z(t)^\top - (Z(t)Z(t)^\top)^\top = 2\imath\mathbb{J}_n$, so that the transformed system is again a physical linear quantum system. Recall that the set of all symplectic matrices of a fixed dimension form a group. In particular, V^{-1} is again symplectic. A similarity transformation with a symplectic matrix V in linear quantum systems corresponds to replacing $G = (S, Kx, (1/2)x^\top Rx)$ with $G' = (S, KV^{-1}x, (1/2)x^\top V^{-T}RV^{-1}x)$. This motivates the following definition:

Definition 3.1 Let $G = (S, Kx, (1/2)x^\top Rx)$ and $G' = (S', K'x, (1/2)x^\top R'x)$ be two linear quantum systems. Then, G' is said to be *transfer function equivalent* to G (or a *transfer function realization* of G) if $S' = S$ and there exists a symplectic matrix V such that $R' = V^{-\top}RV^{-1}$, $K' = KV^{-1}$ (or, equivalently, $R = V^\top R'V$ and $K = K'V$). G is then said to be *transfer function realizable* by G', and vice versa.

Remark 3.4 Two transfer function equivalent systems G and G' need not generate the same input-output dynamics $(\mathcal{A}(t), \mathcal{Y}(t))$ for all $t \geq 0$. This is because they start at a nonzero initial condition $x(0) = (q_1, p_1, q_2, p_2, \ldots, q_n, p_n)^\top$ but have different parameters (S, K, R). Therefore, their transient dynamics may not be the same. However, if the A matrix of G is Hurwitz, the input-output dynamics of G and G' converge asymptotically as $t \to \infty$.

3.3.1 *Pure Cascade Realization of the Transfer Function of Linear Quantum Systems*

Compared to the strict realization, transfer function realization allows the additional freedom of performing a symplectic similarity transformation on the internal vector $x(t)$. Thus, a question that we may ask is, what kind of simplification can we obtain in a transfer function realization of a given system? In particular, are there systems whose transfer functions can be realized by only a cascade of one degree-of-freedom linear quantum systems, without requiring any direct Hamiltonian interactions between them or approximate realizations thereof? An affirmative answer is given here for the latter question, following the treatment in [16]. Moreover, it will be shown that the class of completely passive systems is such a class of systems and their cascade realization will be explicitly given.

3.3.2 Conditions for Realizability by a Pure Cascade Connection

We begin by introducing the following terminology: A square matrix F is said to be lower 2×2 block triangular if it has a lower block triangular form when partitioned into 2×2 blocks:

$$F = \begin{bmatrix} F_{11} & 0_{2\times 2} & 0_{2\times 2} & \dots & 0_{2\times 2} \\ F_{21} & F_{22} & 0_{2\times 2} & \dots & 0_{2\times 2} \\ \vdots & \ddots & \ddots & \ddots & \vdots \\ F_{n1} & F_{n2} & \dots & \dots & F_{nn} \end{bmatrix},$$

where F_{jk}, $j \leq k$, is of dimension 2×2. We first have the following:

Lemma 3.2 *The cascade connection $G_n \lhd G_{n-1} \lhd \cdots \lhd G_1$ of one degree-of-freedom linear quantum systems $G_i = (S_i, K_i x_i, (1/2)x_i^\top R_i x_i)$ $(i = 1, \dots, n)$ realizes a linear quantum system $G = (S, Kx, (1/2)x^\top Rx)$ with $S = S_{n\leftarrow 1}$, $K = \left[\, S_{n\leftarrow 2}K_1 \; S_{n\leftarrow 3}K_2 \; \dots \; K_n \,\right]$, $R = [R_{ij}]_{i,j=1,\dots,n}$, where $R_{jj} = R_j$, $R_{kj} = \Im\{K_k^* S_{k\leftarrow j+1}K_j\}$ whenever $k < j$ and $R_{jk} = R_{kj}^\top$ whenever $j > k$. In particular, $R + \Im\{K^*K\}$ is lower 2×2 block triangular.*

Proof The proof proceeds along the lines of the proof of Theorem 3.1. By the series product formula from Sect. 3.1.1, for the cascade of two one degree-of-freedom linear quantum systems $G_1 = (S_1, K_1 x_1, (1/2)x_1^\top R_1 x_1)$ and $G_2 = (S_2, K_2 x_2, (1/2)x_2^\top R_2 x_2)$, we get the oscillator $G_{(2)} = G_2 \lhd G_1 = (S_2 S_1, S_2 K_1 x_1 + K_2 x_2, (1/2)x_1^\top R_1 x_1 + (1/2)x_2^\top R_2 x_2 + x_2^\top \Im\{K_2^* S_2 K_1\} x_1)$. Letting $x_{(2)} = (x_1^\top, x_2^\top)^\top$, the latter may be compactly written as: $G_{(2)} = (S_{(2)}, K_{(2)}x_{(2)}, (1/2)x_{(2)}^\top R_{(2)} x_{(2)}^\top)$ with $S_{(2)} = S_{2\leftarrow 1} = S_2 S_1$, $K_{(2)} = [\, S_2 K_1 \; K_2 \,]$, and

$$R_{(2)} = \begin{bmatrix} R_1 & \Im\{K_2^* S_{2\leftarrow 2} K_1\}^\top \\ \Im\{K_2^* S_{2\leftarrow 2} K_1\} & R_2 \end{bmatrix}.$$

Repeating the computation for $G_{(k)} = G_k \lhd G_{(k-1)}$ iteratively for $k = 3, \dots, n-1$ and writing $x_{(k)} = (x_1^\top, x_2^\top, \dots, x_k^\top)^\top$ and

$$G_{(k)} = (S_{(k)}, K_{(k)}x_{(k)}, (1/2)x_{(k)}^\top R_{(k)}x_{(k)})$$

at each iteration k, we arrive at the desired result with $G = G_{(n)}$, $S = S_{(n)}$, $K = K_{(n)}$ and $R = R_{(n)}$ as stated in the lemma.

To see that $R + \Im\{K^*K\}$ is lower 2×2 block triangular, we note that K^*K may be expressed as follows:

$$K^*K = \begin{bmatrix} K_1^*K_1 & K_1^*S_2^*K_2 & K_1^*S_{3\twoheadleftarrow 2}^*K_2 & \dots & K_1^*S_{n\twoheadleftarrow 2}^*K_n \\ K_2^*S_2K_1 & K_2^*K_2 & K_2^*S_3^*K_3 & \dots & K_2^*S_{n\twoheadleftarrow 3}^*K_n \\ \vdots & \ddots & \ddots & \ddots & \vdots \\ K_n^*S_{n\twoheadleftarrow 2}K_1 & K_n^*S_{n\twoheadleftarrow 3}K_2 & \dots & K_n^*S_nK_{n-1} & K_n^*K_n \end{bmatrix}.$$

Note that since K^*K is by definition a Hermitian matrix, the 2×2 block elements above the diagonal blocks are the Hermitian transpose of the corresponding elements below the diagonal blocks. It follows therefore that the imaginary part of the block $(K^*K)_{jk}$ at block row j and block column k must satisfy the relation: $\Im\{(K^*K)_{jk}\} = -\Im\{(K^*K)_{kj}\}^\top$. However, from the expression for R derived above and its symmetry, we already have that if $k > j$:

$$R_{jk} = R_{kj}^\top = \Im\{K_k^*S_{k\twoheadleftarrow j+1}K_j\}^\top = \Im\{(K^*K)_{kj}\}^\top.$$

Therefore, the off-diagonal upper block elements of R cancel those of $\Im\{K^*K\}$ when they are summed, and we conclude that the matrix $R + \Im\{K^*K\}$ is lower 2×2 block triangular. □

Recall again the partitioning of R as $R = [R_{jk}]_{j,k=1,\dots,n}$ with $R_{jk} \in \mathbb{R}^{2\times 2}$ and of K as $K = [\, K_1 \; K_2 \; \dots \; K_n \,]$ with $K_k \in \mathbb{C}^{m\times 2}$. We may now state the following result:

Theorem 3.6 *A linear quantum system $G = (S, Kx, (1/2)x^\top Rx)$ with n degrees of freedom is realizable by a pure cascade of n one degree-of-freedom linear quantum systems (without a direct interaction Hamiltonian) if and only if there exists a permutation matrix P such that*

$$Px = (q_{\pi(1)}, p_{\pi(1)}, q_{\pi(2)}, p_{\pi(2)}, \dots, q_{\pi(n)}, p_{\pi(n)})^\top,$$

with π some permutation map of $\{1, 2, \dots, n\}$ to itself, such that the matrix $PAP^\top = 2P\mathbb{J}_n(R + \Im\{K^*K\})P^\top$ is lower 2×2 block triangular. If this condition is satisfied then G can be explicitly constructed as the cascade connection $G_n \lhd G_{n-1} \lhd \dots \lhd G_1$ with $G_1 = (S, K_1'x_1', (1/2)(x_1')^\top R_{11}'x_1')$, and $G_k = (I, K_k'x_k', (1/2)(x_k')^\top R_{kk}'x_k')$ for $k = 2, \dots, n$, where $x_k' = (q_{\pi(k)}, p_{\pi(k)})^\top$, $R' = PRP^\top = [R_{ij}']$, $K' = KP^\top = [\, K_1' \; K_2' \; \dots \; K_n' \,]$, with $R_{ij}' \in \mathbb{R}^{2\times 2}$ and $K_j' \in \mathbb{C}^{m\times 2}$.

Proof The proof of the only if part is a consequence of Lemma 3.2, as follows. If G can be realized by a pure cascade connection of n one degree-of-freedom linear quantum systems, there must exist a permutation map π and an associated permutation matrix P as given in the statement of the theorem such that

$G = (S, K'x', (1/2)(x')^\top R'x')$ and $A' = PAP^\top$ is lower 2×2 block triangular, where $x' = Px$.

Conversely, the if part of the proof can be shown by explicitly constructing a pure cascade connection of n one degree oscillators that realizes G. Suppose there is permutation matrix P such that $A' = PAP^\top$ is lower 2×2 block triangular. Let $G = (S, K'x', (1/2)(x')^\top R'x')$ be as defined in the theorem. Then, the matrix $(1/2)\mathbb{J}_n^{-1}A' = -(1/2)\mathbb{J}_n A' = R' + \Im\{(K')^* K'\}$ is also lower 2×2 block triangular. As we already saw in the proof of Lemma 3.2, this structure implies that $R'_{jk} = \Im\{(K'_j)^* K'_k\}$ whenever $k > j$ and $R'_{kj} = \Im\{(K'_j)^* K'_k\}^\top$ if $k < j$. Now, using the notation of Lemma 3.2, let us define the one degree-of-freedom linear quantum systems G_k for $k = 1, \ldots, n$ as $G_1 = (S, K'_1 x'_1, (1/2)(x'_1)^\top R'_{kk} x'_k)$, and $G_k = (I, K'_k x'_k, (1/2)(x'_k)^\top R'_{kk} x'_k)$ for $k > 1$. It follows from Lemma 3.2 that $G_n \triangleleft G_{n-1} \triangleleft \cdots \triangleleft G_1 = (S, K'x', (1/2)(x')^\top R'x')$. That is, this cascade connection realizes G. □

The following corollary is a direct consequence of Theorem 3.6 on the transfer function realization of a linear quantum system:

Corollary 3.2 *A linear quantum system $G = (S, Kx, (1/2)x^\top Rx)$ is transfer function realizable by a pure cascade connection of one degree-of-freedom linear quantum systems if and only if there is a symplectic transformation matrix V such that the linear quantum system*

$$G' = (S, KV^{-1}x, (1/2)x^\top V^{-\top} RV^{-1}x)$$

has an A matrix which is lower 2×2 block triangular.

Proof By Definition 3.1, G is transfer function realizable by a pure cascade connection if and only if there exists a symplectic matrix V such that $G' = (S, KV^{-1}x, (1/2)$
$x^\top V^{-\top} RV^{-1}x)$ is realizable by a pure cascade connection. But from Theorem 3.6, noting that the matrix P in the theorem can be absorbed into V, this is true if and only if the A matrix associated with G' (i.e., $A = 2\mathbb{J}_n(R' + \Im\{K'^* K'\})$, with $R' = V^{-\top} RV^{-1}$ and $K' = KV^{-1}$) is lower 2×2 block triangular. □

3.3.3 Transfer Function Realization of Completely Passive Linear Quantum Systems

We will now show that any completely passive linear quantum system is transfer function realizable by a cascade connection. Moreover, each component in the cascade will also be completely passive. The first result in this direction was obtained by Petersen [17, 18] who constructively showed that a "generic" subclass of such systems, which satisfy assumptions on the distinctness of the eigenvalues and the invertibility of certain matrices, is transfer function realizable by pure cascading. It will be shown in this section, following [16], that the result in fact holds true for *all* completely passive systems.

Recall from Sect. 2.4.1 that we can express $(1/2)a^*\tilde{R}a$ and $\tilde{K}a$ in the form

$$(1/2)a^*\tilde{R}a = (1/2)x^\top\Re\{\Sigma^*\tilde{R}\Sigma\}x - \frac{1}{4}\sum_{j=1}^{n}\tilde{R}_{jj},$$

and $\tilde{K}a = \tilde{K}\Sigma x$. Therefore, we may set $R = \Re\{\Sigma^*\tilde{R}\Sigma\}$ and $K = \tilde{K}\Sigma$. Also recall that the 2×2 block diagonal elements $\{R_{jj};\ j = 1, \ldots, n\}$ are of the form $R_{jj} = \lambda_j I_2$ for some $\lambda_j \in \mathbb{R}$ for all j. We shall now establish some properties of A and then prove that there exists a unitary symplectic matrix that transforms it into a lower 2×2 block triangular matrix.

Lemma 3.3 $[\,\Sigma^\top\ \Sigma^*\,]^\top A\,[\,\Sigma^*\ \Sigma^\top\,] = \mathrm{diag}(M, M^\#)$, *where*

$$M = (1/2)\Sigma\mathbb{J}_n\Sigma^*(\tilde{R} - \imath\tilde{K}^*\tilde{K}).$$

Proof For the proof, we exploit the identities (2.23) as well as the following easily verified identities: $\Sigma\mathbb{J}_n\Sigma^\top = 0 = \Sigma^\#\mathbb{J}_n\Sigma^*$. Using these identities, we have the following:

$$\begin{aligned}
&[\,\Sigma^\top\ \Sigma^*\,]^\top A\,[\,\Sigma^*\ \Sigma^\top\,]\\
&= 2[\,\Sigma^\top\ \Sigma^*\,]^\top\mathbb{J}_n(R + \Im\{K^*K\})\,[\,\Sigma^*\ \Sigma^\top\,]\\
&= 2[\,\Sigma^\top\ \Sigma^*\,]^\top\mathbb{J}_n(\Re\{\Sigma^*\tilde{R}\Sigma\} + \Im\{\Sigma^*\tilde{K}^*\tilde{K}\Sigma\})\,[\,\Sigma^*\ \Sigma^\top\,]\\
&= 2[\,\Sigma^\top\ \Sigma^*\,]^\top\mathbb{J}_n\Big((1/2)(\Sigma^*\tilde{R}\Sigma + \Sigma^T\tilde{R}^\#\Sigma^\#)\\
&\qquad - \frac{\imath}{2}(\Sigma^*\tilde{K}^*\tilde{K}\Sigma - \Sigma^\top\tilde{K}^\top\tilde{K}^\#\Sigma^\#)\Big)\,[\,\Sigma^*\ \Sigma^\top\,]\\
&= 2[\,\Sigma^\top\ \Sigma^*\,]^\top\mathbb{J}_n\left[\,(1/4)\Sigma^*\tilde{R} - \tfrac{\imath}{4}\Sigma^*\tilde{K}^*\tilde{K}\ \ (1/4)\Sigma^\top\tilde{R}^\# + \tfrac{\imath}{4}\Sigma^\top\tilde{K}^\top\tilde{K}^\#\,\right]\\
&= \mathrm{diag}\Big((1/2)\Sigma\mathbb{J}_n\Sigma^*\tilde{R} - \frac{\imath}{2}\Sigma\mathbb{J}_n\Sigma^*\tilde{K}^*\tilde{K},\ \Sigma^\#\mathbb{J}_n\Sigma^\top\tilde{R}^\# + \frac{\imath}{2}\Sigma^\#\mathbb{J}_n\Sigma^\top\tilde{K}^\top\tilde{K}^\#\Big).
\end{aligned}$$

□

Then, we have the following theorem:

Theorem 3.7 *Let U be the complex unitary matrix in a* Schur decomposition *of the matrix M of Lemma 3.3: $M = U^*\hat{M}U$, where $\hat{M}$ is a lower triangular matrix. Then, the matrix*

$$V = 2\,[\,\Sigma^*\ \Sigma^\top\,]\,\mathrm{diag}(U, U^\#)[\,\Sigma^\top\ \Sigma^*\,]^\top$$

is a real, unitary, *and* symplectic *matrix that transforms A into a lower 2×2 block triangular matrix: $VAV^* = \hat{A}$, where $\hat{A}$ is a real lower 2×2 block triangular matrix.*

Therefore, every completely passive linear quantum system has a transfer function realization by pure cascading and such a realization is obtained by applying the construction of Theorem 3.6 to $G' = (S, KV^\top x, (1/2)x^\top VRV^\top x)$. Moreover, each one degree-of-freedom linear quantum system in the cascade will also be completely passive.

Proof The existence of U is guaranteed by the well-known result that every complex matrix M has a Schur decomposition of the form $M = U^*\hat{M}U$ with $\hat{M}$ lower triangular. Note then that we also have $\hat{M}^\# = U^\# M^\# U^\top$. Let V be as defined in the theorem. Then by Lemma 3.3, the following is true:

$$\begin{aligned} VAV^* &= 4[\,\Sigma^*\ \Sigma^\top\,]\,\mathrm{diag}(\hat{M}, \hat{M}^\#)[\,\Sigma^\top\ \Sigma^*\,]^\top \\ &= 4(\Sigma^*\hat{M}\Sigma + \Sigma^\top\hat{M}^\#\Sigma^\#) \\ &= 8\Re\{\Sigma^*\hat{M}\Sigma\}. \end{aligned}$$

Now, since $\hat{M}$ is a lower triangular matrix, it follows by inspection (using the special structure of Σ) that $\hat{A} = 8\Re\{\Sigma^*\hat{M}\Sigma\}$ is lower 2×2 block triangular, as claimed. That V is real follows from the fact that we may write $V = 2(\Sigma^* U\Sigma + \Sigma^\top U^\#\Sigma^\#) = 4\Re\{\Sigma^* U\Sigma\}$. That it is *unitary* follows from the observation that $\sqrt{2}[\,\Sigma^*\ \Sigma^\top\,]$ and $\sqrt{2}\begin{bmatrix}\Sigma \\ \Sigma^\#\end{bmatrix}$ are unitary (as a consequence of (2.23)) and that $\mathrm{diag}(U, U^\#)$ is also unitary. To see that V is also *symplectic* define $b = Ua$ and $z = Vx$. By the unitarity of U, we have that b and $b^\#$ satisfy the same the commutation relations as a and $a^\#$ (i.e., b is again an annihilation operator). Then, we have

$$\begin{aligned} z = Vx = V2[\,\Sigma^*\ \Sigma^\top\,]\begin{bmatrix} a \\ a^\# \end{bmatrix} &= 2[\,\Sigma^*\ \Sigma^\top\,]\mathrm{diag}(U, U^\#)\begin{bmatrix} a \\ a^\# \end{bmatrix} \\ &= [\,\Sigma^*\ \Sigma^\top\,]\begin{bmatrix} b \\ b^\# \end{bmatrix}. \end{aligned}$$

Now, this implies that z consists of the canonical position and momentum operators associated with the modes in b and satisfies the same CCR as x. But since $z = Vx$ and V is real, preservation of the CCR implies that V is necessarily a symplectic matrix (see, e.g., [19, Sect. III]).

Using the fact $V^{-1} = V^\top = V^*$ established above, it follows from Theorem 3.6 that the completely passive quantum system G is transfer function equivalent to $G' = (S, KV^\top x, (1/2)x^\top VRV^T x)$ whose A matrix is lower 2×2 block triangular. Let $K' = KV^\top = [\,K_1'\ K_2' \ldots K_n'\,]$ and $R' = VRV^\top = [R'_{jk}]_{j,k=1,\ldots,n}$. By Theorem 3.6, we have $G' = G_n \lhd G_{n-1} \lhd \ldots \lhd G_1$ with $G_k = (S_k, K_k' x_k, (1/2)x_k^\top R'_{kk} x_k)$, $S_1 = S$ and $S_k = I$ for $k > 1$. We now show that each G_k is passive. Recall that $K = \tilde{K}\Sigma$ and write $\tilde{K} = [\,\tilde{K}_1 \ldots \tilde{K}_n\,]$ with $\tilde{K}_k \in \mathbb{C}^{m\times 1}$. Using (2.23), we have that $K'x = KV^*x = \tilde{K}U^*a$. By expanding both sides of the equality $K'x = \tilde{K}U^*a$ and collecting and equating terms of the same index, it follows that $K_k' x_k = (\tilde{K}U^*)_k a_k$, where $(\tilde{K}U^*)_k$ is the k-th $\mathbb{C}^{m\times 1}$ block component of $\tilde{K}U^*$.

On the other hand, since G is completely passive we have that $R_{kk} = \lambda_k I_2$ for some $\lambda_k \in \mathbb{R}$, and recalling that $R = \Re\{\Sigma^* \tilde{R} \Sigma\}$, it follows by inspection (after some algebraic manipulations using (2.23)) that $R'_{kk} = \lambda'_k I_2$ for some $\lambda'_k \in \mathbb{R}$. Thus, $x_k^\top R'_{kk} x_k = \lambda'_k a_k^* a_k + c$ for some constant c and we conclude that each G_k is also completely passive. □

We conclude the exposition with an example of a cascade realization for a completely passive linear quantum system.

Example 3.3 Let $G = (I, \tilde{K}a, (1/2)a^* \tilde{R} a + \frac{5}{4})$ be a completely passive system with $\tilde{R} = \begin{bmatrix} 2 & 1+\imath \\ 1-\imath & 3 \end{bmatrix}$ and $\tilde{K} = \begin{bmatrix} 1+0.5\imath & -2+\imath \\ -5-2\imath & 3-4\imath \end{bmatrix}$. Therefore for this system,

$$K = \begin{bmatrix} 0.5+0.25\imath & -0.25+0.5\imath & -1+0.5\imath & -0.5-\imath \\ -2.5-\imath & 1-2.5\imath & 1.5-2\imath & 2+1.5\imath \end{bmatrix},$$

and

$$R = \begin{bmatrix} 0.5 I_2 & 0.25(I_2 - J) \\ 0.25(I_2 + J) & 0.75 I_2 \end{bmatrix}.$$

From Lemma 3.3 and Theorem 3.7, we find that

$$U = \begin{bmatrix} -0.6933 + 0.0039\imath & 0.2244 - 0.6849\imath \\ 0.7204 + 0.0209\imath & 0.2312 - 0.6536\imath \end{bmatrix},$$

and

$$\hat{M} = \begin{bmatrix} -14.8390 - 0.7912\imath & 0 \\ 0.6344 - 0.2225\imath & -0.2235 - 0.4588\imath \end{bmatrix}.$$

Moreover, from the formula for V stated in Theorem 3.7, we have

$$V = \begin{bmatrix} -0.6933 & 0.0039 & 0.2244 & 0.6849 \\ -0.0039 & -0.6933 & -0.6849 & 0.2244 \\ 0.7204 & -0.0209 & 0.2312 & 0.6536 \\ 0.0209 & 0.7204 & -0.6536 & 0.2312 \end{bmatrix}.$$

Let $K' = KV^{-1} = [\, K'_1 \; K'_2 \,]$, then

$$K' = \begin{bmatrix} -0.9144 - 0.7441\imath & 0.7441 - 0.9144\imath & -0.1926 - 0.3684\imath & 0.3684 - 0.1926\imath \\ 3.4433 + 1.2621\imath & -1.2621 + 3.4433\imath & -0.1679 - 0.1501\imath & 0.1501 - 0.1679\imath \end{bmatrix},$$

and

$$R' = VRV^{-1} = \begin{bmatrix} R'_{11} & R'_{12} \\ (R'_{12})^\top & R'_{22} \end{bmatrix} = \begin{bmatrix} 0.7912 & 0 & 0.1113 & 0.3172 \\ 0 & 0.7912 & -0.3172 & 0.1113 \\ 0.1113 & -0.3172 & 0.4588 & 0 \\ 0.3172 & 0.1113 & 0 & 0.4588 \end{bmatrix}.$$

Therefore, G is transfer function realizable by $G' = (I, K'x, (1/2)x^\top R'x)$ as a consequence of Theorem 3.7. It is straightforwardly verified that $R' + \Im\{K'^* K\}$ is lower 2×2 block triangular,

$$R' + \Im\{K'^* K'\} = \begin{bmatrix} 0.7912 & 14.8390 & 0 & 0 \\ -14.8390 & 0.7912 & 0 & 0 \\ 0.2225 & -0.6344 & 0.4588 & 0.2235 \\ 0.6344 & -0.2225 & -0.2235 & 0.4588 \end{bmatrix},$$

from which it follows that G' can be realized by a pure cascade of one degree-of-freedom linear quantum systems. Applying Theorem 3.6 yields $G' = G_2 \triangleleft G_1$ with $G_1 = (I, K_1' x_1, (1/2)x_1^\top R_{11}' x_1)$ and $G_2 = (I, K_2' x_2, (1/2)x_2^\top R_{22}' x_2)$ and inspection then shows that G_1 and G_2 are both completely passive. Figure 3.23 depicts a quantum optical realization of G'.

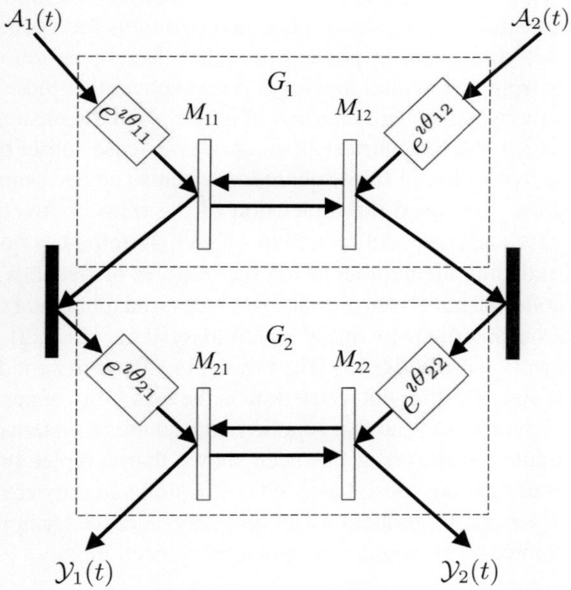

Fig. 3.23 Realization of G' as the cascade connection of G_1 and G_2. G_1 and G_2 are each realized by an optical cavity and a phase shifter. Here, $\theta_{11} = -2.4585$, $\theta_{12} = 0.3513$, $\theta_{21} = -2.0525$, and $\theta_{22} = -2.4121$, and the partially transmitting mirror M_{jk} in the optical cavity G_j, $j, k = 1, 2$, have the coupling coefficients $\gamma_{11} = 1.3898$, $\gamma_{12} = 13.4492$, $\gamma_{21} = 0.1728$ and $\gamma_{22} = 0.0507$, respectively. The resonance frequencies of the optical cavities of G_1 and G_2 have a detuning of $0.7912 - 0.4588 = 0.3324$, with G_1 having the higher resonance frequency. Figure adapted from [16] © 2010 IEEE

3.3.4 Further Reading

A surprising recent result on the transfer realization of linear quantum systems is that it turns out that generic linear quantum systems have a cascade realization, thus including generic *active* systems (those that require an external source of quanta to operate) [20]. This is a result of practical importance as it allows the transfer function of generic linear quantum systems to be realized by a cascade connection of one degree-of-freedom linear quantum systems. Thus, the transfer function realization problem for generic linear quantum systems can be reduced to that of realizing one degree-of-freedom linear quantum systems. Some approaches for the latter realization have been elaborated upon in this chapter.

While the Schur decomposition played a critical role in proving these results for the completely passive case, in the generic case it was shown that there is a symplectic Schur decomposition which decomposes a generic real square matrix A of even dimension into the product $A = VJV^{-1}$, with V symplectic and J lower 2×2 block triangular. This Schur decomposition was based on an algorithm for symplectic QR decomposition, and necessary and sufficient conditions for a generic real square matrix A with even dimension to possess the latter decomposition were obtained. Using techniques from differential topology, it was shown that these conditions are satisfied generically by real square matrices of a fixed even dimension, say $2n \times 2n$, in the sense that it is valid for matrices in an open and dense subset of $\mathbb{R}^{2n \times 2n}$.

In general, the realization of linear quantum systems can be complex with many degrees of freedom. To reduce the dimension of the transfer function realization of linear quantum systems, model reduction of such systems has been considered, with the goal of reducing the number of internal degrees of freedom required in the realization. Methods that have been proposed include an adaptation of the well-known balanced truncation paradigm to linear quantum systems [21, 22] and tangential interpolatory projection methods [23]. The physical realizability condition for linear quantum systems places a stringent restriction on the feasibility of performing model reduction. In particular, it is found that generic linear quantum systems do not possess a balanced realization. However, it has been shown that completely passive linear quantum systems always possess a balanced realization and they are a special class of systems for which model reduction can often be successfully applied, either by balanced truncation or by tangential interpolatory projection.

Appendices

Appendix A: Adiabatic Elimination of Coupled Cavity Modes

In this section, we shall derive formulas for two coupled cavity modes in which one of the cavity has very fast dynamics compared to the other and can be adiabatically eliminated, leaving only the dynamics of the slow cavity mode. The cavities are each coupled to separate bosonic fields and are interacting with one another in a classically pumped nonlinear crystal. A mathematically rigorous theory for the type of adiabatic elimination/singular perturbation that we are interested in here was developed in [12].

The two cavity modes will be denoted by a and b, each defined on two distinct copies of the Hilbert space ℓ^2 of square-integrable sequences ($\mathbb{Z}_+$ denotes the set of all nonnegative integers). Thus, the composite Hilbert space for the two cavity modes is $\mathcal{H} = \ell^2(\mathbb{Z}_+) \otimes l^2(\mathbb{Z}_+)$. The interaction in a nonlinear crystal is given, in some rotating frame, by an interaction Hamiltonian H_{ab} of the form $H_{ab} = \alpha a^* b + \beta a^* b^* + \alpha^* a b^* + \beta^* a b$, for some complex constants α and β. The mode a is coupled to a bosonic field $\mathcal{A}_1$, while b is coupled to the bosonic field $\mathcal{A}_2$, both fields in the vacuum state. The fields $\mathcal{A}_1$ and $\mathcal{A}_2$ live the boson Fock space $\mathcal{F}_2 = \mathcal{F}_1 \otimes \mathcal{F}_1$. We take a to be the slow mode to be retained and b to be the fast mode to be eliminated.

We consider a sequence of linear quantum systems $G_k = (I, \tilde{L}^{(k)}, H_{ab}^{(k)})$ with $\tilde{L}^{(k)} = (\sqrt{\gamma_1} a, k\sqrt{\gamma_2} b)^\top$ and $H_{ab}^{(k)} = \Delta_1 a^* a + k^2 \Delta_2 b^* b + k(\alpha a^* b + \beta a^* b^* + \alpha^* a b^* + \beta^* a b)$ each evolving according to the unitary U_k satisfying the left Hudson–Parthasarathy (H–P) QSDE (as opposed to the conventional right H–P QSDE in (2.2)):

$$dU_k(t) = U_k(t)\left(\tilde{L}^{(k)*}(d\mathcal{A}_1(t), d\mathcal{A}_2(t))^\top - \tilde{L}^{(k)\top}(d\mathcal{A}_1(t), d\mathcal{A}_2(t))^* + \left(\imath H_{ab}^{(k)} - (1/2)\tilde{L}^{(k)*}\tilde{L}^{(k)}\right)dt\right),$$

Here, we are using the left QSDE following the convention used in [12] (see Remark 3 therein) so that the Heisenberg picture dynamics of an operator x is given by $x(t) = U_k(t) x U_k(t)^*$. We shall use the results of [12] to show, in a similar treatment to Sect. 3.2 therein, that in the limit as $k \to \infty$:

$$\lim_{k\to\infty} \sup_{0\le t\le T} \|U_k(t)^*\phi - U(t)^*\phi\| = 0 \;\; \forall \phi \in \mathcal{H}_0 \otimes \mathcal{F}_2 \tag{3.16}$$

for any fixed time $T > 0$, where $\mathcal{H}_0$ is an appropriate Hilbert subspace of $\mathcal{H}$ (to be precisely specified in the next paragraph) for a limiting unitary $U(t)$ (again as a left H–P QSDE) satisfying:

$$\begin{aligned} dU(t) = U(t)\Bigg(&\left(\frac{\imath 2\Delta_2 + \gamma_2}{\imath 2\Delta_2 - \gamma_2} - 1\right) d\Lambda_{22} + \sqrt{\gamma_1} a^* d\mathcal{A}_1(t) - \sqrt{\gamma_1} a\, d\mathcal{A}_1(t)^* \\ &- \imath\sqrt{\gamma_2}(\imath\Delta_2 - \frac{\gamma_2}{2})^{-1}(\alpha a^* + \beta^* a) d\mathcal{A}_2(t) + \\ &\imath\frac{2\sqrt{\gamma_2}}{\imath 2\Delta_2 - \gamma_2}(\alpha^* a + \beta a^*) d\mathcal{A}_2(t)^* + (\imath\Delta_1 - \frac{\gamma_1}{2}) a^* a\, dt + \\ &(\imath\Delta_2 - \frac{\gamma_2}{2})^{-1}(\alpha a^* + \beta^* a)(\alpha^* a + \beta a^*) dt\Bigg), \end{aligned} \tag{3.17}$$

on $\mathcal{H}_0 \otimes \mathcal{F}$. Note that (3.17) is a left H–P QSDE corresponding to the right form in Sect. 2.1.4 by noting that we may write:

$$(\imath \Delta_1 - \frac{\gamma_1}{2})a^*a + (\imath \Delta_2 - \frac{\gamma_2}{2})^{-1}(\alpha a^* + \beta^* a)(\alpha^* a + \beta a^*)$$
$$= \imath \left(\Delta_1 a^* a - \frac{\Delta_2}{\Delta_2^2 + (\frac{\gamma_2}{2})^2}(\alpha a^* + \beta^* a)(\alpha^* a + \beta a^*) \right) - \frac{1}{2}(\tilde{L}_1^* \tilde{L}_1 + \tilde{L}_2^* \tilde{L}_2),$$

with $\tilde{L}_1 = \sqrt{\gamma_1} a$ and $\tilde{L}_2 = \imath \sqrt{\gamma_2}(-\imath \Delta_2 - \frac{\gamma_2}{2})^{-1}(\alpha^* a + \beta a^*)$. As such it satisfies the H–P Condition 1 of [12].

Let $\phi_0, \phi_1, \ldots$ be the standard orthogonal bases of ℓ^2, i.e., ϕ_l is an infinite sequence (indexed starting from 0) of complex numbers with all zeros except a 1 in the lth place. First, let us specify that $\mathcal{H}_0 = \ell^2 \otimes \mathbb{C}\phi_0$, and this is the subspace of $\mathcal{H}$ where the slow dynamics of the system will evolve. Next, we define a dense domain $\mathcal{D} = \text{span}\{\phi_j \otimes \phi_l;\ j, l = 0, 1, 2, \ldots\}$ of $\mathcal{H}$. The strategy is to show that [12, Assumptions 2–3] are satisfied from which the desired result will follow from [12, Theorem 3].

From the definition of $H_{ab}^{(k)}$, $\tilde{L}^{(k)}$ and U_k given above, we can define the operators Y, A, B, G_1, G_2, and W_{jl} $(j, l = 1, 2)$ in [12, Assumption 1] as: $Y = (i\Delta_2 - \frac{\gamma_2}{2})b^*b$, $A = i(\alpha a^* b + \alpha^* a b^* + \beta a^* b^* + \beta^* ab)$, $B = (i\Delta_1 - \frac{\gamma_1}{2})a^*a$, $G_1 = \sqrt{\gamma_1} a^*$, $G_2 = 0$, $F_1 = 0$, $F_2 = \sqrt{\gamma_2} b^*$, $W_{jl} = \delta_{jl}$. Then, we can define the operators $K^{(k)}$, $L_j^{(k)}$ in this assumption as:

$$K^{(k)} = k^2 Y + kA + B;\ L_j^{(k)} = kF_j + G_j\ (j = 1, 2).$$

Let P_0 be the projection operator to $\mathcal{H}_0$. Let us now address Assumption 2. From our definition of $\mathcal{H}_0$, it is clear that we have that (a) $P_0\mathcal{D} \subset \mathcal{D}$. Any element of $P_0\mathcal{D}$ is of the form $f \otimes \phi_0$ for some $f \in \text{span}\{\phi_l;\ l = 0, 1, 2, \ldots\}$; therefore, since $Y = (\imath \Delta_2 - \frac{\gamma_2}{2})b^*b$ and $b\phi_0 = 0$, we find that (b) $YP_0 d = 0$ for all $d \in \mathcal{D}$. Define the operator $\tilde{Y}$ on $\mathcal{D}$ by $\tilde{Y} f \otimes \phi_0 = 0$ and $\tilde{Y} f \otimes \phi_l = l^{-1}(\imath \Delta_2 - \frac{\gamma_2}{2})^{-1} f \otimes \phi_l$ for $l = 1, 2, \ldots$ ($\tilde{Y}$ can then be defined to all of $\mathcal{D}$ by linear extension). From the definition of Y and $\tilde{Y}$, it is easily inspected that (c1) $Y\tilde{Y} f = \tilde{Y} Y f = P_1 f$ for all $f \in \mathcal{D}$, where $P_1 = I - P_0$ (i.e., the projection onto the subspace of $\mathcal{H}$ complementary to $\mathcal{H}_0$). Moreover, because of the simple form of $\tilde{Y}$, it is also readily inspected that (c2) $\tilde{Y}$ has an adjoint $\tilde{Y}^*$ with a dense domain that contains $\mathcal{D}$. Since $F_1 = 0$, we have that (d1) $F_1^* P_0 = 0$ on $\mathcal{D}$, while since $F_2^* f \otimes \phi_0 = \sqrt{\gamma_2} bf \otimes \phi_0 = 0\ \forall f \in \ell^2$, we also have (d2) $F_2^* P_0 = 0$ on $\mathcal{D}$. Finally, from the expression for A and the orthogonality of the bases $\phi_0, \phi_1, \ldots$, a little algebra reveals that (e) $P_0 A P_0 d = 0$ for all $d \in \mathcal{D}$. From (a), (b), (c1–c2), (d1–d2), and (e), we have now verified that Assumption 2 is satisfied.

Finally, we verify that the limiting operator coefficients $K, L_1, L_2, M_1, M_2, N_{jk}$ $(i, j = 1, 2)$ (as operators on $\mathcal{H}_0$) of Assumption 3 coincide with the corresponding coefficients of (3.17). These operator coefficients are defined as $K = P_0(B -$

$A\tilde{Y}A)P_0$, $L_j = P_0(G_j - A\tilde{Y}F_j)P_0$, $M_j = -\sum_{r=1}^{2} P_0 W_{jr}(G_r^* - F_r^*\tilde{Y}A)P_0$ and $N_{jl} = \sum_{r=1}^{2} P_0 W_{jr}(F_r^*\tilde{Y}F_l + \delta_{rl})P_0$. From these definitions and some straightforward algebra, we find that for all $f \in \text{span}\{\phi_l;\ l = 0, 1, 2, \ldots\}$:

$$
\begin{aligned}
Kf \otimes \phi_0 &= \left((\imath\Delta_1 - \frac{\gamma_1}{2})a^*a + (\imath\Delta_2 - \frac{\gamma_2}{2})^{-1}(\alpha a^* + \beta^* a)(\alpha^* a + \beta a^*)\right) f \otimes \phi_0, \\
L_1 f \otimes \phi_0 &= \sqrt{\gamma_1}a^* f \otimes \phi_0, \\
L_2 f \otimes \phi_0 &= -\imath\sqrt{\gamma_2}(\imath\Delta_2 - \frac{\gamma_2}{2})^{-1}(\alpha a^* + \beta^* a) f \otimes \phi_0, \\
M_1 f \otimes \phi_0 &= -\sqrt{\gamma_1}a f \otimes \phi_0, \\
M_2 f \otimes \phi_0 &= \sqrt{\gamma_2}\left(\imath\Delta_2 - \frac{\gamma_2}{2}\right)^{-1}(\alpha^* a + \beta a^*) f \otimes \phi_0,
\end{aligned}
$$

and

$$
\begin{aligned}
N_{11} f \otimes \phi_0 &= f \otimes \phi_0, \ N_{12} f \otimes \phi_0 = 0, \ N_{21} f \otimes \phi_0 = 0, \\
N_{22} f \otimes \phi_0 &= \frac{\gamma_2 + \imath 2\Delta_2}{-\gamma_2 + \imath 2\Delta_2} f \otimes \phi_0.
\end{aligned}
$$

Therefore, we see that $U(t)$ may be written as:

$$
dU(t) = U(t)\left(\sum_{j,l=1}^{2}(N_{jl} - \delta_{jl})d\Lambda_{jl} + \sum_{j=1}^{2} M_j d\mathcal{A}_j^*(t) + \sum_{j=1}^{2} L_j d\mathcal{A}_j(t) + K dt\right).
$$

Since we have already verified that (3.17) is bona fide left QSDE equation, it now follows that Assumption 3 of [12] is satisfied. Now (3.16) follows from [12, Theorem 3] and the proof is complete.

Moreover, we can observe from the derivation above that the coupling of a to $\mathcal{A}_2(t)$ after adiabatic elimination will not change if a is also coupled to other cavities modes $b_3, \ldots, b_m$ via an interaction Hamiltonian of the form $\sum_{i=j}^{m}(\alpha_{j1}ab_j^* + \alpha_{j1}^* a^* b_j + \alpha_{j2}a^* b_j^* + \alpha_{j2}^* ab_j)$, and each additional mode may also be linearly coupled to distinct bosonic fields $\mathcal{A}_3, \ldots, \mathcal{A}_m$, respectively, as long as these other modes are *not* interacting with b and with one another (this amounts to just introducing additional operators F_j, G_j, $j \geq 3$, etc.). Moreover, under these conditions, one can also adiabatically eliminate any of the additional modes and the only effect will be the presence of additional sum terms in $U(t)$ that do not involve b, $\mathcal{A}_1(t)$ and $\mathcal{A}_2(t)$. □

Appendix B: Proof of Theorem 3.4

For this proof, it will be convenient to interchange some rows and columns of the model matrix M to form another model matrix $\tilde{M}$ to avoid complicated bookkeeping and thus reduce unnecessary clutter. Hereby, "rows" and "columns" we mean, respectively, block rows and block columns of M formed with respect to its specified partitioning. This interchange is as follows.

First, we permute rows of M such that the first $2n-1$ rows from top to bottom are the rows labeled (while column labels are kept fixed as they are) s_{00}, s_{12}, s_{21}, s_{13}, s_{31}, ..., s_{1n}, s_{n1}, the next $2(n-2)$ rows, respectively, are the rows labeled s_{23}, s_{32}, s_{24}, s_{42}, ..., s_{2n}, s_{n2}, and so on in the same pattern until we get to the last n rows that are, respectively, those rows labeled s_{11}, s_{22}, ..., s_{nn}. Call the intermediate matrix resulting from this row permutation $\hat{M}$. Then fixing the row labels of $\hat{M}$, we permute its columns such that the first $2n-1$ columns from left to right are, respectively, the columns of $\hat{M}$ labeled r_{00}, r_{12}, r_{21}, r_{13}, r_{31}, ..., r_{1n}, r_{n1}, the next $2(n-2)$ columns are, respectively, the columns labeled r_{23}, r_{32}, r_{24}, r_{42}, ..., r_{2n}, r_{n2}, and so on in the same pattern until the final n columns that are, respectively, the columns labeled r_{11}, r_{22}, ..., r_{nn}. The resulting matrix after this permutation of columns is $\tilde{M}$.

It is important to note here that since the same permutation is applied to the rows and columns, M and $\tilde{M}$ are model matrix representations of the same physical system. That is to say that if M is the model matrix of $G=(S, L, H)$, then $\tilde{M}$ is the model matrix of $\tilde{G}=(PSP^\top, PL, H)$ for some suitable constant real permutation matrix P, while it is clear that G and $\tilde{G}$ are representations of the same physical system. Thus with the same internal connections made, a reduced model matrix for $\tilde{M}$ is also a reduced model matrix for M, up to a possible relabeling of uneliminated ports.

Let $\tilde{L}=PL$ and $\tilde{S}=PSP^\top$. Then, $\tilde{L}$ can be partitioned as $\tilde{L}=(\tilde{L}_\mathrm{i}^\top, \tilde{L}_\mathrm{e}^\top)^\top$, where $\tilde{L}_\mathrm{i}$ is the first $n(n-1)+1$ rows of $\tilde{L}$, while $\tilde{L}_\mathrm{e}$ is the last n rows of $\tilde{L}$. They are of the form:

$$\begin{aligned}\tilde{L}_\mathrm{i} &= (L_{12}^\top, L_{21}^\top, L_{13}^\top, L_{31}^\top, \ldots, L_{1n}^\top, L_{n1}^\top, L_{23}^\top, L_{32}^\top, L_{24}^\top, L_{42}^\top, \ldots, L_{2n}^\top, L_{n2}^\top, \ldots,\\ &\quad L_{(n-2)(n-1)}^\top, L_{(n-1)(n-2)}^\top, L_{(n-2)n}^\top, L_{n(n-2)}^\top, L_{(n-1)n}^\top, L_{n(n-1)}^\top)^\top,\\ \tilde{L}_\mathrm{e} &= (L_{11}^\top, L_{22}^\top, \ldots, L_{nn}^\top)^\top.\end{aligned}$$

Similarly, $\tilde{S}$ can be partitioned as $\tilde{S}=\begin{bmatrix}\tilde{S}_\mathrm{ii} & \tilde{S}_\mathrm{ie}\\ \tilde{S}_\mathrm{ei} & \tilde{S}_\mathrm{ee}\end{bmatrix}$, with $\tilde{S}_\mathrm{ii}$ and $\tilde{S}_\mathrm{ee}$ being block diagonal:

$$\begin{aligned}\tilde{S}_\mathrm{ii} &= \mathrm{diag}(S_{12}, S_{21}, S_{13}, S_{31}, \ldots, S_{1n}, S_{n1}, S_{23}, S_{32}, S_{24}, S_{42}, \ldots, S_{2n}, S_{n2}, \ldots,\\ &\quad S_{(n-2)(n-1)}, S_{(n-1)(n-2)}, S_{(n-2)n}, S_{n(n-2)}, S_{(n-1)n}, S_{n(n-1)}),\\ \tilde{S}_\mathrm{ee} &= \mathrm{diag}(S_{11}, S_{22}, \ldots, S_{nn}),\end{aligned}$$

and $\tilde{S}_\mathrm{ei}$ and $\tilde{S}_\mathrm{ie}$ both being zero matrices. Then, $\tilde{M}$ has a partitioning of the form (3.13) by identifying S_ii, S_ie, S_ei, S_ee, L_i and L_e with $\tilde{S}_\mathrm{ii}$, $\tilde{S}_\mathrm{ie}$, $\tilde{S}_\mathrm{ei}$, $\tilde{S}_\mathrm{ee}$ and $\tilde{L}_\mathrm{i}$, and $\tilde{L}_\mathrm{e}$, respectively. The reduced model matrix resulting from the subsequent simultaneous elimination of all internal edges (s_{jk}, r_{kj}) $j, k=1, \ldots, n$, $j \neq k$, can be conveniently determined by using the adjacency matrix η defined by:

$$\eta = \operatorname{diag}\left(\begin{bmatrix} 0 & I_{c_{12}} \\ I_{c_{21}} & 0 \end{bmatrix}, \begin{bmatrix} 0 & I_{c_{13}} \\ I_{c_{31}} & 0 \end{bmatrix}, \ldots, \begin{bmatrix} 0 & I_{c_{1n}} \\ I_{c_{n1}} & 0 \end{bmatrix}, \begin{bmatrix} 0 & I_{c_{23}} \\ I_{c_{32}} & 0 \end{bmatrix}, \begin{bmatrix} 0 & I_{c_{24}} \\ I_{c_{42}} & 0 \end{bmatrix}, \ldots, \begin{bmatrix} 0 & I_{c_{2n}} \\ I_{c_{n2}} & 0 \end{bmatrix}, \ldots, \begin{bmatrix} 0 & I_{c_{(n-1)n}} \\ I_{c_{(n-1)n}} & 0 \end{bmatrix} \right).$$

(Recall that $c_{jk} = c_{kj}$). Hence, according to Theorem 3.3, the reduced model matrix $\tilde{M}_{\text{red}}$ obtained after elimination of the internal edges $\{(s_{jk}, r_{kj});\ j, k = 1, \ldots, n, j \neq k\}$ has parameters given by (recalling that $\tilde{S}_{\text{ei}}$ and $\tilde{S}_{\text{ie}}$ are zero matrices):

$$\begin{aligned}
\tilde{S}_{\text{red}} &= \tilde{S}_{\text{ee}} + \tilde{S}_{\text{ei}}(\eta - \tilde{S}_{\text{ii}})^{-1}\tilde{S}_{\text{ie}} = \tilde{S}_{\text{ee}}, \\
\tilde{L}_{\text{red}} &= \tilde{L}_{\text{e}} + \tilde{S}_{\text{ei}}(\eta - \tilde{S}_{\text{ii}})^{-1}\tilde{L}_{\text{i}} = \tilde{L}_{\text{e}}, \\
\tilde{H}_{\text{red}} &= \sum_{k=1}^{n} H_k + \sum_{j=\text{i,e}} \Im\{\tilde{L}_j^* \tilde{S}_{j\text{i}}(\eta - \tilde{S}_{\text{ii}})^{-1}\tilde{L}_{\text{i}}\} \\
&= \sum_{k=1}^{n} H_k + \Im\{\tilde{L}_{\text{i}}^* \tilde{S}_{\text{ii}}(\eta - \tilde{S}_{\text{ii}})^{-1}\tilde{L}_{\text{i}}\} \\
&= \sum_{k=1}^{n} H_k + \Im\{\tilde{L}_{\text{i}}^* \eta(\eta - \tilde{S}_{\text{ii}})^{-1}\tilde{L}_{\text{i}}\} \\
&= \sum_{k=1}^{n} H_k + \sum_{j=1}^{n-1}\sum_{k=j+1}^{n} \Im\left\{ [\, L_{jk}^* \; L_{kj}^* \,] \begin{bmatrix} I & -S_{jk} \\ -S_{kj} & I \end{bmatrix}^{-1} \begin{bmatrix} L_{jk} \\ L_{kj} \end{bmatrix} \right\}.
\end{aligned}$$

Since M and $\tilde{M}$ are model matrix representations of the same physical system and the external fields have the same ordering and labeling in both representations, the reduced model matrix of M_{red} and $\tilde{M}_{\text{red}}$ of M and $\tilde{M}$, respectively, after elimination of internal edges (s_{jk}, r_{kj}), coincide. Hence, also the linear quantum stochastic systems G_{red} and $\tilde{G}_{\text{red}}$ associated with M and $\tilde{M}$, respectively, coincide. This completes the proof. □

Appendix C: Proof of Lemma 3.1

We begin by noting that

$$\Im\left\{ \frac{S_{12}}{1 - S_{12}S_{21}} K_1^* K_2 + \frac{S_{21}}{1 - S_{12}S_{21}} K_1^\top K_2^\# \right\} = \frac{1}{2\imath}[\, -K_1^*\Delta^* \; K_1^\top \Delta \,] \begin{bmatrix} K_2 \\ K_2^\# \end{bmatrix},$$

with $\Delta = \frac{S_{21}}{1-S_{21}S_{12}} - \frac{S_{12}^*}{1-S_{21}^*S_{12}^*} = 2\frac{S_{21}-S_{12}^*}{|1-S_{21}S_{12}|^2}$ (exploiting the fact that $S_{12}S_{12}^* = 1 = S_{21}S_{21}^*$). Now, set $K_1 = [\,\kappa\ \imath\kappa\,]$ for an arbitrary nonzero real constant κ, and note that $S_{21}S_{12} \neq 1$ implies that $\Delta \neq 0$ and:

$$\begin{aligned}[\,-K_1^*\Delta^*\ K_1^\top\Delta\,]^{-1} &= \begin{bmatrix} -\kappa\Delta^* & \kappa\Delta \\ \imath\kappa\Delta^* & \imath\kappa\Delta \end{bmatrix}^{-1} \\ &= -\frac{1}{2\imath\kappa^2|\Delta|^2}\begin{bmatrix} \imath\kappa\Delta & -\kappa\Delta \\ -\imath\kappa\Delta^* & -\kappa\Delta^* \end{bmatrix},\end{aligned}$$

and therefore for any real matrix V, $2\imath[\,-K_1^*\Delta^*\ K_1^\top\Delta\,]^{-1}V = \begin{bmatrix} Z \\ Z^\# \end{bmatrix}$ for some complex row vector Z. Therefore, given any R, we see that we may solve the equation

$$[\,-K_1^*\Delta^*\ K_1^\top\Delta\,]\begin{bmatrix} K_2 \\ K_2^\# \end{bmatrix} = 2\imath R,$$

for K_2 and this solution is as given in the statement of the corollary.

Alternatively, we could also have started by setting $K_2 = [\,\kappa\ \imath\kappa\,]$ and analogously solving for K_1 for a given R. It is then an easy exercise that the solution for K_1 in this case is as stated in the corollary. □

Appendix D: Proof of Corollary 3.1

With c_{jk}, S_{jk}, R_{jk}, and K_{jk}, $j,k = 1,\ldots,n$, as defined in the statement of the corollary, from Theorem 3.4 and Lemma 3.1 we have that $S_{\rm red} = I_{nm}$, $L_{\rm red} = (L_{11}^\top, L_{22}^\top, \ldots, L_{nn}^\top)^\top$ with $L_{jj} = K_j x_j$, and

$$H_{\rm red} = \sum_{j=1}^n H_j + \sum_{j=1}^{n-1}\sum_{k=j+1}^n \Im\left\{[\,L_{jk}^*\ L_{kj}^*\,]\begin{bmatrix} 1 & -S_{jk} \\ -S_{kj} & 1 \end{bmatrix}^{-1}\begin{bmatrix} L_{jk} \\ L_{kj} \end{bmatrix}\right\}.$$

Expanding, we have:

$$\begin{aligned}H_{\rm red} &= (1/2)\sum_{j=1}^n x_j^\top R_j x_j + \sum_{j=1}^{n-1}\sum_{k=j+1}^n \Im\left\{\frac{1}{1-S_{jk}S_{kj}}\right. \\ &\quad \left.\times(L_{jk}^*L_{jk} + S_{jk}L_{jk}^*L_{kj} + S_{kj}L_{kj}^*L_{jk} + L_{kj}^*L_{kj})\right\} \\ &= (1/2)\sum_{j=1}^n x_j^\top\left(R_j + 2{\rm sym}\left(\sum_{k=1,k\neq j}^n \Im\{\frac{K_{jk}^*K_{jk}}{1-S_{jk}S_{kj}}\}\right)\right)x_j\end{aligned}$$

$$+\sum_{j=1}^{n-1}\sum_{k=j+1}^{n} x_j^\top \Im\{\frac{S_{jk}}{1-S_{jk}S_{kj}}K_{jk}^* K_{kj} + \frac{S_{kj}}{1-S_{jk}S_{kj}}K_{jk}^\top K_{kj}^\#\}x_k$$

$$= (1/2)\sum_{j=1}^{n} x_j^\top R_{jj} x_j + \sum_{j=1}^{n-1}\sum_{k=j+1}^{n} x_j^\top \big(R_{jk} - \Im\{K_j^\top K_k^\#\}\big)x_k$$

$$= (1/2)x^\top R x - \sum_{j=1}^{n-1}\sum_{k=j+1}^{n} x_j^\top \Im\{K_j^\top K_k^\#\}x_k,$$

where $\mathrm{sym}(A) = (1/2)(A + A^\top)$, and $R = [R_{jk}]_{j,k=1,\ldots,n}$ and $R_{kj} = R_{jk}^\top$. From this, it is clear that using the concatenation product we can decompose G_{red} as $G_{\mathrm{red}} = (0, 0, H_{\mathrm{red}}) \boxplus \boxplus_{j=1}^{n}(I_m, L_{jj}, 0)$. Let $G_{\mathrm{red},0} = (0, 0, H_{\mathrm{red}})$ and $G_{\mathrm{red},j} = (I_m, L_{jj}, 0)$, $j = 1, \ldots, n$. Now, using the series product rule, we easily compute that

$$\begin{aligned} G_{\mathrm{net}} &= G_{\mathrm{red},0} \boxplus (G_{\mathrm{red},n} \lhd \ldots \lhd G_{\mathrm{red},2} \lhd G_{\mathrm{red},1}) \\ &= \left(0, 0, (1/2)x^\top R x - \sum_{j=1}^{n-1}\sum_{k=j+1}^{n} x_j^\top \Im\{K_j^\top K_k^\#\}x_k\right) \boxplus \\ &\quad \left(I_m, [\,K_1\ K_2\ \ldots\ K_n\,]x, \sum_{j=1}^{n-1}\sum_{k=j+1}^{n} x_j^\top \Im\{K_j^\top K_k^\#\}x_k\right) \\ &= \left(I_m, [\,K_1\ K_2\ \ldots\ K_n\,]x, (1/2)x^\top R x\right). \end{aligned}$$

Therefore, G_{net} realizes a linear quantum stochastic system with parameters S_{net}, L_{net} and H_{net}, as claimed. □

References

1. J.C. Willems, J.W. Polderman, *Introduction to Mathematical Systems Theory: A Behavioral Approach*. Texts in Applied Mathematics, vol. 26 (Springer, Berlin, 1998)
2. B.D.O. Anderson, S. Vongpanitlerd, *Network Analysis and Synthesis: A Modern Systems Theory Approach*. Networks Series (Prentice-Hall Inc, Upper Saddle River, 1973)
3. H.I. Nurdin, M.R. James, A.C. Doherty, Network synthesis of linear dynamical quantum stochastic systems. SIAM J. Control Optim. **48**(4), 2686–2718 (2009), © 2009 Society for Industrial and Applied Mathematics. Reprinted with permission. All rights reserved
4. U. Leonhardt, Quantum physics of simple optical instruments. Rep. Prog. Phys. **66**, 1207–1249 (2003)
5. J. Gough, M.R. James, The series product and its application to quantum feedforward and feedback networks. IEEE Trans. Autom. Control **54**(11), 2530–2544 (2009)
6. J. Gough, M.R. James, Quantum feedback networks: Hamiltonian formulation. Commun. Math. Phys. **287**, 1109–1132 (2009)
7. C.W. Gardiner, P. Zoller, *Quantum Noise: A Handbook of Markovian and Non-Markovian Quantum Stochastic Methods with Applications to Quantum Optics*, 3rd edn. (Springer, Berlin, 2004)

8. C.M. Caves, B.L. Schumaker, New formalism for two-photon quantum optics. I. Quadrature phases and squeezed states. Phys. Rev. A **31**, 3068–3092 (1985)
9. B. Schumaker, C.M. Caves, New formalism for two-photon quantum optics. ii. Mathematical foundation and compact notation. Phys. Rev. A **31**, 3093–3111 (1985)
10. U. Leonhardt, A. Neumaier, Explicit effective Hamiltonians for linear quantum-optical networks. J. Opt. B **6**, L1–L4 (2004)
11. H.M. Wiseman, G.J. Milburn, Quantum theory of field-quadrature measurements. Phys. Rev. A **47**(1), 642–663 (1993)
12. L. Bouten, R. van Handel, A. Silberfarb, Approximation and limit theorems for quantum stochastic models with unbounded coefficients. J. Funct. Anal. **254**, 3123–3147 (2008)
13. H.I. Nurdin, Synthesis of linear quantum stochastic systems via quantum feedback networks. IEEE Trans. Autom. Control **55**(4), 1008–1013 (2010). Reprinted, with permission, © 2010 IEEE. Extended preprint version available at arXiv:0905.0802
14. M.R. James, H.I. Nurdin, I.R. Petersen, H^{∞} control of linear quantum stochastic systems. IEEE Trans. Autom. Control **53**(8), 1787–1803 (2008)
15. H.I. Nurdin, M.R. James, I.R. Petersen, Coherent quantum LQG control. Automatica **45**, 1837–1846 (2009)
16. H.I. Nurdin, On synthesis of linear quantum stochastic systems by pure cascading. IEEE Trans. Autom. Control **55**(10), 2439–2444 (2010) © 2010 IEEE
17. I.R. Petersen, Cascade cavity realization for a class of complex transfer functions arising in coherent quantum feedback control, in *Proceedings of the 2009 European Control Conference (Budapest, Hungary, 23–26 August 2009)*, pp. 190–195
18. I.R. Petersen, Cascade cavity realization for a class of complex transfer functions arising in coherent quantum feedback control. Automatica **47**(8), 1757–1763 (2011)
19. G. Linblad, Brownian motion of harmonic oscillators: existence of a subdynamics. J. Math. Phys. **39**(5), 2763–2780 (1998)
20. H.I. Nurdin, S. Grivopoulos, I.R. Petersen, The transfer function of generic linear quantum stochastic systems has a pure cascade realization. *Automatica*, pp. 324–333 (2016)
21. H.I. Nurdin, Structures and transformations for model reduction of linear quantum stochastic systems. IEEE Trans. Autom. Control **59**(9), 2413–2425 (2014)
22. O. Techakesari, H.I. Nurdin, On the quasi-balanceable class of linear quantum stochastic systems. Syst. Control Lett. **78**, 25–31 (2015)
23. O. Techakesari, H.I. Nurdin, Tangential interpolatory projection for model reduction of linear quantum stochastic systems. IEEE Trans. Autom. Control **62**(1), 5–17 (2017)
24. H.I. Nurdin, On synthesis of linear quantum stochastic systems by pure cascading (2010). arXiv:0905.0802 (arXiv preprint)

Chapter 4
Quantum Filtering for Linear Dynamical Quantum Systems

Abstract This chapter introduces quantum filtering theory as a quantum analogue of stochastic filtering theory. When applied to linear quantum systems, this leads to a quantum version of the Kalman filter. Concepts from quantum probability theory that are relevant to quantum filtering theory are introduced, including a suitable notion of quantum conditional expectations. The chapter also briefly introduces robust observers for linear quantum systems to deal with modeling uncertainties for the purpose of estimation.

As discussed in Sect. 1.2.2, for classical linear stochastic systems with partial observation, we can construct a dynamical estimate called the Kalman filter, which generates an optimal estimator for the system's state vector in the least mean square sense. In order to emphasize the importance of the Kalman filter, we highlight the fact that it can be used for constructing LQG controllers based on the separation principle, as illustrated in Sect. 1.2.2. This chapter discusses the quantum analogue of the Kalman filtering theory for linear dynamical quantum systems, as well as a more general setting of the quantum filtering theory developed by Belavkin (see Chap. 1).

This chapter is structured as follows. First, we begin with a general overview of a version of quantum conditional expectations, which gives a least mean square estimate as in the classical case. Then, we use the quantum conditional expectation to derive a special case of the quantum filtering equation. The Schrödinger picture representation of this equation will be given, which is called the stochastic master equation or the quantum trajectory equation in the physics literature. Next, the general theory is applied to the special linear setup where the state is limited to coherent states, and the quantum version of the Kalman filter is given. Finally, we consider a practical aspect of estimation for linear quantum systems. Quantum filtering is a model-based approach for estimation; as a consequence, the optimal estimate can be constructed only under the assumption that the system dynamics (more precisely the system's (A, B, C, D) matrices) are perfectly known. Of course in any realistic situation, the model is only an approximation. Hence, we need a robust filter that can estimate the

---

H.I. Nurdin and N. Yamamoto, *Linear Dynamical Quantum Systems*,
Communications and Control Engineering, DOI 10.1007/978-3-319-55201-9_4

system's variable, even when the model is imperfect. A quantum version of robust filters will be discussed.

4.1 Quantum Conditional Expectations

4.1.1 Quantum Probability Space

In the framework of quantum filtering, we will only consider a restricted notion of quantum conditional expectations, as linear projective mappings from the space of observables to an algebra of commutative observables. Hence, we first need to specify such an algebra of observables and the notion of a state on this algebra. For the purpose of illustration and to emphasize the basic underlying idea, in the following we discuss only the case of finite-dimensional systems, wherein the dimension of the Hilbert space is finite-dimensional and an observable can be represented by a complex matrix. Quantum harmonic oscillators are thus, strictly speaking, excluded because they are defined on an infinite-dimensional Hilbert space. Nonetheless, the notions and results below have an infinite-dimensional analogue, which readers can find in, e.g., [1] and the references cited therein.

The content we describe here partly follows [2]. First recall that a quantum mechanical random variable is represented by a linear self-adjoint operator on a Hilbert space. This axiom means that quantum random variables in general do not commute with one another, and thus, the conventional classical probability space $(\Omega, \mathcal{F}, \mathbf{P})$ needs to be replaced by a *quantum probability space*, which is composed of a $*$-algebra and a state defined as follows.

Definition 4.1 ($*$*-algebra*) Let $\mathcal{H}$ be a finite-dimensional complex Hilbert space. A $*$*-algebra* $\mathcal{A}$ is a set of linear operators $\mathcal{H} \to \mathcal{H}$ such that $I, \alpha A + \beta B, AB, A^* \in \mathcal{A}$ for any $A, B \in \mathcal{A}$ and $\alpha, \beta \in \mathbb{C}$. $\mathcal{A}$ is called commutative if $[A, B] = 0$ for any $A, B \in \mathcal{A}$.

In the above definition, the adjoint A^* is defined as the unique operator satisfying $\langle A^*x, y\rangle = \langle x, Ay\rangle,\ \forall x, y \in \mathcal{H}$. The $*$-algebra introduced above corresponds to a σ-algebra $\mathcal{F}$ in a classical probability space. The next is the quantum version of the expectation $\mathbf{E}$, which is referred to as a "state"; this notion has a one-to-one correspondence with the standard notion of a state in quantum mechanics, which is represented by a ket vector or a density matrix.

Definition 4.2 (*State*) A *state* on $\mathcal{A}$ is a linear map $\mathbb{P} : \mathcal{A} \to \mathbb{C}$ that is positive: $\mathbb{P}(A^*A) \geq 0,\ \forall A \in \mathcal{A}$, and normalized $\mathbb{P}(I) = 1$.

In the finite-dimensional context, the state can be given explicitly as follows. Let $(e_1, \ldots, e_d)$ be an orthonormal basis of the d-dimensional Hilbert space $\mathcal{H}$. Also recall that the trace can be defined as $\mathrm{Tr}(A) = \sum_{i=1}^{d} \langle e_i, Ae_i\rangle$ for all $A \in \mathcal{A}$; note that $\mathrm{Tr}(A)$ does not depend on the choice of basis. Then, the state $\mathbb{P}$ can be expressed in terms of the trace and an associated matrix. That is, there exists a unique matrix ρ that satisfies

$$\mathbb{P}(A) = \mathrm{Tr}(\rho A), \quad \rho = \rho^* \geq 0, \quad \mathrm{Tr}(\rho) = 1. \tag{4.1}$$

ρ is called the *density matrix*; due to the equivalence between ρ and $\mathbb{P}$, we sometimes also refer to ρ as the state. Note that, from the latter two conditions in (4.1), $\mathbb{P}(A^*A) \geq 0 \; \forall A \in \mathcal{A}$ and $\mathbb{P}(I) = 1$ are satisfied. Also especially when ρ is a *pure state*, meaning that ρ is a rank-one projection matrix $\rho = |b\rangle\langle b|$, $|b\rangle \in \mathcal{H}$, then $\mathbb{P}(A)$ can be represented as

$$\mathbb{P}(A) = \mathrm{Tr}(|b\rangle\langle b|A) = \langle b|A|b\rangle. \tag{4.2}$$

Now a quantum probability space can be defined as follows.

Definition 4.3 (*Quantum probability space*) Let $\mathcal{A}$ be a $*$-algebra of operators on a finite-dimensional complex Hilbert space $\mathcal{H}$ and $\mathbb{P}$ be a state on $\mathcal{A}$. Then, $(\mathcal{A}, \mathbb{P})$ is called a (finite-dimensional) quantum probability space.

For a quantum probability space $(\mathcal{A}, \mathbb{P})$, if $A \in \mathcal{A}$ is self-adjoint, then we call A a quantum random variable or an observable; this is the definition of observables in the scenario of quantum probability theory. Now, let us consider a commutative $*$-algebra $\mathcal{A}$, in which all quantum random variables in $\mathcal{A}$ commute with each other as in the classical case. In this case, $(\mathcal{A}, \mathbb{P})$ is called a commutative quantum probability space. Now, we can state the well-known *Spectral Theorem* (Theorem 4.1 below), emphasizing that a commutative quantum probability space is equivalent to a classical one. This is based on the following notion:

Definition 4.4 (*$*$-isomorphism*) Let Ω be a set and let $\mathcal{F}$ be a σ-algebra on Ω. A *$*$-isomorphism* between a commutative $*$-algebra $\mathcal{C}$ and the set of bounded $\mathcal{F}$-measurable functions $\ell^\infty(\mathcal{F})$ on Ω is a linear bijection $\iota : \mathcal{C} \to \ell^\infty(\mathcal{F})$ such that $\iota(A^*)(i) = \iota(A)(i)^*$ and $\iota(AB)(i) = \iota(A)(i)\iota(B)(i)$ for all $A, B \in \mathcal{C}$ and $i \in \Omega$.

Theorem 4.1 (Spectral Theorem) *Let $(\mathcal{C}, \mathbb{P})$ be a finite-dimensional commutative quantum probability space. Then, there exists a classical probability space $(\Omega, \mathcal{F}, \mathbf{P})$ and a $*$-isomorphism $\iota : \mathcal{C} \to \ell^\infty(\mathcal{F})$ such that $\mathbb{P}(A) = \mathbf{E}_\mathbf{P}[\iota(A)]$, $\forall A \in \mathcal{C}$.*

Proof The theorem can be proven by construction. First, let $\mathcal{H} = \mathbb{C}^n$ and $\Omega = \{1, \ldots, n\}$. Since $[A, B] = 0$ and $[A, A^*] = 0 \; \forall A, B \in \mathcal{C}$, all the elements in $\mathcal{C}$ can be diagonalized simultaneously. Hence, we can set $A = \mathrm{diag}\{a_1, \ldots, a_n\}$ and define a classical random variable $\iota(A) : \Omega \to \mathbb{C}$ by $\iota(A)(i) = a_i$. Let P be a projection in $\mathcal{C}$, i.e., $P = P^* = P^2$, then $\iota(P)$ is the indicator function of a subset S_P of Ω. We define $\mathcal{F}$ as the set of subsets S_P of Ω where P runs through the projections in $\mathcal{C}$. Furthermore, we define a probability measure $\mathbf{P}$ on $\mathcal{F}$ by $\mathbf{P}(S_P) = \mathbb{P}(P)$, $\forall P \in \mathcal{C}$. As a result, we have constructed a classical probability space $(\Omega, \mathcal{F}, \mathbf{P})$. It is then easy to verify $\mathbf{E}_\mathbf{P}[\iota(A)] = \mathbb{P}(A)$. □

Thanks to the above spectral theorem, a *single* observable A can *always* be represented as a classical random variable $\iota(A)$ on a classical probability space $(\Omega, \mathcal{F}, \mathbf{P})$, where the measure **P** is determined from the state. Then, a *measurement* on A generates one of the values that $\iota(A)$ can take, with probability according to **P**. Here, we pose a very important remark; if *two* observables do not commute with each other, then these observables cannot be represented as classical random variables on the *same* probability space. Such observables are called *incompatible*, meaning that they cannot be measured simultaneously in a single realization of an experiment.

Example 4.1 Let us now see an example, a quantum two-level system, like a single atom with two distinguishable two energy levels. The Hilbert space is $\mathcal{H} = \mathbb{C}^2$. The quantum probability space of a *two-level system* is composed of the $*$-algebra of 2×2 complex matrices $\mathcal{M}$ and a state ψ on $\mathcal{M}$. The state ψ can be represented as $\psi(A) = \mathrm{Tr}(\rho A)$, $\forall A \in \mathcal{M}$ for some matrix ρ satisfying $\rho = \rho^* \geq 0$ and $\mathrm{Tr}(\rho) = 1$. Let us then study a commutative $*$-subalgebra $\mathcal{D} = \{D = \mathrm{diag}(d_1, d_2) : d_1, d_2 \in \mathbb{R}\} \subset \mathcal{M}$. From Theorem 4.1, there always exists a classical probability space that is in one-to-one correspondence with $(\mathcal{D}, \psi)$. Such a classical space is composed of the sample space $\Omega = \{1, 2\}$, the set of events $\mathcal{F} = \{\emptyset, \{1\}, \{2\}, \Omega\}$, and the probability measure constructed as follows. Let d_1 and d_2 be eigenvalues of an observable D in $\mathcal{D}$ and define a classical random variable $\iota(D)$ on $(\Omega, \mathcal{F})$ as $\iota(D)(1) = d_1$ and $\iota(D)(2) = d_2$. Now, $D \in \mathcal{D}$ has a spectral decomposition $D = \sum d_i P_i$ with the projection matrices $P_1 = \mathrm{diag}(1, 0)$ and $P_2 = \mathrm{diag}(0, 1)$. Hence, according to the proof of Theorem 4.1, a probability distribution for $\iota(D)$ can be defined through $\mathbf{P}(\{1\}) = \psi(P_1) = \mathrm{Tr}(\rho P_1) = \rho_{11}$ and $\mathbf{P}(\{2\}) = \rho_{22}$.

Lastly, let us define the *composite quantum probability space* of two quantum probability spaces $(\mathcal{A}_1, \mathbb{P}_1)$ and $(\mathcal{A}_2, \mathbb{P}_2)$, which are defined on the finite-dimensional Hilbert spaces $\mathcal{H}_1$ and $\mathcal{H}_2$, respectively. First, for two vectors $|a_1\rangle \in \mathcal{H}_1$ and $|a_2\rangle \in \mathcal{H}_2$, we define the tensor (Kronecker) product $|a_1\rangle \otimes |a_2\rangle$. Then, the Hilbert space $\mathcal{H}_1 \otimes \mathcal{H}_2$ can be introduced by defining an inner product $\langle a_1 \otimes a_2, b_1 \otimes b_2\rangle := \langle a_1, b_1\rangle\langle a_2, b_2\rangle$. Now, the composite quantum probability space $(\mathcal{A}_1 \otimes \mathcal{A}_2, \mathbb{P}_1 \otimes \mathbb{P}_2)$ can be defined on $\mathcal{H}_1 \otimes \mathcal{H}_2$ as follows. First, an arbitrary element of $\mathcal{A}_1 \otimes \mathcal{A}_2$ is represented as a linear combination of product-type elements of the form $A_1 \otimes A_2 \in \mathcal{A}_1 \otimes \mathcal{A}_2$. Second, the state $\mathbb{P}_1 \otimes \mathbb{P}_2$ is introduced through the relation $(\mathbb{P}_1 \otimes \mathbb{P}_2)(A_1 \otimes A_2) = \mathbb{P}_1(A_1)\mathbb{P}_2(A_2)$ and its linear extension to all of $\mathcal{A}_1 \otimes \mathcal{A}_2$.

4.1.2 Conditional Expectations

Let us consider a quantum probability space $(\mathcal{A}, \mathbb{P})$ and two commuting self-adjoint elements A and B on $\mathcal{A}$. From the Spectral Theorem (Theorem 4.1), A and B can be represented as classical random variables $\iota(A)$ and $\iota(B)$ on a single classical probability space $(\Omega, \mathcal{F}, \mathbf{P})$. This allows us to specify the classical conditional expectation $\mathbf{E}_{\mathbf{P}}[\iota(A) \mid \iota(B)]$. Then, the *quantum conditional expectation* $\mathbb{P}(A \mid B)$ can be defined as

$$\mathbb{P}(A \mid B) = \iota^{-1}\Big(\mathbf{E}_{\mathbf{P}}[\iota(A) \mid \iota(B)]\Big).$$

Next, we aim to extend the above discussion to the case where the variable A can be conditioned on a commutative $*$-subalgebra $\mathcal{C}$ of $\mathcal{A}$, rather than a single variable B. In fact, as long as A commutes with every element in $\mathcal{C}$, Theorem 4.1 is applicable to the commutative $*$-algebra generated from A and $\mathcal{C}$. In this case, we can define

$$\mathbb{P}(A \mid \mathcal{C}) = \iota^{-1}\Big(\mathbf{E}_{\mathbf{P}}[\iota(A) \mid \sigma(\iota(\mathcal{C}))]\Big), \tag{4.3}$$

where $\sigma(\iota(\mathcal{C}))$ denotes the σ-algebra generated from $\iota(K)$, $K \in \mathcal{C}$. Moreover, this argument means that the quantum conditional expectation onto $\mathcal{C}$ can be defined for every self-adjoint element in the *commutant* of $\mathcal{C}$: for a commutative $*$-subalgebra $\mathcal{C}$, the commutant of $\mathcal{C}$ is given by

$$\mathcal{C}' := \{A \in \mathcal{A} \ : \ [A, C] = 0, \ \forall C \in \mathcal{C}\}.$$

Then, we have the formal definition of the conditional expectation as shown below (note that the general treatment of quantum conditional expectations based on operator algebras can be found in, e.g., [3]).

Definition 4.5 (*Quantum conditional expectation*) Let $(\mathcal{A}, \mathbb{P})$ be a quantum probability space, and let $\mathcal{C}$ be a commutative $*$-subalgebra of $\mathcal{A}$. Then, the map $\mathbb{P}(\,\cdot \mid \mathcal{C}) : \mathcal{C}' \to \mathcal{C}$ is called (a version of) the quantum conditional expectation from $\mathcal{C}'$ to $\mathcal{C}$ if $\mathbb{P}(\mathbb{P}(A \mid \mathcal{C})K) = \mathbb{P}(AK)$, $\forall A \in \mathcal{C}'$, $\forall K \in \mathcal{C}$.

Note that an arbitrary element A in the commutant can be uniquely expressed as $A = A_1 + \imath A_2$ with A_1 and A_2 self-adjoint. Hence, the conditional expectation of A onto $\mathcal{C}$ can be defined as simply a linear extension of (4.3). Finally, we provide some basic properties of both the classical and quantum conditional expectations as follows: (i) $\mathbb{P}(A \mid \mathcal{C})$ is unique with probability one, (ii) $\mathbb{P}(\mathbb{P}(A \mid \mathcal{C})) = \mathbb{P}(A)$, (iii) $\mathbb{P}(CA \mid \mathcal{C}) = C\mathbb{P}(A \mid \mathcal{C})$ if $C \in \mathcal{C}$ and $A \in \mathcal{C}'$ (module property), and (iv) $\mathbb{P}(\mathbb{P}(A \mid \mathcal{B}) \mid \mathcal{C}) = \mathbb{P}(A \mid \mathcal{C})$ if $\mathcal{C} \subset \mathcal{B}$ (tower property). Note from the tower property that $\mathbb{P}(\,\cdot \mid \mathcal{C})$ is a projection. A particularly important fact deduced from the above properties is that, as in the classical case, $\mathbb{P}(A \mid \mathcal{C})$ is a least mean square estimate of A given $\mathcal{C}$ in the sense that

$$\begin{aligned}\|A - \mathbb{P}(A \mid \mathcal{C})\|_{\mathbb{P}} &\leq \|A - \mathbb{P}(A \mid \mathcal{C})\|_{\mathbb{P}} + \|\mathbb{P}(A \mid \mathcal{C}) - B\|_{\mathbb{P}} \\ &= \|A - B\|_{\mathbb{P}}, \ \ \forall B \in \mathcal{C},\end{aligned} \tag{4.4}$$

where we have defined $\|X\|_{\mathbb{P}}^2 := \mathbb{P}(X^*X)$.

4.2 Quantum Filtering Theory

In this section, in the setting of the quantum probability theory elaborated in the previous section, we illustrate a dynamical estimation method for open quantum systems via quantum filtering.

4.2.1 Quantum Filtering: The Idea

We begin by describing the idea of quantum filtering in a simple setup where the system is only coupled to a single probe field, which is in the vacuum state. The case of multiple probe fields is discussed in the following subsection.

The system dynamics considered here is generated by the following special class of Hudson–Parthasarathy QSDEs (2.2):

$$dU(t) = \Big(\big(-\iota H - \frac{1}{2}L^*L\big)dt + Ld\mathcal{A}(t)^* - L^*d\mathcal{A}(t)\Big)U(t), \quad U(0) = I, \quad (4.5)$$

where again H is the system Hamiltonian and L represents the coupling to the probe field with annihilation process $\mathcal{A}(t)$. For simplicity, we further assume that the internal scattering process is identity: $S = I$. Also recall that the operator $\mathcal{A}(t)$ satisfies the quantum Itō rule $d\mathcal{A}(t)d\mathcal{A}(t)^* = dt$ and all other products between $d\mathcal{A}(t), d\mathcal{A}(t)^*$, and dt being zero. Now, let us see the time evolution of an observable X, i.e., $j_t(X) := U(t)^*XU(t)$. Using the quantum Itō calculus, we end up with

$$dj_t(X) = j_t(\mathcal{L}X)dt + j_t([L^*, X])d\mathcal{A}(t) + j_t([X, L])d\mathcal{A}(t)^*, \quad (4.6)$$

where

$$\mathcal{L}X := \iota[H, X] + L^*XL - \frac{1}{2}L^*LX - \frac{1}{2}XL^*L.$$

The quantum Langevin equation (1.18) derived for an oscillator is a special class of (4.6), though 1.18 is represented in the Stratonovich form with white noise operators. Note also that the quantum probability space of interest is generated by $j_t(X)$ with fixed X and $\mathbb{P}$, the latter now being the composite of the initial state of the system and the vacuum state of the field.

Next, let us discuss the measurement process. Here, we assume that the real part of the output field operator is measured by a homodyne detector. This means that we measure the output field operator $y_{\mathrm{m}}(t) := \mathcal{Y}(t) + \mathcal{Y}(t)^* = U(t)^*(\mathcal{A}(t) + \mathcal{A}(t)^*)U(t)$. Again by using the quantum Itō rule, the time evolution of $y_{\mathrm{m}}(t)$ can be obtained as

$$dy_{\mathrm{m}}(t) = j_t(L + L^*)dt + d\mathcal{A}(t) + d\mathcal{A}(t)^*. \quad (4.7)$$

Recall from Sect. 2.1.4 that the output observable $y_{\mathrm{m}}(t)$ has the *self-non-demolition* property:

$$[y_{\mathrm{m}}(s), y_{\mathrm{m}}(t)] = 0, \quad \forall s, t. \quad (4.8)$$

This means that we find a commutative $*$-algebra (the von Neumann algebra) generated by $y_{\mathrm{m}}(s)$, denoted by $\mathscr{Y}_{\mathrm{m},t} = \mathrm{vN}\{y_{\mathrm{m}}(s), 0 \le s \le t\}$. Therefore, there exists a $*$-isomorphism ι defined in Definition 4.4 such that $\mathscr{Y}_{\mathrm{m},t}$ is equivalent to a classical stochastic process representing the observations. Moreover, from Sect. 2.1.4 we also

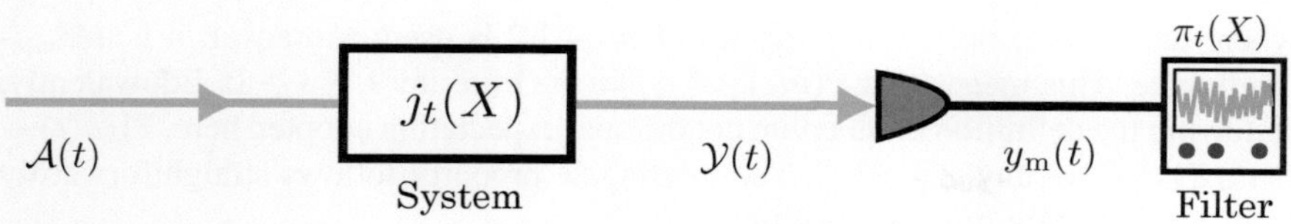

Fig. 4.1 Setup of quantum filtering

have that $y_{\rm m}(t)$ satisfies the *quantum non-demolition (QND)* condition,

$$[y_{\rm m}(s), j_t(X)] = 0, \quad \forall s \le t, \tag{4.9}$$

for all system observables X.

Our goal is to construct the best estimate of the system observable $j_t(X)$ based on the observations $\mathscr{Y}_{{\rm m},t}$; see Fig. 4.1 for the setup. In particular, if the criterion is given by the least mean-square error, $\mathbb{P}[(j_t(X) - Z)^2]$ with Z an element of $\mathscr{Y}_{{\rm m},t}$, we can have an explicit solution; that is, because now $\mathscr{Y}_{{\rm m},t}$ is a commutative algebra and $j_t(X)$ lives in the commutant of $\mathscr{Y}_{{\rm m},t}$, from (4.4) we find that the mean-square error is minimized when Z is given by the quantum conditional expectation:

$$\pi_t(X) = \operatorname*{argmin}_{Z\in\mathscr{Y}_{{\rm m},t}} \mathbb{P}\big[(j_t(X) - Z)^2\big] = \mathbb{P}(j_t(X) \mid \mathscr{Y}_{{\rm m},t}).$$

Consequently, the optimal filter for the system dynamics (4.6) is given by the time evolution of $\pi_t(X)$ as follows:

$$\begin{aligned} d\pi_t(X) &= \pi_t(\mathcal{L}X)dt \\ &\quad + \big(\pi_t(XL + L^*X) - \pi_t(X)\pi_t(L + L^*)\big)\big(dy_{\rm m}(t) - \pi_t(L + L^*)dt\big). \end{aligned} \tag{4.10}$$

This is called the *quantum filtering* or *Belavkin equation*. A brief explanation about how to derive (4.10) is given at the end of this subsection. Note that (4.10) is an equation on $\mathscr{Y}_{{\rm m},t}$, hence it is essentially a classical stochastic differential equation on the equivalent classical σ-algebra. In particular, as in the classical case, the last term $d\nu(t) := dy_{\rm m}(t) - \pi_t(L + L^*)dt$ is a classical Wiener increment and is called the *innovations process* or simply the innovation. As in the classical case, it is zero-mean;

$$\begin{aligned} \mathbb{P}(\nu(t)) &= \mathbb{P}\Big(\int_0^t dy_{\rm m}(s)\Big) - \mathbb{P}\Big(\int_0^t \pi_s(L + L^*)ds\Big) \\ &= \mathbb{P}\Big[\int_0^t j_s(L + L^*)ds + d\mathcal{A}(s) + d\mathcal{A}(s)^*\Big] - \mathbb{P}\Big[\int_0^t \mathbb{P}\big(j_s(L + L^*) \mid \mathscr{Y}_{{\rm m},s}\big)ds\Big] \\ &= \mathbb{P}\Big(\int_0^t j_s(L + L^*)ds\Big) - \mathbb{P}\Big(\int_0^t j_s(L + L^*)ds\Big) = 0, \end{aligned}$$

where the tower property mentioned in Sect. 4.1.2 is used. Moreover, it is a $\mathscr{Y}_{\mathrm{m},t}$-martingale. This means that $\mathbb{P}(\nu(t) \mid \mathscr{Y}_{\mathrm{m},s}) = \nu(s)$ for any $t \geq s \geq 0$. Equivalently, following the definition of quantum conditional expectation adopted here, $\mathbb{P}[(\nu(t) - \nu(s))Z] = 0$ for any $Z \in \mathscr{Y}_{\mathrm{m},s}$. The martingale property follows straightforwardly from the definition of $\nu(t)$ as follows:

$$\begin{aligned}\mathbb{P}\left[(\nu(t) - \nu(s))Z\right] &= \mathbb{P}\left[\left(y_{\mathrm{m}}(t) - y_{\mathrm{m}}(s) + \int_s^t \pi_\tau(L + L^*)d\tau\right) Z\right] \\ &= \mathbb{P}\Big[\left(\mathcal{A}(t) - \mathcal{A}(s)\right) Z + \int_s^t \big(j_\tau(L + L^*) - \pi_\tau(L + L^*)\big) Z d\tau\Big] \\ &= \mathbb{P}\left[\left(\mathcal{A}(t) - \mathcal{A}(s)\right) Z\right] = \mathbb{P}\left[\mathcal{A}(t) - \mathcal{A}(s)\right] \mathbb{P}(Z) = 0,\end{aligned}$$

where we have used the fact that

$$\mathbb{P}\left(\pi_\tau(L + L^*)Z\right) = \mathbb{P}\left(j_\tau(L + L^*)Z\right),$$

for every $\tau \in [s, t]$ and for every $Z \in \mathscr{Y}_{\mathrm{m},t}$ in going from the second to third line. Also note that all elements of $\mathscr{Y}_{\mathrm{m},s}$ act non-trivially only on $\mathfrak{h} \otimes \mathcal{F}_1(s])$, where $\mathfrak{h}$ denotes the Hilbert space of the system and, from Chap. 2, $\mathcal{F}_1(s])$ is the factor of the Fock space of multipliticy one up to time s, while $\mathcal{A}(t) - \mathcal{A}(s)$ acts non-trivially only on $\mathcal{F}_1([s, t])$ (in particular commuting with any element of $\mathscr{Y}_{\mathrm{m},s}$). The martingale property implies that the increment $\nu(t) - \nu(s)$ is independent of any element of $\mathscr{Y}_{\mathrm{m},s}$ for any $t > s$. That is, roughly speaking, the increment $\nu(t) - \nu(s)$ is independent of any function of $\{y_{\mathrm{m}}(\tau),\ 0 \leq \tau \leq s\}$.

There are several approaches to deriving the quantum filtering equation (4.10), including the reference probability approach [1, 4, 5] and the martingale method [1]. Here, we will take the simplest route using a technique employed in [6]. We begin with the ansatz that the quantum filter is of the form

$$d\pi_t(X) = \mathbb{H}_1 dt + \mathbb{H}_2 dy_{\mathrm{m}}(t), \tag{4.11}$$

where $\mathbb{H}_1$ and $\mathbb{H}_2$ are functions of $\mathscr{Y}_{\mathrm{m},t}$. The goal is to derive explicit forms of $\mathbb{H}_1$ and $\mathbb{H}_2$. Now, we introduce

$$\bar{h}(t) = \exp\Big(\int_0^t h(s) dy_{\mathrm{m}}(s) - \frac{1}{2}\int_0^t h(s)^2 ds\Big),$$

where $h(t) \in \mathscr{Y}_{\mathrm{m},t}$ and thus $\bar{h}(t) \in \mathscr{Y}_{\mathrm{m},t}$ as well. Using the Itō rule we find that $d\bar{h}(t) = \bar{h}(t)h(t)dy_{\mathrm{m}}(t)$. To determine $\mathbb{H}_1$ and $\mathbb{H}_2$, we calculate $d\mathbb{P}(\bar{h}(t) j_t(X))$ in two ways as follows. First, from (4.6) we have that

$$
\begin{aligned}
d\mathbb{P}\Big(\bar{h}(t)j_t(X)\Big) &= \mathbb{P}\Big(d\bar{h}(t)j_t(X) + \bar{h}(t)dj_t(X) + d\bar{h}(t)dj_t(X)\Big) \\
&= \mathbb{P}\Big(\bar{h}(t)h(t)j_t(L+L^*)dt \cdot j_t(X) + \bar{h}(t)j_t(\mathcal{L}X)dt \\
&\qquad + \bar{h}(t)h(t)d\mathcal{A}(t) \cdot j_t([X,L])d\mathcal{A}(t)^*\Big) \\
&= \mathbb{P}\Big(\bar{h}(t)h(t)j_t(LX + L^*X) + \bar{h}(t)j_t(\mathcal{L}X) + \bar{h}(t)h(t)j_t([X,L])\Big)dt \\
&= \mathbb{P}\Big(\bar{h}(t)h(t)j_t(XL + L^*X) + \bar{h}(t)j_t(\mathcal{L}X)\Big)dt,
\end{aligned}
$$

where $\mathbb{P}[d\mathcal{A}(t)] = \mathbb{P}[d\mathcal{A}(t)^*] = 0$ and the quantum Itō rule $d\mathcal{A}(t)d\mathcal{A}(t)^* = dt$ are used. Now, from the tower property of the quantum conditional expectation, the above equation can be further calculated as

$$
\begin{aligned}
\frac{d}{dt}\mathbb{P}\Big(\bar{h}(t)j_t(X)\Big) &= \mathbb{P}\Big[\mathbb{P}\Big(\bar{h}(t)h(t)j_t(XL + L^*X) + \bar{h}(t)j_t(\mathcal{L}X) \,\Big|\, \mathscr{Y}_{\mathrm{m},t}\Big)\Big] \\
&= \mathbb{P}\Big[\bar{h}(t)h(t)\mathbb{P}\Big(j_t(XL + L^*X) \,\Big|\, \mathscr{Y}_{\mathrm{m},t}\Big) + \bar{h}(t)\mathbb{P}\Big(j_t(\mathcal{L}X) \,\Big|\, \mathscr{Y}_{\mathrm{m},t}\Big)\Big] \\
&= \mathbb{P}\Big(\bar{h}(t)h(t)\pi_t(XL + L^*X) + \bar{h}(t)\pi_t(\mathcal{L}X)\Big).
\end{aligned}
$$

In the second way, we first apply the tower property and have

$$
\begin{aligned}
d\mathbb{P}\Big(\bar{h}(t)j_t(X)\Big) &= d\mathbb{P}\Big[\mathbb{P}\Big(\bar{h}(t)j_t(X) \,\Big|\, \mathscr{Y}_{\mathrm{m},t}\Big)\Big] = d\mathbb{P}\Big(\bar{h}(t)\mathbb{P}(j_t(X) \mid \mathscr{Y}_{\mathrm{m},t})\Big) \\
&= d\mathbb{P}\Big(\bar{h}(t)\pi_t(X)\Big) = \mathbb{P}\Big(d\bar{h}(t)\pi_t(X) + \bar{h}(t)d\pi_t(X) + d\bar{h}(t)d\pi_t(X)\Big) \\
&= \mathbb{P}\Big[\bar{h}(t)h(t)\Big(\pi_t(L+L^*)\pi_t(X) + \mathbb{H}_2\Big) + \bar{h}(t)\Big(\mathbb{H}_1 + \mathbb{H}_2\pi_t(L+L^*)\Big)\Big]dt.
\end{aligned}
$$

Comparing the above two equations and noting that $h(t)$ can be chosen arbitrarily in $\mathscr{Y}_{\mathrm{m},t}$, we end up with

$$
\pi_t(L+L^*)\pi_t(X) + \mathbb{H}_2 = \pi_t(XL + L^*X), \quad \mathbb{H}_1 + \mathbb{H}_2\pi_t(L+L^*) = \pi_t(\mathcal{L}X),
$$

which from (4.11) leads to (4.10) by some straightforward rearrangement and regrouping of terms.

4.2.2 *Quantum Filter: Multiple-Input Multiple-Output Case*

Now, we consider a general quantum system coupled to m bosonic fields $\mathcal{A} = (\mathcal{A}_1, \ldots, \mathcal{A}_m)^\top$ in the vacuum state, with Hamiltonian H and coupling operators $L = (L_1, \ldots, L_m)^\top$, and scattering matrix $S \in \mathbb{C}^{m\times m}$. (Note that this setup is not the most general, as explained later.) When the system is driven by coherent fields, then, as elaborated in Chap. 2, the Hudson–Parthasarathy QSDE is given by

$$dU(t) = \Big[\mathrm{Tr}\Big((S-I)^\top d\Lambda(t)\Big) + d\mathcal{A}(t)^* L(t) - L(t)^* S d\mathcal{A}(t) \\ - \Big(\imath H(t) + \frac{1}{2}L(t)^* L(t)\Big)dt\Big]U(t), \tag{4.12}$$

with initial condition $U(0) = I$. Here we have

$$L(t) = L + Sf(t), \quad H(t) = H + \Im\{L^* Sf(t)\},$$

where $f(t)$ is the time-dependent amplitude of the coherent input fields, taking a value in $\mathbb{C}^m$ (see Sect. 2.7.5). Recall that the time evolution of a bounded operator X in the Heisenberg picture, i.e., $X(t) = j_t(X) = U(t)^* X U(t)$, is given by

$$dj_t(X) = j_t\Big(\imath[H(t), X] + L(t)^* X L(t) - \frac{1}{2}L(t)^* L(t) X - \frac{1}{2} X L(t)^* L(t)\Big)dt \\ + d\mathcal{A}(t)^* j_t(S^*[X, L(t)]) + j_t([L(t)^*, X]S)d\mathcal{A}(t),$$

with initial condition $j_0(X) = X$ [5]. Also, the field output defined by $\mathcal{Y}(t) = U(t)^* \mathcal{A}(t) U(t)$ obeys the following equation:

$$d\mathcal{Y}(t) = j_t(L(t))dt + Sd\mathcal{A}(t). \tag{4.13}$$

Next, we describe the measurement process. In the previous section, we consider the measurement of the real part of $\mathcal{Y}(t)$, which leads to the measurement output equation (4.7), but there are in fact infinitely many choices. For instance, we could also measure the imaginary part of $\mathcal{Y}(t)$. Those output fields can be represented in the quadrature form as

$$y(t) = 2\Big(\Re\{\mathcal{Y}_1(t)\}, \Im\{\mathcal{Y}_1(t)\}, \ldots, \Re\{\mathcal{Y}_m(t)\}, \Im\{\mathcal{Y}_m(t)\}\Big)^\top,$$

as defined in Sect. 2.3.2. We consider a measurement $y_{\mathrm{m}}(t)$ of the form $y_{\mathrm{m}}(t) = F_{\mathrm{m}} y(t)$, with $F_{\mathrm{m}} \in \mathbb{R}^{p\times 2m}$. We require F_{m} to be full row rank and such that

$$[y_{\mathrm{m}}(t), y_{\mathrm{m}}(s)^\top] = y_{\mathrm{m}}(t) y_{\mathrm{m}}(s)^\top - (y_{\mathrm{m}}(s) y_{\mathrm{m}}(t)^\top)^\top = 0, \quad \forall s, t \geq 0.$$

Let $\mathscr{Y}_{\mathrm{m},t} = \mathrm{vN}\{y_{\mathrm{m}}(s), 0 \leq s \leq t\}$ denote the von Neumann algebra generated by $y_{\mathrm{m},j}(s)$, $j = 1, 2, \ldots, p$ and $0 \leq s \leq t$. Here $y_{\mathrm{m},j}$ is the j-th element of the vector y_{m}. As we saw above, the assumption on F_{m} ensures that $\mathscr{Y}_{\mathrm{m},t} = \mathrm{vN}\{y_{\mathrm{m}}(s),\ 0 \leq s \leq t\}$ is a commutative von Neumann algebra. Moreover, since the output $y(t)$ satisfies $[y(s), j_t(X)] = 0$ for all $0 \leq s \leq t$ (recall Sect. 2.1.4), we also have that $[y_{\mathrm{m}}(s), j_t(X)] = F_{\mathrm{m}}[y(s), j_t(X)] = 0$ for all $0 \leq s \leq t$ and for any bounded system operator X. Therefore, $y_{\mathrm{m}}(t)$ is isomorphic to a classical stochastic process by an infinite dimensional version of Theorem 4.1, and the quantum conditional expectation $\pi_t(x) = \mathbb{P}(j_t(X) \mid \mathscr{Y}_{m,t})$ is well defined (recall that $\mathbb{P}$ denotes the quantum expectation with respect to the initial state of the system and field). To simplify the exposition in what will follow, we impose an additional structure on $F_{\rm m}$. Namely, we require that $p \leq m$ and $F_{\rm m}$ can be decomposed as,

$$F_{\rm m} = M[\, I_p \; 0_{p\times(2m-p)} \,][\, V \; V^{\#} \,]\Gamma_m, \tag{4.14}$$

with M some invertible real $p \times p$ matrix, V some complex $m \times m$ unitary matrix, and Γ_m is as defined in Sect. 2.3.2. What these conditions mean is that effectively, we can write

$$\begin{aligned} y_{\rm m}(t) &= M[\, I_p \; 0_{p\times(2m-p)} \,](\tilde{\mathcal{Y}}(t) + \tilde{\mathcal{Y}}(t)^{\#}) \\ &= M[\, I_p \; 0_{p\times(2m-p)} \,]U(t)^*(\tilde{\mathcal{A}}(t) + \tilde{\mathcal{A}}(t)^{\#})U(t) \end{aligned}$$

for an input field annihilation process $\tilde{\mathcal{A}}(t) = V\mathcal{A}(t)$ and the corresponding output field annihilation process $\tilde{\mathcal{Y}}(t) = V\mathcal{Y}(t)$. Recall that $\mathcal{A}(t)$ has been taken to be in the vacuum state. Since V is unitary, $\tilde{\mathcal{A}}(t)$ is also in the vacuum state and satisfies the same vacuum Itō products as $\mathcal{A}(t)$. Moreover, if we define $\tilde{\Lambda}(t) = V^{\#}\Lambda(t)V^{\top}$ then $\tilde{\Lambda}(t)$ satisfies the same vacuum Itō products as $\Lambda(t)$. The imposed structure simplifies the derivation of the associated quantum filtering equation, using the results of [5] extended to the multiple-input multiple-output case. However, we emphasize that the additional conditions on $F_{\rm m}$ can be relaxed and more general linear measurements performed on $y(t)$ can be analyzed, but requiring more laborious algebraic manipulations. For details, see [7].

Let

$$\tilde{y}_{\rm m}(t) = M^{-1}y_{\rm m}(t) = [\, I_p \; 0_{p\times(2m-p)} \,](\tilde{\mathcal{Y}}(t) + \tilde{\mathcal{Y}}(t)^{\#})$$

be another measurement process. Since $\tilde{y}_{\rm m}$ is linearly related to $y_{\rm m}$ and M is invertible, $\tilde{y}_{\rm m}$ and $y_{\rm m}$ generate the same von Neumann algebra, we have that

$$\pi_t(X) = \mathbb{P}(j_t(X) \mid {\rm vN}\{y_{\rm m}(s), 0 \leq s \leq t\}) = \mathbb{P}(j_t(X) \mid {\rm vN}\{\tilde{y}_{\rm m}(s), 0 \leq s \leq t\}).$$

Then, define $\tilde{L} = VL$ and $\tilde{S} = VSV^*$, and the operators

$$\tilde{L}(t) = \tilde{L} + \tilde{S}f(t), \quad \tilde{H}(t) = H + \Im\{\tilde{L}^*\tilde{S}f(t)\}. \tag{4.15}$$

Note that $\tilde{H}(t) = H(t)$. In terms of these operators and the processes $\tilde{\mathcal{A}}(t)$, $\tilde{\mathcal{A}}(t)^*$ and $\tilde{\Lambda}(t)$ defined above, the QSDE (4.12) can be rewritten as:

$$\begin{aligned} dU(t) = \Big[&{\rm Tr}\Big((\tilde{S} - I)^{\top} d\tilde{\Lambda}(t)\Big) + d\tilde{\mathcal{A}}(t)^*\tilde{L}(t) - \tilde{L}(t)^*\tilde{S}d\tilde{\mathcal{A}}(t) \\ &-\Big(\imath\tilde{H}(t) + \frac{1}{2}\tilde{L}(t)^*\tilde{L}(t)\Big)dt\Big]U(t). \end{aligned} \tag{4.16}$$

It then follows, by a straightforward extension of the results of [5] to multiple inputs and multiple outputs, that the quantum filtering equation is given by

$$\begin{aligned} & d\pi_t(X) \\ &= \pi_t\Big[\imath[\tilde{H}(t), X] + \sum_{j=1}^{m} \frac{1}{2}\Big(\tilde{L}_j(t)^*[X, \tilde{L}_j(t)] + [\tilde{L}_j(t)^*, X]\tilde{L}_j(t)\Big)\Big]dt \\ &\quad + \sum_{j=1}^{p}\Big(\pi_t(X\tilde{L}_j(t) + \tilde{L}_j(t)^*X) - \pi_t(\tilde{L}_j(t) + \tilde{L}_j(t)^*)\pi_t(X)\Big)\, d\tilde{\nu}_j(t), \end{aligned} \tag{4.17}$$

where $\tilde{\nu}_j(t) = \tilde{y}_{\mathrm{m},j}(t) - \int_0^t \pi_s(\tilde{L}_j(s) + \tilde{L}_j(s)^*)ds$ is the j-th component of the innovations process $\tilde{\nu}(t) = \tilde{y}_{\mathrm{m}}(t) - [\, I_p \ 0\,] \int_0^t \pi_s(\tilde{L}(s) + \tilde{L}(s)^*)ds$. The process $\tilde{\nu}(t)$ is a $\mathscr{Y}_{\mathrm{m},t}$-martingale Wiener process [5]. This can be shown by an analogous argument to the proof of the martingale property of the innovations process for a system coupled to a single-input field given in Sect. 4.2.1. In particular, $\tilde{\nu}(t)$ is zero-mean and is independent of any function of $\mathscr{Y}_{\mathrm{m},t}$.

4.2.3 *Stochastic Master Equation*

Let us return to the simple setup where the system is coupled to a single probe field and also the internal system scattering process is identity (i.e., $S = I$). First, we remark that the quantum filter (4.10) describes the time evolution of an observable, which is projected onto the commutative algebra; that is, (4.10) is the Heisenberg picture representation of the estimate. Here, we switch to the Schrödinger picture to see the time evolution of the conditional state as an estimate of the state. To do this, let us recall that, by defining ι to be a $*$-isomorphism from $\mathscr{Y}_{\mathrm{m},t}$ to the σ-algebra corresponding to the classical measurement record, we have $\iota(\pi_t(X)) = \iota(\mathbb{P}(U(t)^*XU(t) \mid \mathscr{Y}_{\mathrm{m},t}))$; this means that $\iota(\pi_t(\cdot))$ is a nonnegative, linear, and identity-preserving functional of X. Thus, $\iota(\pi_t(\cdot))$ is a state possessing the properties of Definition 4.2 almost surely. Hence, we can connect $\iota(\pi_t(X))$ with a stochastic operator $\rho(t)$ as $\iota(\pi_t(X)) = \mathrm{Tr}(X\rho(t))$. As described in (4.1), then $\rho(t)$ belongs to the following convex set:

$$\mathcal{S} := \{\rho \,:\, \rho = \rho^* \geq 0, \ \ \mathrm{Tr}\rho = 1\}. \tag{4.18}$$

The state $\rho(t)$ is called a *conditional state*, because it is essentially the quantum analogue of a classical conditional probability density. Applying the relation $\iota(\pi_t(X)) = \mathrm{Tr}(X\rho(t))$ to (4.10), we obtain the time evolution of $\rho(t)$ as follows:

$$d\rho = \mathcal{L}^*\rho dt + \Big(L\rho + \rho L^* - \mathrm{Tr}(L\rho + \rho L^*)\rho\Big)\Big(dy_{\mathrm{m}}(t) - \mathrm{Tr}(L\rho + \rho L^*)dt\Big),$$
$$\mathcal{L}^*\rho := -\imath[H, \rho] + L\rho L^* - \frac{1}{2}L^*L\rho - \frac{1}{2}\rho L^*L, \tag{4.19}$$

where we identify $\iota(y_{\mathrm{m}}(t))$ with $y_{\mathrm{m}}(t)$ for simplicity. This equation is called the *stochastic master equation (SME)*. It is the quantum version of the classical Kushner–Stratonovich equation, which is the stochastic partial differential equation describing the time evolution of a conditional probability density of a stochastic system. Note again that (4.19) is a (matrix-valued or operator-valued) classical stochastic differential equation driven by the zero-mean Wiener process $d\nu(t) = dy_{\mathrm{m}}(t) - \mathrm{Tr}(L\rho(t) + \rho(t)L^*)dt$. We also point out that this ideal SME (4.19) preserves a pure state; that is, $\mathrm{Tr}(\rho(t)^2) = \mathrm{Tr}(\rho(0)^2)$ holds if $\rho(0)$ is a pure state.

Now, by averaging the SME and exploiting the fact that the innovation is a martingale, we obtain the dynamics of $\bar{\rho} := \mathbf{E}(\rho)$:

$$\frac{d\bar{\rho}}{dt} = \mathcal{L}^*\bar{\rho}dt = -\imath[H, \bar{\rho}] + L\bar{\rho}L^* - \frac{1}{2}L^*L\bar{\rho} - \frac{1}{2}\bar{\rho}L^*L. \tag{4.20}$$

This is called the quantum *master equation (ME)*, which describes the unconditional time evolution of the quantum system interacting with a coherent field via the coupling operator L. Note that (4.20) can be derived directly from the QSDE (4.6) as follows. First, the expectation of $j_t(X)$, i.e., $\mathbb{P}[j_t(X)]$, yields $d\mathbb{P}[j_t(X)]/dt = \mathbb{P}[j_t(\mathcal{L}X)]$, due to $\mathbb{P}[d\mathcal{A}(t)] = \mathbb{P}[d\mathcal{A}(t)^*] = 0$. Second, because $\mathbb{P}[j_t(X)] = \mathbb{P}[U(t)^*XU(t)]$ is a nonnegative, linear, and identity-preserving functional of X, it can be expressed as $\mathbb{P}[j_t(X)] = \mathrm{Tr}(X\bar{\rho}(t))$. Combining these two relations, we end up with (4.20).

Finally, it is straightforward to generalize the SME and ME to the multi-input setup discussed in Sect. 4.2.2. That is, if the system couples m bosonic input fields and the first p output fields are measured by homodyne detectors, then the time evolution of the conditional state $\rho(t)$ is given by the following SME:

$$\begin{aligned} d\rho = \Big\{ &-\imath[\tilde{H}(t), \rho] + \sum_{j=1}^{m} \Big(\tilde{L}_j(t)\rho\tilde{L}_j(t)^* - \frac{1}{2}\tilde{L}_j(t)^*\tilde{L}_j(t)\rho - \frac{1}{2}\rho\tilde{L}_j(t)^*\tilde{L}_j(t)\Big)\Big\}dt \\ &+ \sum_{j=1}^{p} \Big(\tilde{L}_j(t)\rho + \rho\tilde{L}_j(t)^* - \mathrm{Tr}(\tilde{L}_j(t)\rho + \rho\tilde{L}_j(t)^*)\rho\Big) \\ &\times \Big(dy_{\mathrm{m},j}(t) - \mathrm{Tr}(\tilde{L}_j(t)\rho + \rho\tilde{L}_j(t)^*)dt\Big), \end{aligned}$$

where $\tilde{H}(t)$ and $\tilde{L}(t)$ are defined in (4.15), and $y_{m,j}(t)$ represents the measurement record corresponding to the j-th output field. Also, the ensemble average of the measurement results yields the following ME:

$$\frac{d\bar{\rho}}{dt} = -\imath[\tilde{H}(t), \bar{\rho}] + \sum_{j=1}^{m} \Big(\tilde{L}_j(t)\bar{\rho}\tilde{L}_j(t)^* - \frac{1}{2}\tilde{L}_j(t)^*\tilde{L}_j(t)\bar{\rho} - \frac{1}{2}\bar{\rho}\tilde{L}_j(t)^*\tilde{L}_j(t)\Big).$$

4.2.4 QND Interaction and the Projection Postulate

As an important application of the quantum filter, here we study a special case where the coupling operator L is Hermitian, i.e., $L = L^*$. For instance, the Faraday interaction for an atomic ensemble described in Sect. 1.5.5 satisfies this assumption, where $L = \sqrt{M} J_z$ with M the coupling strength. We further assume that the system Hamiltonian is zero, $H = 0$; this is because we are working in a rotating frame with respect to H satisfying $[H, L] = 0$. The point of this setting is that, from (4.6), the time evolution of the observable L is given by

$$dj_t(L) = j_t\Big(L^*LL - \frac{1}{2}LL^*L - \frac{1}{2}L^*LL\Big)dt + j_t([L^*, L])d\mathcal{A} + j_t([L, L])d\mathcal{A}^* = 0.$$

That is, the system observable L does not change in time, i.e., $j_t(L) = j_0(L) = L$; in other words, this special interaction does not disturb the physical property of L, hence it is called a quantum non-demolition (QND) interaction. On the other hand, when the amplitude quadrature of $\mathcal{Y}(t)$ is measured by a homodyne detector, then the measurement output equation is given by

$$dy_{\rm m} = j_t(L + L^*)dt + d\mathcal{A} + d\mathcal{A}^* = 2j_t(L)dt + d\mathcal{A} + d\mathcal{A}^* = 2Ldt + d\mathcal{A} + d\mathcal{A}^*.$$

Therefore, by measuring this quadrature component of the output probe field, we can extract the information about L without disturbing this observable.

The statistics of the information we obtain can be explicitly evaluated by examining the corresponding filtering equation,

$$d\pi_t(X) = \pi_t\Big(LXL - \frac{1}{2}L^2X - \frac{1}{2}XL^2\Big)dt + \big[\pi_t(XL + LX) - 2\pi_t(X)\pi_t(L)\big]d\nu.$$

It is then immediate to see that

$$d\pi_t(L) = 2\big[\pi_t(L^2) - \pi_t(L)^2\big]d\nu = 2\langle(L - \langle L\rangle_c)^2\rangle_c d\nu, \tag{4.21}$$

where $\langle\cdot\rangle_c := \pi_t(\cdot)$. That is, the conditional expectation is a pure diffusion process weighted by the conditional variance $\langle\Delta L^2\rangle_c = \langle(L - \langle L\rangle_c)^2\rangle_c$, thus $\mathbb{P}[d\pi_t(L)] = 0$ and so $\mathbb{P}[\pi_t(L)] = \mathbb{P}[\pi_0(L)] = \mathbb{P}(L)$, which is consistent with the claim above that the QND interaction does not disturb L. Moreover, this conditional variance obeys

$$\begin{aligned} d\langle\Delta L^2\rangle_c &= d\big[\pi_t(L^2) - \pi_t(L)^2\big] \\ &= 2\big[\pi_t(L^3) - \pi_t(L^2)\pi_t(L)\big]d\nu - 2\pi_t(L)d\pi_t(L) - [d\pi_t(L)]^2 \\ &= 2\big[\pi_t(L^3) - 3\pi_t(L^2)\pi_t(L) + 2\pi_t(L)^3\big]d\nu - 4\langle\Delta L^2\rangle_c^2 dt \\ &= 2\langle\Delta L^3\rangle_c d\nu - 4\langle\Delta L^2\rangle_c^2 dt, \end{aligned} \tag{4.22}$$

where we have used the Itō rule $d\nu^2 = dt$. This leads to the following important relation:

$$\frac{d}{dt}\mathbb{P}[\langle \Delta L^2 \rangle_c] = -4\mathbb{P}[\langle \Delta L^2 \rangle_c^2].$$

In general, it is known from classical stochastic stability theory that, if $d\mathbf{E}(x)/dt \le 0$ for a nonnegative random variable $x(t)$ contained in a finite set, then $\lim_{t\to\infty} d\mathbf{E}(x)/dt = 0$ holds almost surely. Hence, in our case, we have $\mathbb{P}[\langle \Delta L^2 \rangle_c^2] \to 0$ almost surely as $t \to \infty$. This means that the conditional state converges to one of the eigenvectors of L, which is referred to as *quantum state reduction*. In particular for the finite-dimensional case (i.e., the case where L is a Hermitian matrix), very importantly, the probability of this event can be exactly calculated as follows. Let $L = \sum_k \ell_k |\ell_k\rangle\langle \ell_k|$ be the spectral decomposition of L, with ℓ_k the kth eigenvalue and $|\ell_k\rangle$ the corresponding eigenvector. Then, as in the above calculation, we have $d\pi_t(|\ell_k\rangle\langle\ell_k|) = 2\pi_t(|\ell_k\rangle\langle\ell_k|)[\ell_k - \pi_t(L)]d\nu$, leading to $\mathbb{P}[\pi_\infty(|\ell_k\rangle\langle\ell_k|)] = \mathbb{P}[\pi_0(|\ell_k\rangle\langle\ell_k|)] = \langle \ell_k|\rho(0)|\ell_k\rangle$. Now, we know from the above result that the conditional state converges into one of the eigenvectors of L, and thus, the left-hand side of this equation is exactly the probability for the state $\rho(0)$ to converge to $|\ell_k\rangle\langle\ell_k|$. As a result, we have

$$\mathbf{P}(\{\rho(0) \to |\ell_k\rangle\langle\ell_k|\}) = \langle \ell_k|\rho(0)|\ell_k\rangle.$$

This is the renowned Born's projection postulate: continuous measurement with QND interaction realizes the Born's projection postulate.

4.3 Quantum Kalman Filter for Gaussian Linear Quantum Systems

Following the theme of this monograph, we will now focus on linear dynamical quantum systems. First, a simple example is provided to show the idea behind quantum Kalman filtering and how it is used. Then, based on that intuition, a general Kalman filter dynamics is given for a class of linear measurements applied to the system.

4.3.1 Example in Quantum Optics

The quantum Kalman filter provides the best estimate of observables of a linear Gaussian quantum system (i.e., a linear quantum system where all the noise and the initial state is Gaussian). A very nice structure of the Kalman filter is that it enables an exact computation of the estimate, while in the general case such an exact computation is not possible because infinitely many conditional moments need to be computed. In fact, thanks to the linear and Gaussian nature of the system, the conditional mean and variance are enough to compute the quantum Kalman filtering equation. Here, we focus on a special example taken from quantum optics and illustrate how the Kalman filter can be used.

The example we consider here is a degenerate parametric amplifier (DPA) discussed in Sect. 1.5.3. This system is characterized by Hamiltonian $H = \imath\epsilon(a^{*2} - a^2)/2$ (in a rotating frame) and coupling operator $L = \sqrt{\gamma}a$. Here, we ignore the optical loss. The QSDE of this system is given by

$$dj_t(a) = -\frac{\gamma}{2}j_t(a)dt + \frac{\epsilon}{2}j_t(a^*)dt - \sqrt{\gamma}d\mathcal{A}(t), \quad d\mathcal{Y}(t) = \sqrt{\gamma}j_t(a)dt + d\mathcal{A}(t),$$

which is also given by (1.37) with $\kappa = 0$ in the white noise form. Here, we consider the measurement of the amplitude quadrature of the output field $\mathcal{Y}(t)$ by a homodyne detector, which generates the measurement output $y_{\mathrm{m}}(t)$ satisfying

$$dy_{\mathrm{m}}(t) = \sqrt{\gamma}j_t(a + a^*)dt + d\mathcal{A}(t) + d\mathcal{A}(t)^*.$$

As described before, this is identified as a classical stochastic signal $\iota(y_{\mathrm{m}}(t))$ and is processed by the quantum filter, which yields the best estimate of both the quadratures $j_t(q) = j_t(a + a^*)$ and $j_t(p) = j_t(-\imath(a - a^*))$. The entire setup is depicted in Fig. 4.2. Note that in the scenario of quantum optics, we usually do not intend to estimate the internal cavity modes, but rather we are interested in the external output field and its squeezing level. But this simple example can be used to model several important systems such as an opto-mechanical oscillator, in which case estimating those internal degrees of freedom (such as the position and momentum) is important. Also from this example, we will see the necessity of feedback control to produce an unconditional squeezed state, which will be discussed in Sect. 6.6.2.

The filtering equation of $j_t(q)$ and $j_t(p)$ can be derived directly by substituting $X = q$ and $X = p$ to (4.10). Let us begin by deriving the equation for $\pi_t(q)$. First, it can be seen immediately that $\pi_t(\mathcal{L}q) = (\epsilon - \gamma)\pi_t(q)/2$ and $\pi_t(L + L^*) = \sqrt{\gamma}\pi_t(q)$. The remaining part (coefficient of the innovation part) can be calculated as follows;

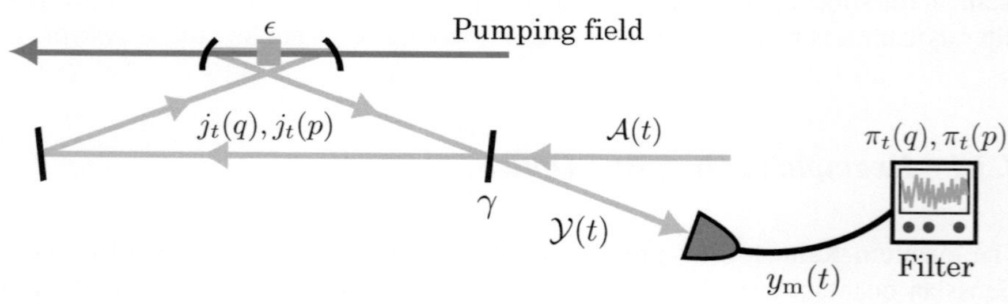

Fig. 4.2 Schematic of the quantum filter for an DPA. The homodyne detector placed at the output of the DPA measures the amplitude quadrature of the output field $\mathcal{Y}(t)$ and generates the measurement output $y_{\mathrm{m}}(t)$, which is used for producing the estimate $\pi_t(x) = (\pi_t(q), \pi_t(p))^\top$

$$\begin{aligned}\pi_t(qL + L^*q) - \pi_t(q)\pi_t(L + L^*) &= \frac{\sqrt{\gamma}}{2}\pi_t(2q^2 + \imath[q, p]) - \pi_t(q) \cdot \sqrt{\gamma}\pi_t(q)\\ &= \sqrt{\gamma}\pi_t(\Delta q^2) - \sqrt{\gamma},\end{aligned}$$

where $\Delta q = q - \pi_t(q)$. Hence the filtering equation for q is

$$d\pi_t(q) = \frac{\epsilon - \gamma}{2}\pi_t(q)dt + \sqrt{\gamma}\big(\pi_t(\Delta q^2) - 1\big)(dy_{\rm m} - \sqrt{\gamma}\pi_t(q)dt). \tag{4.23}$$

This implies that we need to have an equation to compute $\pi_t(\Delta q^2)$. To obtain this equation, here we show only the critical part: the coefficient of the innovation part of the filtering equation for q^2 can be calculated to be

$$\pi_t(q^2L + L^*q^2) - \pi_t(q^2)\pi_t(L + L^*) = \sqrt{\kappa}\Big(\pi_t(q^3) - \pi_t(q^2)\pi_t(q) - 2\pi_t(q)\Big),$$

suggesting that the third-order moment may be necessary. However, with the assumption that the initial state is a quantum Gaussian state such as a coherent or squeezed state, the third moment is actually zero, i.e., $\pi_t(\Delta q^3) = 0$ (further discussions in relation to this will be given in Sect. 4.3.2), which leads to $\pi_t(q^3) - 3\pi_t(q^2)\pi_t(q) + 2\pi_t(q)^3 = 0$. Using this relation, we end up with the equation of $\pi_t(\Delta q^2)$ as follows:

$$d\pi_t(\Delta q^2) = \frac{\epsilon + \gamma}{2}\pi_t(\Delta q^2)dt - \gamma\pi_t(\Delta q^2)^2 dt. \tag{4.24}$$

That is, this is an ordinary differential equation in the variable $\pi_t(\Delta q^2)$, hence it can be solved numerically if not explicitly. The filtering equation of p can also be derived in the same way:

$$d\pi_t(p) = -\frac{\epsilon + \gamma}{2}\pi_t(p)dt + \sqrt{\gamma}\pi_t\Big(\frac{\Delta q\Delta p + \Delta p\Delta q}{2}\Big)(dy_{\rm m} - \sqrt{\gamma}\pi_t(q)dt). \tag{4.25}$$

Also the conditional variance of p obeys

$$d\pi_t(\Delta p^2) = -\frac{\epsilon + \gamma}{2}\pi_t(\Delta p^2)dt + \gamma dt. \tag{4.26}$$

The above set of Eqs. (4.23)–(4.26), can be summarized as follows:

$$d\pi_t(x) = A\pi_t(x)dt + (P(t) - I)C^\top(dy_{\rm m}(t) - C\pi_t(x)dt), \tag{4.27}$$

$$\frac{dP(t)}{dt} = AP(t) + P(t)A^\top + \gamma I - (P(t) - I)C^\top C(P(t) - I), \tag{4.28}$$

where

$$\pi_t(x) = \begin{bmatrix} \pi_t(q) \\ \pi_t(p) \end{bmatrix}, \quad A = \begin{bmatrix} (\epsilon - \gamma)/2 & 0 \\ 0 & -(\epsilon + \gamma)/2 \end{bmatrix}, \quad C = [\sqrt{\gamma} \;\; 0]$$

and $P(t)$ is the estimation error symmetrized covariance matrix defined by

$$P(t) = \begin{bmatrix} \pi_t(\Delta q^2) & \pi_t(\Delta q \Delta p + \Delta p \Delta q)/2 \\ \pi_t(\Delta q \Delta p + \Delta p \Delta q)/2 & \pi_t(\Delta p^2) \end{bmatrix}.$$

The set of Eqs. (4.27) and (4.28) is called the quantum Kalman filtering equations (for the DPA). A notable feature of these equations is that (4.28) is a deterministic differential equation, implying that the estimation error does not depend on the observation record; (4.28) is called the *Ricatti differential equation*, and this deterministic property holds more generally, as will be outlined in the next subsection.

Now, let us examine our filtering equation under the assumption that the initial state is set to a pure coherent state with mean $\langle x \rangle = (2, 2)^\top$ and covariance matrix $\langle \Delta x \Delta x^\top + (\Delta x \Delta x^\top)^\top \rangle / 2 = I$ with $\Delta x = x - \langle x \rangle$. We set $\pi_0(x) = \langle x \rangle$ and $P(0) = I$ in the filtering equation. The time evolution of the filtering equation (4.27) and (4.28) is illustrated in Fig. 4.3. From this, we observe several interesting properties of the estimate for the DPA system, which are summarized as follows.

- $\pi_t(q)$ changes probabilistically, while $\pi_t(p)$ obeys a deterministic time evolution and converges to zero. This is because the off-diagonal elements of $P(t)$ is always zero due to $P(0) = I$. This fact implies that we do not obtain any information about p from the observation; in fact, the dynamics of $\pi_t(p)$ is now $d\pi_t(p)/dt = (-\epsilon - \gamma)\pi_t(p)/2$, and this is exactly the same as the unconditional time evolution of p, i.e., $\mathbb{P}(p)$, which is generated from the master equation (4.20).

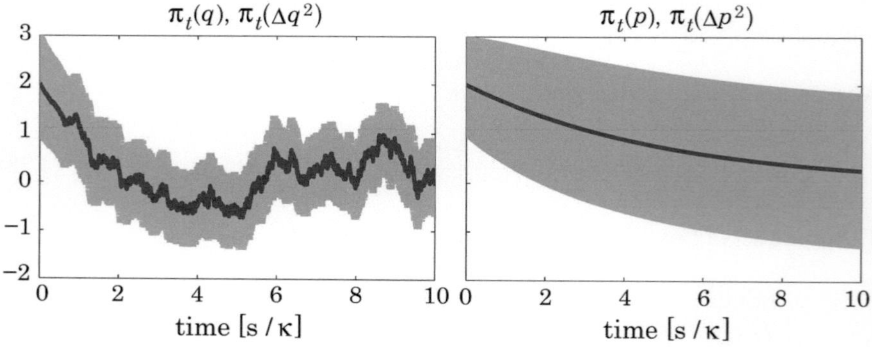

Fig. 4.3 The time evolution of $\pi_t(q)$ (*left*) and $\pi_t(p)$ (*right*). The estimation errors $\pi_t(\Delta q^2)$ and $\pi_t(\Delta p^2)$ are depicted in *blue* along with the trajectories of $\pi_t(q)$ and $\pi_t(p)$, respectively. The squeezing strength is chosen as $\epsilon = -0.6\gamma$, which guarantees the stability of the system (i.e., A is Hurwitz) (color figure online)

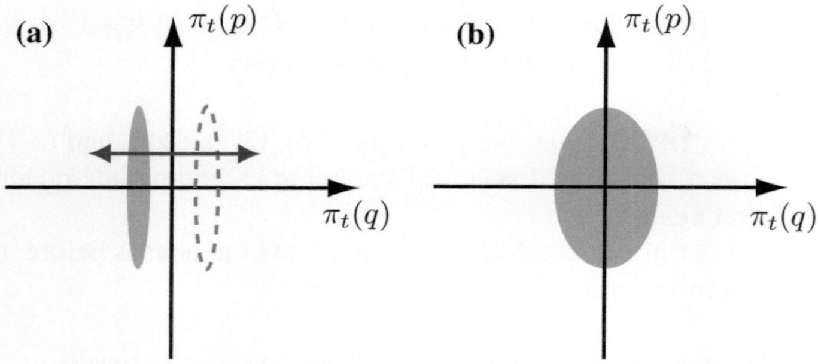

Fig. 4.4 Phase-space picture of the **a** conditional and **b** unconditional states of the DPA. The conditional state fluctuates along the $\pi_t(q)$ axis. The ensemble average of this fluctuation yields the unconditional state, which is squeezed but not pure

- In this case, the steady-state solution of (4.28) can be explicitly obtained as

$$\pi_\infty(\Delta q^2) = \frac{\epsilon+\gamma}{\gamma}, \quad \pi_\infty(\Delta p^2) = \frac{\gamma}{\epsilon+\gamma}.$$

  Hence $\pi_\infty(\Delta q^2)\pi_\infty(\Delta p^2) = 1$ holds. Thus, if $\epsilon \neq 0$, the conditional state is a *pure* squeezed state; in fact as discussed in Sect. 4.2.3, the equivalent SME preserves the purity of the conditional state. In particular, in the simulation, we chose $\epsilon = -0.6\gamma$, in which case $\pi_\infty(\Delta q^2) = 2/5$ and $\pi_\infty(\Delta q^2) = 5/2$, hence it is a q-squeezed state.

From the second observation above, the conditional state (when $t \gg 1$) can be illustrated in the $(\pi_t(q), \pi_t(p))$ phase space. As shown in Fig. 4.4a, it is a pure squeezed state fluctuating along the $\pi_t(q)$ axis. However, due to this probabilistic nature, the ensemble average of this conditional squeezed state is not a pure state anymore, as illustrated in Fig. 4.4b. In fact, the unconditional variance of q at the steady state can be calculated as $\mathbb{P}(\Delta q^2) = \langle \pi_\infty(\Delta q^2)\rangle = \gamma/(\gamma-\epsilon)$; hence together with the fact $\mathbb{P}(\Delta p^2) = \pi_\infty(\Delta p^2) = \gamma/(\epsilon+\gamma)$ we have that $\mathbb{P}(\Delta q^2)\mathbb{P}(\Delta p^2) = \gamma^2/(\gamma^2-\epsilon^2) > 1$; hence it is not a pure squeezed state. Therefore, to obtain a pure squeezed state in a deterministic way, we need to compensate the fluctuation along the $\pi_t(q)$ axis, and this can be carried out by feedback. This topic will be discussed in Sect. 6.6.2.

4.3.2 *Quantum Filtering: Multiple-Input and Multiple-Output Case*

We now turn to the case of linear quantum systems with multiple inputs and outputs, with coherent input fields, with dynamics given by

$$
\begin{aligned}
dx(t) &= Ax(t)dt + B(f(t)dt + dw(t));\ x(0) = x. \\
dy(t) &= Cx(t)dt + D(f(t)dt + dw(t)),
\end{aligned} \tag{4.29}
$$

where the matrices (A, B, C, D) are given by (2.6), (2.7), (2.8), and (2.7), respectively, and $f(t)$ is a $\mathbb{R}^{2m}$-valued vector of the phase and amplitude quadratures of the coherent input fields.

We follow the setting of Sect. 4.2.1 and restrict the treatment as before to the case where the measurement process $y_{\mathrm{m}}(t)$ is of the form

$$
dy_{\mathrm{m}}(t) = F_{\mathrm{m}}dy(t) = C_{\mathrm{m}}x(t)dt + D_{\mathrm{m}}(f(t)dt + dw(t)), \tag{4.30}
$$

where $C_{\mathrm{m}} = F_{\mathrm{m}}C$, $D_{\mathrm{m}} = F_{\mathrm{m}}D$, and F_{m} is as given by (4.14). Now, with $x = (q_1, p_1, q_2, p_2, \ldots, q_n, p_n)^\top$ as before, let

$$
\pi_t(x) = (\pi_t(q_1), \pi_t(p_1), \pi_t(q_2), \pi_t(p_2), \ldots, \pi_t(q_n), \pi_t(p_n))^\top.
$$

We now also restrict ourselves to the case where the state of the oscillators to be jointly Gaussian, while the input fields, being coherent states, are already Gaussian. This will be required in the following to obtain the general form of the quantum Kalman filtering equation for linear Gaussian quantum systems with multiple inputs and outputs. The linear evolution of $x(t)$ and the Gaussian state of the system imply that $x(t)$ is a Gaussian observable in sense that,

$$
\begin{aligned}
\mathbb{P}\left(e^{\imath\lambda^\top x(t)}\right) &= \mathbb{P}\left(e^{\imath\lambda^\top\left(\Phi(t)x+\int_0^t \Phi(t-\tau)B(f(\tau)d\tau+dw(\tau))\right)}\right) \\
&= \langle e^{\imath\lambda^\top\left(\Phi(t)x+\int_0^t \Phi(t-\tau)Bf(\tau)d\tau\right)}\rangle\langle e^{\imath\lambda^\top\int_0^t \Phi(t-\tau)Bdw(\tau)}\rangle \\
&= e^{\imath\lambda^\top(\Phi(t)\langle x\rangle+\int_0^t \Phi(t-\tau)Bf(\tau)d\tau)-(1/2)\lambda^\top R(t)\lambda},
\end{aligned}
$$

for all $\lambda \in \mathbb{R}^n$, where $\Phi(t) = e^{At}$ and

$$
R(t) = \Phi(t)\left(\langle(x-\langle x\rangle)(x-\langle x\rangle)^\top\rangle + \int_0^t \Phi(-\tau)BB^\top\Phi(-\tau)^\top d\tau\right)\Phi(t)^\top.
$$

In the second line above, the first quantum expectation $\langle\cdot\rangle$ on the right is with respect to the oscillators, while the second quantum expectation is with respect to the fields. For any integer $k > 0$, let $0 \leq s_1 < s_2 < \cdots < s_k$. By a similar calculation, $y_{\mathrm{m}}(s_1), y_{\mathrm{m}}(s_2), \ldots, y_{\mathrm{m}}(s_k)$ are jointly Gaussian observables in the sense that,

$$
\mathbb{P}\left(e^{\imath\mathrm{Tr}(\chi^\top y_{\mathrm{m}}(s_1,\ldots,s_k))}\right) = e^{\imath\left(\mathrm{vec}(\chi)^\top\mu(s_1,\ldots,s_k)\right)-(1/2)\mathrm{vec}(\chi)^\top Q(s_1,\ldots,s_k)\mathrm{vec}(\chi)},
$$

for any $\chi \in \mathbb{R}^{p\times k}$, where $y_{\mathrm{m}}(s_1, \ldots, s_k) = [\, y_{\mathrm{m}}(s_1) \ \ldots \ y_{\mathrm{m}}(s_k)\,]$, and $\mathrm{vec}(\chi) \in \mathbb{R}^{p\times k}$ is a column vector formed by stacking the columns of χ from left to right,

$$\mu(s_1, s_2, \ldots, s_k) = \begin{bmatrix} \int_0^{s_1} \left(D_{\rm m} f(\tau) + C_{\rm m} \int_0^{\tau} \Phi(\tau - \tau') B f(\tau') d\tau'\right) d\tau \\ \int_0^{s_2} \left(D_{\rm m} f(\tau) + C_{\rm m} \int_0^{\tau} \Phi(\tau - \tau') B f(\tau') d\tau'\right) d\tau \\ \vdots \\ \int_0^{s_k} \left(D_{\rm m} f(\tau) + C_{\rm m} \int_0^{\tau} \Phi(\tau - \tau') B f(\tau') d\tau'\right) d\tau \end{bmatrix},$$

and

$$\begin{aligned} &Q(s_1, s_2, \ldots, s_k) \\ &= \Bigg[C_{\rm m} \int_0^{s_i} \Phi(\tau) d\tau \langle (x - \langle x \rangle)(x - \langle x \rangle)^\top \int_0^{s_j} \Phi(\tau)^\top d\tau C_{\rm m}^\top \\ &\quad + \int_0^{\infty} \left(D_{\rm m} 1_{[0,s_i]}(\tau') + C_{\rm m} \int_{\tau'}^{s_i} 1_{[0,s_i]}(\tau) \Phi(\tau - \tau') B d\tau \right) \\ &\quad \times \left(D_{\rm m} 1_{[0,s_j]}(\tau') + C_{\rm m} \int_{\tau'}^{s_j} 1_{[0,s_j]}(\tau) \Phi(\tau - \tau') B d\tau \right)^\top d\tau' \Bigg]_{i,j=1,2,\ldots,k} . \end{aligned}$$

Extrapolating from the above, it is easy to see that the observables $x_j(t)$ ($j = 1, 2, \ldots, n$) and $y_{\rm m}(s_1)$, …, $y_{\rm m}(s_k)$ are jointly Gaussian for any integer $k > 1$ and any $0 \leq s_1 < s_2 < \cdots < s_k \leq t$, in the sense that any linear combination of these observables is a Gaussian observable (extending the definition above in the obvious way).

We know that $\lambda^\top x(t)$ commutes with any element of the von Neuman algebra $\mathscr{Y}_{{\rm m},t}$ for any $\lambda \in \mathbb{R}^n$, and the conditional expectation $\pi_t(\lambda^\top x) = \lambda^\top \pi_t(x)$ is an element of this algebra. Thus, by the Spectral Theorem, $\lambda^\top \pi_t(x)$ and $\{y_{\rm m}(s); 0 \leq s \leq t\}$ can be viewed as classical random variables that can be realized on the same classical probability space. Even more than that, since $\lambda^\top x(t)$ and $\{y_{\rm m}(s); 0 \leq s \leq t\}$ are jointly Gaussian observables, as discussed in the preceding paragraph, it follows that the conditional expectation $\lambda^\top \pi_t(x)$ must be (equivalent to) a classical Gaussian random variable for any $\lambda \in \mathbb{R}^n$. In particular, $\{\pi_t(x_j); j = 1, 2, \ldots, n\}$ are essentially jointly Gaussian random variables.

Now, using the definitions of $\tilde{H}(t)$ and $\tilde{L}(t)$ to explicitly calculate all commutators on the right-hand side of (4.17), and taking the initial state of the system to be Gaussian, so that the conditional expectations are jointly Gaussian random variables and their third-order covariances vanish, we have

$$\begin{aligned} d\pi_t(x) &= A\pi_t(x)dt + Bf(t)dt \\ &\quad + [P(t)C^\top F_{\rm m}^\top + B\Sigma](D_{\rm m} D_{\rm m}^\top)^{-1}[dy_{\rm m}(t) - F_{\rm m} C\pi_t(x)dt] \\ &= A\pi_t(x)dt + Bf(t)dt + [P(t)C^\top F_{\rm m}^\top + B\Sigma](D_{\rm m} D_{\rm m}^\top)^{-1} d\nu(t), \end{aligned} \tag{4.31}$$

with initial condition $\pi_0(x) = \langle x \rangle$, where $\Sigma = D^\top F_{\rm m}^\top$. Also $\nu(t)$ is the innovations process $\nu(t) = y_{\rm m}(t) - F_{\rm m} \int_0^t (C\pi_s(x) + Du(s))ds$, and $P(t)$ is the symmetrized covariance matrix,

$$P(t) = \frac{1}{2}\pi_t\Big[(x - \pi_t(x))(x - \pi_t(x))^\top + ((x - \pi_t(x))(x - \pi_t(x))^\top)^\top\Big],$$

satisfying the deterministic matrix Riccati differential equation

$$\begin{aligned}\dot{P}(t) &= AP(t) + P(t)A^\top + BB^\top \\ &\quad - (P(t)C^\top F_{\mathrm{m}}^\top + B\Sigma)(D_{\mathrm{m}} D_{\mathrm{m}}^\top)^{-1}(P(t)C^\top F_{\mathrm{m}}^\top + B\Sigma)^\top, \qquad (4.32)\end{aligned}$$

with initial condition

$$P(0) = \frac{1}{2}\langle \Delta x \Delta x^\top + (\Delta x \Delta x^\top)^\top \rangle, \qquad (4.33)$$

with $\Delta x = x - \langle x \rangle$.

4.4 Robust Linear Quantum Observers

A central problem of the quantum filter is that, because it is a model-based estimator, the exact system parameters must be known for the optimal estimation to be carried out (in the Kalman filtering case, this means that all the matrices (A, B, C, D) must be exactly known). In fact, a nominal filter with the wrong parameters is fragile in the sense that the estimation error can be large or even diverge. Of course, this is not a problem appearing only in the quantum case, but rather it is a long-standing problem that has been extensively investigated in classical systems and control theory. Actually, in the literature for classical systems, we find numerous treatments of *robust filtering*, which generates a guaranteed estimate even under system uncertainties. This section is devoted to discussing a quantum version of these classical robust filters [8].

4.4.1 Guaranteed-Error Robust Observer

For simplicity, let us deal with a general single-input and single-output linear quantum system with single-mode variable $x(t) = (q(t), p(t))^\top = (j_t(q), j_t(p))^\top$. The system couples to a vacuum field, represented by the coupling operator $L = Kx$ with $x = (q, p)^\top$. Here, we consider the case where the system Hamiltonian H contains uncertainty of the following form:

$$H = \frac{1}{2}x^\top(R + \Delta R(t))x - x^\top \mathbb{J} F u(t), \quad \Delta R(t)^2 \leq gI.$$

$u(t)$ is a control input, which can be a function of the measurement result (i.e., a feedback control input). The real symmetric matrix $\Delta R(t)$ represents a time-varying uncertainty bounded by the known constant $g > 0$. The other matrices, R, F, and K, are known. By defining $\Delta A(t) := \mathbb{J}\Delta R(t)$, the dynamics of $x(t)$ is given by

$$dx(t) = (A + \Delta A)x(t)dt + Fu(t)dt + \imath\mathbb{J}[K^\top d\mathcal{A}^*(t) - K^* d\mathcal{A}(t)]. \tag{4.34}$$

Also here, we consider measuring the amplitude quadrature by homodyne detection, generating the measurement output

$$dy_{\mathrm{m}}(t) = (K + K^*)x(t)dt + d\mathcal{A}(t) + d\mathcal{A}(t)^*. \tag{4.35}$$

Note that we can consider a more general case where K also contains some uncertainty; see [8].

Now, motivated by the structure of the Kalman filter (4.31), we aim to design a *classical* linear observer of the form

$$dx'(t) = A'x'(t)dt + Fu(t)dt + B'dy_{\mathrm{m}}(t), \tag{4.36}$$

where A' and B' are a matrix and vector to be determined so that the variance of the estimation error is guaranteed to be within a certain bound. The vector of *classical* random variables, $x'(t) = (q'(t), p'(t))^\top \in \mathbb{R}^2$, represents an estimate of $x(t)$, which is not necessarily an optimal estimate of $x(t)$. (Hence, we refer to a system with dynamics (4.36) as an "observer" rather than a "filter".) Furthermore, the control input $u(t)$ is fixed to a linear function of the observer variable, i.e., $u(t) = \ell^\top x'(t)$, where ℓ is a fixed real column vector with dimension 2. We now offer two important remarks about the setting considered above. First, note that we are now considering a hybrid system composed of the quantum system (4.34) and the classical system (4.36). This could lead to some difficulty in making sense of the system of differential equations that describe the two systems simultaneously, especially when they are interacting and there is a measurement involved, the latter causing the state of the quantum part to collapse. This kind of hybrid system can be given a consistent interpretation by embedding the classical linear dynamics as a commutative subdynamics of an appropriate linear quantum system. Details of this technique of embedding classical dynamics into a quantum system, along with some examples, can be found in [9, Appendix, pp. 2542–2543] and, for the linear case, [10, Sect. 3], [11–13]. The second remark is regarding the feedback control input $u(t)$. It is not an essential point of the theory that will be described; the robust observer method introduced below holds even if $\ell = 0$.

Now, an explicit form of (A', B') that enjoys a guaranteed estimation error bound is provided in the following theorem.

Theorem 4.2 *Suppose there exist positive scalars δ_1, δ_2, and ϵ such that the following two coupled Riccati equations have positive definite solutions $P_1 > 0$ and $P_2 > 0$:*

$$(A + F\ell^\top)P_1 + P_1(A + F\ell^\top)^\top + \mathbb{D} + \frac{1}{\epsilon}P_1^2 + (\epsilon g + \delta_1)I = 0, \tag{4.37}$$

$$\begin{aligned} &\mathbb{A}P_2 + P_2\mathbb{A}^\top + \mathbb{D} - (P_2\mathbb{F}^\top + \mathbb{J}^\top\Im(K)^\top)(\mathbb{F}P_2 + \Im(K)\mathbb{J}) \\ &\quad - P_2(\ell F^\top P_1^{-1} + P_1^{-1}F\ell^\top)P_2 + (\epsilon g + \delta_2)I = 0, \end{aligned} \tag{4.38}$$

where the matrices $\mathbb{A}$, $\mathbb{D}$ *and the vector* $\mathbb{F}$ *are defined by*

$$\mathbb{A} = A + (\mathbb{D} + \epsilon g I)P_1^{-1}, \quad \mathbb{D} = \mathbb{J}\Re(K^*K)\mathbb{J}^\top, \quad \mathbb{F} = K + K^* + \Im(K)\mathbb{J}P_1^{-1}.$$

Then, the observer

$$\begin{aligned} dx'(t) &= (\mathbb{A} - P_2\ell F^\top P_1^{-1})x'(t)dt + Fu(t)dt \\ &\quad + \left(P_2\mathbb{F}^\top + \mathbb{J}^\top\Im(K)^\top\right)(dy_{\mathrm{m}}(t) - \mathbb{F}x'(t)dt) \end{aligned} \tag{4.39}$$

generates the estimate $x'(t) = (q'(t), p'(t))^\top$ *that satisfies*

$$\lim_{t\to\infty} \langle (q(t) - q'(t))^2 + (p(t) - p'(t))^2 \rangle \le \mathrm{Tr}(P_2), \tag{4.40}$$

for all admissible uncertainties contained in $\Delta R(t)$.

Proof We consider the augmented variable $\bar{z}(t) = (x(t), x(t) - x'(t))^\top$, where $x(t)$ and $x'(t)$ satisfy (4.34) and (4.36), respectively. Then, $\bar{z}(t)$ obeys the following linear QSDE:

$$d\bar{z}(t) = (\bar{A} + \Delta\bar{A}(t))\bar{z}(t)dt + \bar{b}d\mathcal{A}(t)^* + \bar{b}^*d\mathcal{A}(t),$$

where

$$\begin{aligned} \bar{A} &= \begin{bmatrix} A + F\ell^\top & -F\ell^\top \\ A - A' - B'(K + K^*) & A' \end{bmatrix}, \quad \bar{b} = \begin{bmatrix} \imath\mathbb{J}K^\top \\ \imath\mathbb{J}K^\top - B' \end{bmatrix}, \\ \Delta\bar{A}(t) &= \begin{bmatrix} \Delta A(t) \ 0 \\ \Delta A(t) \ 0 \end{bmatrix} = \begin{bmatrix} \mathbb{J} \\ \mathbb{J} \end{bmatrix} \Delta R(t)[I \ 0] =: \bar{M}\Delta R(t)\bar{N}. \end{aligned}$$

Now, the symmetrized covariance matrix of $\bar{z}$ is given by $\bar{V}_{nm} = \langle \bar{z}_n\bar{z}_m + \bar{z}_m\bar{z}_n \rangle/2$, $(n, m = 1, \ldots, 4)$. This satisfies the following generalized uncertain relation:

$$\langle \bar{z}(t)\bar{z}(t)^\top \rangle = \bar{V}(t) + \frac{\imath}{2}\bar{\mathbb{J}} \ge 0, \quad \bar{\mathbb{J}} = \begin{bmatrix} \mathbb{J} \ \mathbb{J} \\ \mathbb{J} \ \mathbb{J} \end{bmatrix},$$

and the time evolution of $\bar{V}(t)$ is calculated as

$$\begin{aligned}\frac{d}{dt}\bar{V}(t) &= (\bar{A}+\Delta\bar{A}(t))\langle\bar{z}(t)\bar{z}(t)^\top\rangle + \langle\bar{z}(t)\bar{z}(t)^\top\rangle(\bar{A}+\Delta\bar{A}(t))^\top + \bar{b}^*\bar{b}^\top \\ &= (\bar{A}+\Delta\bar{A}(t))\Big(\bar{V}(t)+\frac{\imath}{2}\bar{\mathbb{J}}\Big) + \Big(\bar{V}(t)+\frac{\imath}{2}\bar{\mathbb{J}}\Big)(\bar{A}+\Delta\bar{A}(t))^\top + \bar{b}^*\bar{b}^\top \\ &= (\bar{A}+\Delta\bar{A}(t))\bar{V}(t) + \bar{V}(t)(\bar{A}+\Delta\bar{A}(t))^\top + \bar{\mathbb{D}}, \qquad (4.41)\end{aligned}$$

where $\bar{\mathbb{D}}$ is given by

$$\bar{\mathbb{D}} = \begin{bmatrix}\mathbb{D} & \mathbb{D}\\ \mathbb{D} & \mathbb{D}\end{bmatrix} - \left[\begin{array}{c|c} 0 & [B'\Im(K)\mathbb{J}]^\top \\ \hline B'\Im(K)\mathbb{J} & B'\Im(K)\mathbb{J} + [B'\Im(K)\mathbb{J}]^\top - B'B'^\top\end{array}\right].$$

Our goal is to design A' and B' such that the matrix inequality

$$\exists \bar{X} > 0 \ \text{ s.t. } \ (\bar{A}+\Delta\bar{A}(t))\bar{X} + \bar{X}(\bar{A}+\Delta\bar{A}(t))^\top + \bar{\mathbb{D}} < 0 \qquad (4.42)$$

is satisfied for all admissible uncertainties: then, it is immediate to see that (4.41) yields $\lim_{t\to\infty}\bar{V}(t) \le \bar{X}$. In order to obtain a sufficient condition for (4.42) that does not involve $\Delta\bar{A}(t)$, we use the following relation:

$$\Big(\frac{1}{\sqrt{\epsilon}}\bar{X}\bar{N}^\top - \sqrt{\epsilon}\bar{M}\Delta R(t)\Big)\Big(\frac{1}{\sqrt{\epsilon}}\bar{X}\bar{N}^\top - \sqrt{\epsilon}\bar{M}\Delta R(t)\Big)^\top \ge 0,$$

where $\epsilon > 0$ is a free parameter. Then, combining the above relation with the assumption on the uncertainty, $\Delta R(t)^2 \le gI$, we obtain

$$\begin{aligned}\Delta\bar{A}(t)\bar{X} + \bar{X}\Delta\bar{A}(t)^\top &= (\bar{M}\Delta R(t))(\bar{N}\bar{X}) + (\bar{N}\bar{X})^\top(\bar{M}\Delta R(t))^\top \\ &\le \frac{1}{\epsilon}\bar{X}\bar{N}^\top\bar{N}\bar{X} + \epsilon\bar{M}(\Delta R(t))^2\bar{M}^\top \\ &\le \frac{1}{\epsilon}\bar{X}\bar{N}^\top\bar{N}\bar{X} + \epsilon g\bar{M}\bar{M}^\top.\end{aligned}$$

Therefore, the condition (4.42) holds for all admissible uncertainties if there exists a positive definite matrix $\bar{X} > 0$ such that the following Riccati inequality holds:

$$\bar{\Psi} := \bar{A}\bar{X} + \bar{X}\bar{A}^\top + \bar{\mathbb{D}} + \epsilon g\bar{M}\bar{M}^\top + \frac{1}{\epsilon}\bar{X}\bar{N}^\top\bar{N}\bar{X} < 0.$$

Especially here we aim to find a solution of the form $\bar{X} = \mathrm{diag}(P_1, P_2)$ with P_1 and P_2 denoting 2×2 positive definite matrices. Then, partitioning the 4×4 matrix $\bar{\Psi}$ into $\bar{\Psi} = (\Psi_{ij})$ with 2×2 matrices Ψ_{ij}, we obtain

$$\Psi_{11} = (A + F\ell^\top)P_1 + P_1(A + F\ell^\top)^\top + \mathbb{D} + \epsilon g I + P_1^2/\epsilon,$$
$$\Psi_{21} = (A - A' - B'K - B'K^*)P_1 + \mathbb{D} + \epsilon g I - B'\Im(K)\mathbb{J} - P_2\ell F^\top,$$
$$\Psi_{22} = A'P_2 + P_2A'^\top + \mathbb{D} + \epsilon g I - B'\Im(K)\mathbb{J} - [B'\Im(K)\mathbb{J}]^\top + B'B'^\top.$$

Let us now assume that the Riccati equation (4.37), which is equal to $\Psi_{11} = -\delta_1 I < 0$, has a solution $P_1 > 0$. Then, the equality $\Psi_{21} = 0$ yields $A' = \mathbb{A} - B'\mathbb{F} - P_2\ell F^\top P_1^{-1}$. Moreover, Ψ_{22} is then calculated as

$$\begin{aligned}\Psi_{22} &= \mathbb{A}P_2 + P_2\mathbb{A}^\top + \mathbb{D}\\ &\quad + \Big(B' - P_2\mathbb{F}^\top - \mathbb{J}^\top\Im(K)^\top\Big)\Big(B'^\top - \mathbb{F}P_2 - \Im(K)\mathbb{J}\Big)\\ &\quad - (P_2\mathbb{F}^\top + \mathbb{J}^\top\Im(K)^\top)(\mathbb{F}P_2 + \Im(K)\mathbb{J})\\ &\quad - P_2(\ell F^\top P_1^{-1} + P_1^{-1}F\ell^\top)P_2 + \epsilon g I.\end{aligned}$$

Hence, the optimal B' that minimizes the maximum eigenvalue of Ψ_{22} is given by

$$B' = P_2\mathbb{F}^\top + \mathbb{J}^\top\Im(K)^\top.$$

Then, the existence of a solution $P_2 > 0$ in (4.38) directly implies $\Psi_{22} = -\delta_2 I < 0$. As a result, we obtain $\bar{\Psi} = \mathrm{diag}(-\delta_1 I, -\delta_2 I) < 0$, which leads to the objective condition (4.42). Therefore, as mentioned above, we have $\lim_{t\to\infty}\bar{V}(t) \le \bar{X}$. Then, as the third and fourth diagonal elements of the matrix $\bar{V}(t)$ are, respectively, given by $\bar{V}_{33} = \langle\bar{z}_3^2\rangle = \langle(q(t) - q'(t))^2\rangle$ and $\bar{V}_{44} = \langle\bar{z}_4^2\rangle = \langle(p(t) - p'(t))^2\rangle$, we obtain the assertion (4.40). □

The quantum robust observer constructed above has the following important property, which one would expect it to have: if the uncertainties are small (or zero), the robust observer is close (or identical) to the optimal quantum Kalman filter. The formal statement of this fact is given as follows (for the proof, see [8]).

Proposition 4.1 *Consider the case where the uncertainties converge to zero: i.e.,* $\Delta R(t) \to 0$. *Then, there exist parameters* δ_1, δ_2, *and* ϵ *such that the robust observer* (4.39) *converges to the stationary Kalman filter associated with the system* (4.34) *with* $\Delta A(t) = 0$.

This proposition also states that we can find the parameters $(\delta_1, \delta_2, \epsilon)$ such that the robust observer (4.39) approximates the stationary Kalman filter when the uncertainties are small, because the solutions of the Riccati equations (4.37) and (4.38) are continuous with respect to the above parameters.

4.4.2 Example

Here, we will demonstrate that, in the presence of uncertainty, the robust observer actually outperforms the nominal Kalman filter, which is no longer optimal for uncertain systems. The system considered here is a single particle with position q and momentum p, which is trapped around the origin of the antiharmonic potential field $V(q) = -0.05q^2$. For instance, this can be interpreted as an approximate potential for the double-well potential $V(q) = q^4 - 0.05q^2$. Then, the corresponding Hamiltonian is given by $H^{\text{free}} = 2p^2 - 0.05q^2 = x^\top Rx$ with $R = \text{diag}(-0.05, 2)$. Apart from this, the system is driven by the control Hamiltonian

$$H^{\text{control}} = -u(t)q, \quad \text{i.e.,} \quad F = \begin{bmatrix} 0 \\ 1 \end{bmatrix}, \tag{4.43}$$

where $u(t)$ is the control input that will be described later. In addition, the system is subjected to an uncertain Hamiltonian $\Delta H = -\sqrt{d(t)}q^2$, where $d(t)$ is an unknown time-varying parameter bounded by the known constant $g \geq 0$, i.e., $d(t) \in [0, g]$ for all t. For this system, we consider the continuous QND measurement of q, corresponding to $L = q$ or equivalently $K = [1\ 0]$.

With the above setup, we perform a numerical comparison between the developed robust observer and the nominal Kalman filter. For the robust observer, we evaluate the guaranteed upper bound of the estimation error $\text{Tr}(P_2)$ in (4.40). To calculate $\text{Tr}(P_2)$, we set the parameters δ_1 and δ_2 to be both 0.1 and also chose ϵ that minimizes $\text{Tr}(P_2)$. For the nominal Kalman filter, on the other hand, we evaluate the stationary mean-square error between the "true" system and the estimator for the "nominal" system corresponding to $d(t) = 0$ (see Appendix C of [8]). As for the control part, here we take the feedback of the form $u(t) = \ell^\top x'(t)$ in the robust observer case and $u(t) = \ell^\top \pi_t(x)$ in the nominal Kalman filter case. The control gain ℓ is set to that of the stationary LQG controller for the nominal system (see Chap. 5 for details of LQG control for linear quantum systems).

Let us describe the simulation results. In Table 4.1, the two estimation errors mentioned above are provided for several values of g. The uncertainty $d(t)$ is now set to the worst case value $d(t) = g$ for each g. In the first row of the table, "N/A" means that the nominal Kalman filter fails in the estimation. That is, the error dynamics between the true system and the nominal Kalman filter diverges, and it has no stationary solution. On the other hand, the second row in Table 4.1 shows that the robust observer is not very sensitive to the uncertain parameter g and eventually works as a tolerant estimator.

Table 4.1 The stationary estimation error of the nominal Kalman filter, denoted by "KAL", and the guaranteed upper bound of the estimation error of the robust observer (i.e., $\text{Tr}(P_2)$), denoted by "ROB". Note that the robust observer does not coincide with the nominal Kalman filter even when $g = 0$, because the parameter δ_2 is now set to a non-zero value. Reprinted with permission from [8] © (2006) by the American Physical Society

g	0.00	0.20	0.38	0.60	0.80	0.97
KAL	1.43	2.38	40.88	N/A	N/A	N/A
ROB	1.73	3.32	4.74	7.04	10.12	14.13

4.4.3 Further Reading

We have discussed a robust estimation method that guarantees the estimation error to be within a known bound. A general framework in this direction is to design an estimator or controller so that a given cost is guaranteed to be bounded by a known constant even under unknown perturbations. In particular, it is important to guarantee stability of the system under unknown perturbations, and there are actually several methods to design such robust estimators/controllers; see e.g., [14]. Another direction pursuing the robustness property is the risk-sensitive approach, which has also been studied extensively in the classical case [15, 16]; in this method, rather than minimizing a cost function $V(x) \geq 0$ with x a system variable, we consider the problem minimizing $e^{V(x)}$. In fact, we find that several quantum versions of these risk-sensitive estimators or controllers have been developed [2, 8, 17, 18].

References

1. L. Bouten, R. van Handel, M.R. James, An introduction to quantum filtering. SIAM J. Control Optim. **46**, 2199–2241 (2007)
2. N. Yamamoto, L. Bouten, Quantum risk-sensitive filtering and robustness. IEEE Trans. Autom. Control **54**(1), 92–107. Reprinted, with permission, © 2009 IEEE (2009)
3. S. Sakai, *C*-Algebras and W*-Algebras* (Springer, Berlin, 1998)
4. L. Bouten, R. van Handel, Quantum filtering: a reference probability approach (2006). arXiv preprint, http://arXiv.org/abs/math-ph/0508006
5. L. Bouten, R. van Handel, On the separation principle of quantum control, in *Quantum Stochastics and Information: Statistics, Filtering and Control (University of Nottingham, UK, 15–22 July 2006)*, ed. by V.P. Belavkin, M. Guta (World Scientific, Singapore, 2008), pp. 206–238
6. R. van Handel, J.K. Stockton, H. Mabuchi, Feedback control of quantum state reduction. IEEE Trans. Autom. Control **50**(6), 768–780 (2005)
7. H.I. Nurdin, Quantum filtering for multiple input multiple output systems driven by arbitrary zero-mean jointly Gaussian input fields. Russ. J. Math. Phys. **21**(3), 386–398 (2014). Reprinted, with permission, © (2006) by the American Physical Society
8. N. Yamamoto, Robust observer for uncertain linear quantum systems. Phys. Rev. A **74**, 032107. Reprinted, with permission, Copyright (2006) by the American Physical Society (2006)
9. J. Gough, M.R. James, The series product and its application to quantum feedforward and feedback networks. IEEE Trans. Autom. Control **54**(11), 2530–2544 (2009)
10. M.R. James, H.I. Nurdin, I.R. Petersen, H^{∞} control of linear quantum stochastic systems. IEEE Trans. Autom. Control **53**(8), 1787–1803 (2008)
11. S. Wang, H.I. Nurdin, G. Zhang, M.R. James, Quantum optical realization of classical linear stochastic systems. Automatica **49**(10), 3090–3096 (2013)
12. S. Wang, H.I. Nurdin, G. Zhang, M.R. James, *Network Synthesis for a Class of Mixed Quantum-Classical Linear Stochastic Systems*. arXiv: 1403.6928 (2014, arXiv preprint)
13. S. Wang, H.I. Nurdin, G. Zhang, M.R. James, Synthesis and structure of mixed quantum-classical linear systems, in *Proceedings of the 51st IEEE Conference on Decision and Control (Maui, Hawaii, Dec 10–13, 2012)* (2012), pp. 1093–1098
14. I.R. Petersen, V. Ugrinovskii, M.R. James, Robust stability of uncertain linear quantum systems. Philos. Trans. R. Soc. A **370**, 5354–5363 (2012)
15. P. Dupuis, M. James, I. Petersen, Robust properties of risk-sensitive control. Math. Control Syst. Signals **13**, 318–332 (2000)

16. R.K. Boel, M.R. James, I.R. Petersen, Robustness and risk sensitive filtering. IEEE Trans. Autom. Control **47**(3), 451–461 (2002)
17. M.R. James, A quantum Langevin formulation of risk-sensitive optimal control. J. Opt. B: Quantum Semiclass. Opt. **7**, 198 (2005)
18. C. D'Helon, A.C. Doherty, M.R. James, S.D. Wilson, Quantum risk-sensitive control, in *Proceedings of the 45th IEEE Conference on Decision and Control (CDC)* (2006), pp. 3132 – 3137

Chapter 5
Feedback Control of Linear Dynamical Quantum Systems

Abstract This chapter introduces and treats the topic of quantum feedback control on linear quantum systems. Two distinct approaches to feedback control of quantum systems are considered: measurement-based feedback control and coherent feedback control. For measurement-based feedback control, the notions of controlled Hudson–Parthasarathy QSDEs and controlled quantum filtering equations are introduced. Three control problems are formulated and solved: measurement-based linear quadratic Gaussian control, coherent linear quadratic Gaussian control, and coherent feedback H^∞ control. Examples are provided to illustrate each control method.

In this chapter, we shall formulate and discuss the solutions to three classes of quantum feedback control problems for linear quantum systems and illustrate physical settings where these control problems are relevant. They are the measurement-based (quantum) linear quadratic Gaussian (LQG) control problem, coherent quantum LQG problem, and coherent feedback H^∞ control problem. The exposition in this chapter will build upon the theoretical foundations that have been developed in the previous sections, especially the use of quantum filters in the measurement-based LQG control problem.

5.1 Measurement-Based Quantum Feedback Control

5.1.1 Controlled Quantum Evolution and Quantum Filter

We first set up the theoretical framework for measurement-based (quantum) feedback control. In this type of control, continuous linear measurement is performed on the output $\mathcal{Y}(t)$ of a linear quantum system to gain information about the system that is then used to drive a classical controller. The controller processes the measurement to produce a control signal that is fed back to the system to control its evolution.

Section 5.2 contains materials reprinted from Automatica [12], with permission from Elsevier. Section 5.3 and the associated appendices contain materials reprinted, with permission, from [34] © 2008 IEEE.

H.I. Nurdin and N. Yamamoto, *Linear Dynamical Quantum Systems*,
Communications and Control Engineering, DOI 10.1007/978-3-319-55201-9_5

Describing a quantum system undergoing measurement and feedback is non-trivial, and to properly describe such feedback in the quantum setting requires setting up a modified Hudson–Parthasarathy QSDE that depends on the control signals, the so-called Hudson–Parthasarathy QSDE for controlled quantum Markov flows [1, 2]. We shall introduce this QSDE in this chapter and derive the solution to the measurement-based LQG control problem, following the treatment in [2].

Following Sect. 4.2.2, we consider an n-degree of freedom linear quantum system coupled with m bosonic fields, with quadratic Hamiltonian $H = (1/2)x^\top Rx$, coupling operator $L = Kx$ ($K \in \mathbb{C}^{m\times 2n}$), and scattering matrix $S \in \mathbb{C}^{m\times m}$. We will reuse the notation introduced in Sect. 4.2.2. Thus, let $y_{\rm m}(t)$ be the output of a linear measurement performed on the quantum output $y(t)$ of a linear quantum system in real quadrature form (2.5), with the same number of inputs and outputs m. We take the measurement $y_{\rm m}$ to be of the form $y_{\rm m}(t) = F_{\rm m}y(t)$, with $F_{\rm m} \in \mathbb{R}^{p\times 2m}$ satisfying the conditions set out in Sect. 4.2.2. We assume that we have two control signals $u_1(t) \in \mathbb{R}^{c_1}$ and $u_2(t) \in \mathbb{R}^{c_2}$ (with c_1 and c_2 non-negative integers, and the convention that $u_j(t)$ is dropped if $c_j = 0$), which at each time $t \geq 0$ are functions of $\mathbb{R}^p$-valued stochastic processes on the time interval $[0, t]$. That is, at each time $t \geq 0$, $u_j(t) \triangleq u_j(\{d(s),\ 0 \leq s \leq t\})$ for some $\mathbb{R}^p$-valued stochastic process $d(t)$. We assume that control can be applied in two ways:

1. Adding a control Hamiltonian term of the form $H_c(t) = x^\top N_1 u_1(t)$ to H, that depends on the $\mathbb{R}^{p_1}$-valued control signal $u_1(t)$ with $N_1 \in \mathbb{R}^{2n\times c_1}$.
2. Driving the system with coherent input fields with amplitudes given by the elements of the vector $N_2 u_2(t)$, with a given $N_2 \in \mathbb{C}^{m\times c_2}$ and $u(t)$ an $\mathbb{R}^{c_2}$-valued control signal for each t.

The measurement-based feedback control loop is depicted in Fig. 5.1.

Let $z(t) = F_{\rm m}w(t)$. Since $w(t)$ satisfies the same differential commutation relation as $y(t)$, we have that $[z(t), z(s)^\top] = 0$ for all $s, t \geq 0$. Thus, z is isomorphic to some classical stochastic process, and the definition $u_j(\{z(s),\ 0 \leq s \leq t\})$ makes sense. To reduce notational clutter, let us introduce the notation $z_{[0,t]} = \{z(s),\ 0 \leq$

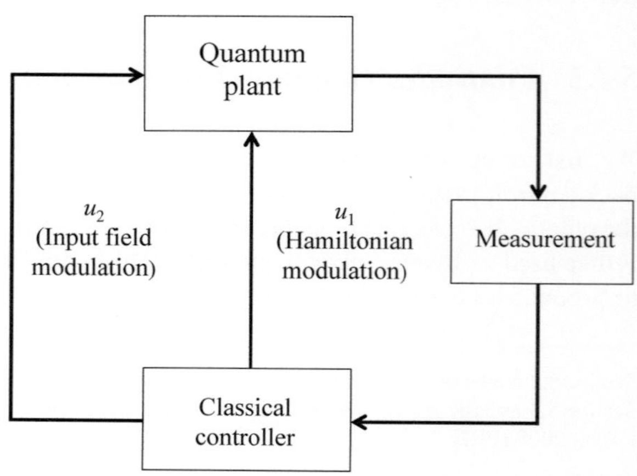

Fig. 5.1 Measurement-based feedback control loop

$s \leq t\}$ and write $u_j(\{z(s),\ 0 \leq s \leq t\})$ simply as $u_j(z_{[0,t]})$. We can now define the *controlled* Hudson–Parthasarathy QSDE $U(t)$,

$$\begin{aligned} dU(t) = \big(&\mathrm{Tr}((S-I)^\top d\Lambda(t)) + d\mathcal{A}^*(t)(L + SN_2u_2(z_{[0,t]})) \\ &- (S^*L + N_2u_2(z_{[0,t]}))d\mathcal{A}(t) \\ &- \big(\imath\big(H + H_c(t) + \Im\{L^*SN_2u_2(z_{[0,t]})\}\big) \\ &+ {}^1\!/_2(L + N_2u_2(z_{[0,t]}))^*(L + N_2u_2(z_{[0,t]}))\, dt\big)\, U(t), \end{aligned} \tag{5.1}$$

with initial condition $U(0) = I$. Since u_j acts on a commutative process z, it is important to note that $U(t)^*u_j\big(z_{[0,t]}\big)U(t) = u_j\big(U(t)^*z_{[0,t]}U(t)\big) = u_j\big((y_{\mathrm{m}})_{[0,t]}\big)$ (here, following the convention above, $(y_{\mathrm{m}})_{[0,t]} = \{y_{\mathrm{m}}(s),\ 0 \leq s \leq t\}$). That is, in the Heisenberg picture, u_j becomes a function of the measurement y_{m}, as it should be.

Define the time-dependent operators,

$$L(t) = L + SN_2u_2(z_{[0,t]}) \text{ and } H(t) = H + H_c(t) + \Im\{L^*SN_2u_2(z_{[0,t]})\}.$$

For any bounded operator X of the quantum system, let $X(t) = j_t(X) = U(t)^*XU(t)$ be the time evolution of X in the Heisenberg picture. Then by an application of the quantum Itō rule, we have that $j_t(X)$ satisfies the QSDE [2, 3],

$$\begin{aligned} &dj_t(X) \\ &\quad= j_t\Big(\imath[H(t), X] + L(t)^*XL(t) - {}^1\!/_2(L(t)^*L(t)X + XL(t)^*L(t))\Big)dt \\ &\quad\quad + d\mathcal{A}(t)^*\, j_t(S^*[X, L(t)]) + j_t([L(t)^*, X]S)d\mathcal{A}(t) \end{aligned}$$

with initial condition $j_0(X) = X$. Constructing the vector $x(t)$ of the quadrature form of the linear quantum system as in Sect. 2.3, we have the equations for the *controlled* linear quantum dynamics,

$$dx(t) = Ax(t)dt + B_cu(t)dt + Bdw(t), \tag{5.2}$$

with A and B as given by (2.6) and (2.7), respectively,

$$u(t) = \begin{bmatrix} u_1\big((y_{\mathrm{m}})_{[0,t]}\big) \\ u_2\big((y_{\mathrm{m}})_{[0,t]}\big) \end{bmatrix},$$

and $B_c = \begin{bmatrix} B_{c,1} & B_{c,2} \end{bmatrix}$ with $B_{c,1} \in \mathbb{R}^{2n\times c_1}$ and $B_{c,2} \in \mathbb{R}^{2n\times c_2}$ given by

$$\begin{aligned} B_{c,1} &= 2\mathbb{J}_nN_1, \\ B_{c,2} &= -4\mathbb{J}_n\Re\{\imath K^*SN_2\}. \end{aligned}$$

We also have the linear QSDE for the measurement y_{m},

$$\begin{aligned} dy_{\mathrm{m}}(t) &= F_{\mathrm{m}}dy(t) \\ &= F_{\mathrm{m}}(Cx(t)dt + D_c u(t)dt + Ddw(t)), \end{aligned} \tag{5.3}$$

with C and D given by (2.8) and (2.9), respectively, and

$$D_c = P_m^\top \begin{bmatrix} 0_{m\times c_1} & 2\Re\{SN_2\} \\ 0_{m\times c_1} & 2\Im\{SN_2\} \end{bmatrix}.$$

Let V, $\tilde{\mathcal{A}}(t) = V\mathcal{A}(t)$, $\tilde{\mathcal{Y}}(t) = V\mathcal{Y}(t) = VU(t)^*\mathcal{A}(t)U(t)$, and $\tilde{\Lambda}(t) = V^\#\Lambda(t)V^\top$ be as in Sect. 4.2.2, and

$$\tilde{y}_{\mathrm{m}}(t) = M^{-1}y_{\mathrm{m}}(t) = [\, I_p \; 0_{p\times(2m-p)}\,](\tilde{\mathcal{Y}}(t) + \tilde{\mathcal{Y}}(t)^\#),$$

be another measurement process. Recall the notation $\mathscr{Y}_{\mathrm{m},t} = \mathrm{vN}\{y_{\mathrm{m}}(s), 0 \le s \le t\}$ from Chap. 4 for the von Neumann algebra generated by $y_{\mathrm{m},j}(s)$, $j = 1, 2, \ldots, p$ and $0 \le s \le t$, where $y_{\mathrm{m},j}$ is the j-th element of the vector y_{m}. We similarly let $\tilde{\mathscr{Y}}_{\mathrm{m},t} = \mathrm{vN}\{\tilde{y}_{\mathrm{m}}(s), 0 \le s \le t\}$ denote the von Neumann algebra generated by the components of $\tilde{y}_{\mathrm{m}}(s), 0 \le s \le t$. Since $\tilde{y}_{\mathrm{m}}$ is linearly related to y_{m} and M is invertible, they generate the same von Neumann algebra. Therefore, $\pi_t(X) = \mathbb{P}(j_t(X) \mid \mathscr{Y}_{\mathrm{m},t}) = \mathbb{P}(j_t(X) \mid \tilde{\mathscr{Y}}_{\mathrm{m},t})$. Define $\tilde{L} = VL$, $\tilde{N}_2 = VN_2$, and $\tilde{S} = VSV^*$, and the operators

$$\tilde{L}(t) = \tilde{L} + \tilde{S}\tilde{N}_2 u_2(z_{[0,t]}) \text{ and } \tilde{H}(t) = H + H_c(t) + \Im\{\tilde{L}^*\tilde{S}\tilde{N}_2 u_2(z_{[0,t]})\}.$$

Note that $\tilde{H}(t) = H(t)$. In terms of these operators and the processes $\tilde{\mathcal{A}}(t)$, $\tilde{\mathcal{A}}(t)^*$, and $\tilde{\Lambda}(t)$, the controlled QSDE can be rewritten as:

$$\begin{aligned} dU(t) = \Big(&\mathrm{Tr}((\tilde{S} - I)^\top d\tilde{\Lambda}(t)) + d\tilde{\mathcal{A}}(t)^*\tilde{L}(t) - \tilde{L}(t)^*\tilde{S}d\tilde{\mathcal{A}}(t) \\ &- \imath\left(\tilde{H}(t) + 1/2\tilde{L}(t)^*\tilde{L}(t)\right)dt\Big)U(t). \end{aligned} \tag{5.4}$$

It then follows by a straightforward extension of the results of [2] to multiple inputs and multiple outputs that the controlled quantum filtering equation is given by

$$\begin{aligned} &d\pi_t(X) \\ &= \pi_t\left(\imath[\tilde{H}(t), X] + \sum_{j=1}^{m} 1/2\left(\tilde{L}_j(t)^*[X, \tilde{L}_j(t)] + [\tilde{L}_j(t)^*, X]\tilde{L}_j(t)\right)\right)dt \\ &\quad + \sum_{j=1}^{p}\left(\pi_t(X\tilde{L}_j(t) + \tilde{L}_j(t)^*X) - \pi_t(\tilde{L}_j(t) + \tilde{L}_j(t)^*)\pi_t(X)\right)d\tilde{\nu}_j(t), \end{aligned} \tag{5.5}$$

where $\tilde{\nu}_j(t) = \tilde{y}_{\mathrm{m},j}(t) - \int_0^t \pi_s(\tilde{L}_j(s) + \tilde{L}_j(s)^*)ds$ is the j-th component of the innovations process $\tilde{\nu}(t) = \tilde{y}_{\mathrm{m}}(t) - [\, I_p \; 0\,]\int_0^t \pi_s(\tilde{L}(s) + \tilde{L}(s)^*)ds$. The process $\tilde{\nu}(t)$

is a $\mathscr{Y}_{\mathrm{m},t}$-martingale zero-mean Wiener process [2], following the same line of proof sketched earlier in Chap. 4. Thus, $\tilde{\nu}(t) - \tilde{\nu}(\tau)$ is independent of any function of $\mathscr{Y}_{\mathrm{m},\tau}$ for any $t \geq \tau \geq 0$.

With $x = (q_1, p_1, q_2, p_2, \ldots, q_n, p_n)^\top$ as before, let

$$\hat{x}(t) = \pi_t(x) = (\pi_t(q_1), \pi_t(p_1), \pi_t(q_2), \pi_t(p_2), \ldots, \pi_t(q_n), \pi_t(p_n))^\top.$$

Then, using the definitions of $\tilde{H}(t)$ and $\tilde{L}(t)$, explicitly calculating all commutators on the right-hand side of (5.5), and taking the initial state of the system to be Gaussian (analogous to the calculations leading to (4.31) and (4.32) in Chap. 4), yields:

$$\begin{aligned} d\hat{x}(t) &= A\hat{x}(t)dt + B_c u(t)dt + (P(t)C^\top F_\mathrm{m}^\top + B\Sigma) \\ &\quad \times (D_m D_\mathrm{m}^\top)^{-1}(dy_\mathrm{m}(t) - F_\mathrm{m}(C\hat{x}(t) + D_c u(t))dt) \\ &= A\hat{x}(t)dt + B_c u(t)dt \\ &\quad + (P(t)C^\top F_\mathrm{m}^\top + B\Sigma)(D_m D_\mathrm{m}^\top)^{-1} d\nu(t). \end{aligned} \tag{5.6}$$

with initial condition $\hat{x}(0) = \langle x \rangle$, where $\Sigma = D^\top F_\mathrm{m}^\top$ and $D_\mathrm{m} = F_\mathrm{m} D$, $\nu(t)$ is the innovations process $\nu(t) = y_\mathrm{m}(t) - F_\mathrm{m} \int_0^t (C\hat{x}(t) + D_c u(s))ds$, and $P(t)$ is the symmetrized covariance matrix,

$$P(t) = \frac{1}{2}\pi_t\left((x - \hat{x}(t))(x - \hat{x}(t))^\top + ((x - \hat{x}(t))(x - \hat{x}(t))^\top)^\top\right),$$

satisfying the deterministic matrix Riccati differential equation

$$\begin{aligned} \dot{P}(t) &= AP(t) + P(t)A^\top + BB^\top \\ &\quad - (P(t)C^\top F_\mathrm{m}^\top + B\Sigma)(D_\mathrm{m} D_\mathrm{m}^\top)^{-1}(P(t)C^\top F_\mathrm{m}^\top + B\Sigma)^\top, \end{aligned} \tag{5.7}$$

with initial condition

$$P(0) = \frac{1}{2}\mathbb{P}\left(\Delta x \Delta x^\top + (\Delta x \Delta x^\top)^\top\right), \tag{5.8}$$

with $\Delta x = x - \langle x \rangle$.

In treatment of the measurement-based LQG control, coherent quantum LQG control, and coherent feedback H^∞ control that will now follow, the initial state of the system will always be assumed to be a Gaussian state and the bosonic fields coupled to the quantum control system are also Gaussian.

5.1.2 Measurement-Based LQG Control

In the measurement-based LQG control problem, we wish to find a control law $u(t)$ of dimension $c = c_1 + c_2$, with components belonging in $\mathscr{Y}_{\mathrm{m},t}$ for each $t \geq 0$, to minimize the quadratic cost function

$$J_T(u) = \mathbb{P}\Bigg(\int_0^T (C_x x(t) + D_u u(t))^\top (C_x x(t) + D_u u(t))dt \\ + x(T)^\top Q x(T) \Bigg), \tag{5.9}$$

for a fixed terminal time $T > 0$, where C_x, D_u, and Q are some prespecified constant matrices in $\mathbb{R}^{k\times 2n}$, $\mathbb{R}^{k\times c}$, and $\mathbb{R}^{2n\times 2n}$, with k some positive integer and $Q = Q^\top \geq 0$. Problems for linear quantum systems with quadratic costs of this type are commonly encountered for *cooling and confinement*; see, e.g., [4–6] and the references therein. To be concrete, take, for instance, a single quantum harmonic oscillator ($n = 1$), $C_x = [\, I_2\ 0\,]^\top$, $D_u = [\,0\ I_c\,]^\top$ and $Q = I$, then we have that

$$\begin{aligned} &\mathbb{P}((C_x x(t) + D_u u(t))^\top (C_x x(t) + D_u u(t)) + x(T)^\top Q x(T)) \\ &= \mathbb{P}(q(t)^2 + p(t)^2 + u(t)^\top u(t) + q(T)^2 + p(T)^2). \end{aligned}$$

For this particular cost, cooling is quantified as minimizing the cumulative kinetic energy $\int_0^T \mathbb{P}(p(t)^2)dt$ of the system over the interval $[0, T]$ and its mean kinetic energy $\mathbb{P}(p(T)^2)$ at the terminal time T, and confinement is quantified as minimizing the cumulative displacement of the system $\int_0^T \mathbb{P}(q(t)^2)dt$ away from the origin over the interval $[0, T]$ and its mean displacement from the origin $\mathbb{P}(q(T)^2)$ at the terminal time T. The remaining term $\int_0^T \mathbb{P}(u(t)^\top u(t))dt$ represents a penalty on the cumulative control effort required to achieve the cooling and confinement task.

Using the properties of the quantum conditional expectation, the cost function can be expressed as:

$$\begin{aligned} &J_T(u) \\ &= \mathbb{P}\left(\int_0^T \mathbb{P}\left(C_x x(t) + D_u u(t))^\top (C_x x(t) + D_u u(t)) \,\middle|\, \mathscr{Y}_{\mathrm{m},t} \right) dt \right) \\ &\quad + \mathbb{P}\left(\mathbb{P}\left(x(T)^\top Q x(T) \,\middle|\, \mathscr{Y}_{\mathrm{m},t}\right)\right) \\ &= \mathbb{P}\left(\int_0^T \pi_t \left(C_x x(t) + D_u u(t))^\top (C_x x(t) + D_u u(t))\right) dt \right) \\ &\quad + \mathbb{P}\left(\pi_t\left(x(T)^\top Q x(T)\right)\right) \\ &= \mathbb{P}\Bigg(\int_0^T \left(\pi_t\left(x(t)^\top C_x C_x^\top x(t)\right) + \pi_t(x)^\top C_x^\top D_u u(t) \right. \\ &\quad \left. + u(t)^\top D_u^\top C_x \pi_t(x) + u(t)^\top D_u^\top D_u u(t)\right) dt \Bigg) \\ &\quad + \mathbb{P}\left(\pi_t\left(x(T)^\top Q x(T)\right)\right). \end{aligned}$$

Since the innovations process $\nu(t) = y_{\mathrm{m}}(t) - F_{\mathrm{m}} \int_0^t (C\hat{x}(s) + D_c u(s))ds$ is a $\mathscr{Y}_{\mathrm{m},t}$-martingale zero-mean Wiener process, it follows that (5.6) can be viewed as essentially a controlled fully observable classical linear stochastic system with deterministic initial condition $\hat{x}(0) = \langle x \rangle$ driven by a Wiener noise process $\nu(t)$. Moreover, we note that (5.6)–(5.8) are precisely the Kalman filtering equation that would be obtained if one ignores the quantum mechanical origin of (5.2) and viewing them as describing a classical linear stochastic state-space model with state vector $x(t)$ [7, 8]. The upshot of this observation is that via the quantum conditional expectation, the measurement-based LQG control problem can be mapped to an equivalent classical LQG control problem, and so the solution of the former problem is of an analogous form to the latter. That is, the measurement-based LQG control law is of the form $u_{\rm opt}(t) = -K(t)\hat{x}(t)$, where $K(t) \in \mathbb{R}^{c\times 2n}$ is an optimal time-dependent gain such that $u_{\rm opt}$ solves the deterministic linear quadratic regulator (LQR) problem of minimizing the deterministic quadratic cost [9, 10]

$$J_{{\rm LQR},T}(u) = \int_0^T (C_x x'(t) + D_u u(t))^\top (C_x x'(t) + D_u u(t))dt + x'(T)^\top Q x'(T),$$

with $x'(t)$ a classical state vector satisfying the deterministic state-space equation,

$$\dot{x}'(t) = Ax'(t) + B_c u(t).$$

Using this analogy, the solution to the infinite horizon measurement-based LQG problem with $T \to \infty$, $Q = 0$, and infinite horizon cost function

$$\begin{aligned} J_\infty(u) = \limsup_{T\to\infty} \frac{1}{T}\mathbb{P}\Bigg(\int_0^T (C_x x(t) + D_u u(t))^\top (C_x x(t) + D_u u(t))dt \\ + x(T)^\top Q x(T)\Bigg), \end{aligned} \tag{5.10}$$

can also be obtained from the solution of the analogous infinite horizon classical LQG problem.

In the above, we have taken a shortcut to the solution of the measurement-based LQG problem by exploiting the linearity of the dynamics and quadratic forms of the cost function to set up an analogy with the classical LQG control problem. The problem can also be solved in a general framework for measurement-based optimal control of quantum systems. One can formulate the quadratic cost-to-go, pose the measurement-based LQG problem as a dynamic programming problem, and write down the associated Bellman equation that must be satisfied by the associated value function. As in the classical setting, for a wide range of problems a *separated law* holds that is a functional of the stochastic density operator $\rho(t)$ of the quantum filter, defined via the identity $\mathrm{Tr}(\rho(t)X) = \iota(\pi_t(X))$ for all bounded system operators X. Roughly speaking, if there is a separated law satisfying the associated Bellman's equation, then this policy will be optimal. This is the case for the measurement-based LQG control problem. For the complete details, we refer the reader to [2, 11], with [11] focusing on the measurement-based quantum LQG problem. This framework can be applied to a wide range of optimal control problems for measurement-based feedback control of quantum systems, beyond the quantum LQG control problem.

We conclude the exposition of measurement-based LQG control with a discussion of an example application to a trapped atom system, as proposed in [5].

Example 5.1 Consider the off-resonant interaction between a two-level atom and an optical cavity in which it is trapped. The interaction between the atom and the optical cavity is given by the Hamiltonian,

$$H = H_a - \hbar \frac{g_0^2}{\Delta} a^* a \cos(k_0 q_{\mathrm{at}})^2,$$

where a is the annihilation operator of the cavity mode, ω_0 is the resonance frequency of the cavity mode, $k_0 = \omega_0/c$ is the wave number of the cavity mode (here, c denotes the speed of light), Δ the detuning between the cavity mode and the two-level atom (i.e., $\Delta = \omega_a - \omega_0$, where ω_a is the atomic transition frequency), q_{at} the atomic position operator, g_0 the cavity-QED coupling constant giving the strength of the interaction between the cavity mode and the atom, and H_a is the Hamiltonian for the mechanical motion of the atom.

The work [5] is interested in estimating the position of the atom in the cavity based on continuous measurement of the phase quadrature of the output light leaking out of the cavity (i.e., $-\imath\mathcal{Y}(t) + \imath\mathcal{Y}(t)^*$) via homodyne detection. In order to do this, it is assumed that the atom is trapped in a region of a size which is small compared to the wavelength of the light, about a region halfway between a node and an antinode of the standing field inside the cavity, so that the following approximation can be made:

$$\cos(k_0 q_{\mathrm{at}})^2 = \cos(k_0 q_{\mathrm{at},0} + k_0 q'_{\mathrm{at}})^2 \approx 1/2 + k_0 q'_{\mathrm{at}}.$$

Renaming q'_{at} as q_{at}, the Hamiltonian H then takes the approximate form,

$$H = H_a - \hbar \frac{g_0^2}{2\Delta} a^* a - \hbar k_0 \frac{g_0^2}{\Delta} a^* a q_{\mathrm{at}},$$

The optical cavity is driven by a field in a coherent state $|f\rangle$ with f a constant function equal to $\alpha = 2\hbar E$ for all times, where $E = \sqrt{\frac{\gamma P}{\hbar\omega_0}}$ and P are the driving laser power. The decay rate of the cavity is taken to be γ, so that the coupling operator of the cavity to the laser is $L_1 = \sqrt{\gamma} a$. Measurement will be performed on the output field from the cavity; however, this measurement may be lossy (photons are lost in the measurement process), and this is modeled by mixing the output from the cavity at a beam splitter with an auxiliary vacuum field to which photons will be lost. The beam splitter is assumed to have transmission rate of $\epsilon \geq 0$ and is represented by the unitary matrix

$$S = \begin{bmatrix} \sqrt{\epsilon} & \sqrt{1-\epsilon} \\ -\sqrt{1-\epsilon} & \sqrt{\epsilon} \end{bmatrix}.$$

If $b_{\text{in},1}$ is the input to the first port of the beam splitter and $b_{\text{in},2}$ the input to the second port, then the corresponding outputs $b_{\text{out},1}$ and $b_{\text{out},2}$, respectively, would be given by

$$\begin{bmatrix} b_{\text{out},1} \\ b_{\text{out},2} \end{bmatrix} = S \begin{bmatrix} b_{\text{in},1} \\ b_{\text{in},2} \end{bmatrix} = \begin{bmatrix} \sqrt{\epsilon} & \sqrt{1-\epsilon} \\ -\sqrt{1-\epsilon} & \sqrt{\epsilon} \end{bmatrix} \begin{bmatrix} b_{\text{in},1} \\ b_{\text{in},2} \end{bmatrix}.$$

In our case, $b_{\text{in},1}$ will come from the output field of the cavity, and measurements will be performed on the output $b_{\text{out},1}$. Overall, using the (S, L, H) notation and series product operation from Chap. 3, the coherently driven atom and cavity system followed by the beam splitter is given by

$$\begin{aligned} G &= (S, 0, 0) \lhd \left(I, \begin{bmatrix} \sqrt{\gamma} a + \alpha \\ 0 \end{bmatrix}, H + \Im\{\alpha/2(a - a^*)\} \right) \\ &= \left(\begin{bmatrix} \sqrt{\epsilon} & \sqrt{1-\epsilon} \\ -\sqrt{1-\epsilon} & \sqrt{\epsilon} \end{bmatrix}, \begin{bmatrix} \sqrt{\epsilon\gamma}\, a + \sqrt{\epsilon}\alpha \\ -\sqrt{(1-\epsilon)\gamma}\, a - \sqrt{1-\epsilon}\,\alpha \end{bmatrix}, \right. \\ &\qquad \left. H + \Im\{\alpha/2(a - a^*)\} \right) \end{aligned}$$

The measurement will be that of the phase quadrature of the first output of the beam splitter and takes the form (using the fact that α is real)

$$dy_{\text{m}}(t) = \beta\sqrt{\epsilon\gamma}\left(\sqrt{\epsilon\gamma}(-\imath a(t) + \imath a(t)^*)dt - \imath d\mathcal{B}(t) + \imath d\mathcal{B}^*(t)\right),$$

where β is a real constant relating to the strength of the local oscillator and the reflectivity of the beam splitter in the homodyne detection setup, and $\mathcal{B}(t) = \sqrt{\epsilon}\mathcal{A}_1(t) + \sqrt{1-\epsilon}\mathcal{A}_2(t)$. Here, $\mathcal{A}_1(t)$ is the vacuum annihilation operator for the field coupled to the cavity, and $\mathcal{A}_2(t)$ is the vacuum annihilation operator for the field entering the second port of the beam splitter. In the context of the theory leading to this example, we have that $p = 1$, $V = \text{diag}(-\imath, 1)$ and $M = \beta\sqrt{\epsilon\gamma}$. Let $L_1 = \sqrt{\epsilon\gamma}\, a + \sqrt{\epsilon}\alpha$, $L_2 = -\sqrt{(1-\epsilon)\gamma}\, a - \sqrt{1-\epsilon}\,\alpha$. The quantum filtering equation for any atomic operator X is given by

$$\begin{aligned} d\pi_t(X) &= \pi_t\left(\imath/\hbar[H, X] + \sum_{j=1}^{2} 1/2\left(L_j^*[X, L_j] + [L_j^*, X]L_j\right) \right) dt \\ &\quad + \left(\pi_t(-\imath X L_1 + \imath L_1^* X) - \pi_t(-\imath L_1 + \imath L_1^*)\pi_t(X)\right) d\nu_1(t) \\ &= \pi_t\left(\imath[H, X] + \gamma/2\left(a^*[X, a] + [a^*, X]a\right)\right) dt \\ &\quad + \sqrt{\epsilon\gamma}\left(\pi_t(-\imath X a + \imath a^* X) - \pi_t(-\imath a + \imath a^*)\pi_t(X)\right)(\beta\sqrt{\epsilon\gamma})^{-1} d\nu_1(t), \end{aligned}$$

where $\nu_1(t)$ is the innovations process given by,

$$\begin{aligned}\nu_1(t) &= y_{\mathrm{m}}(t) - \beta\sqrt{\epsilon\gamma}\int_0^t \pi_s(-\imath L_1 + \imath L_1^*)ds \\ &= y_{\mathrm{m}}(t) - \beta\epsilon\gamma\int_0^t \pi_s(-\imath a + \imath a^*)ds.\end{aligned}$$

In [5] the quantum filtering equation is simplified further by transforming to a "displacement picture," using the fact that without the atom the cavity state would converge to the coherent state $\left|-2\sqrt{\frac{P}{\hbar\gamma\omega_0}}\right\rangle$, where the state of the optical cavity becomes close to the vacuum, and making the assumption

$$\frac{k_0 g_0^2(|-2E/\gamma|^2+1)|\langle x\rangle|}{\Delta\gamma} = \delta \ll 1.$$

Under these assumptions, the cavity has very fast dynamics and goes to a steady state very quickly, relative to the timescale of the dynamics of the atom. The cavity can thus be adiabatically eliminated. It is shown that the quantum filtering equation can be approximated as,

$$\begin{aligned}d\pi_t(X) &= \pi_t\left(\imath/\hbar[H', X] + \kappa\left(q_{\mathrm{at}}[X, q_{\mathrm{at}}] + [q_{\mathrm{at}}, X]q_{\mathrm{at}}\right)\right)dt \\ &\quad + \sqrt{2\epsilon\kappa}\left(\pi_t(Xq_{\mathrm{at}} + q_{\mathrm{at}}X) - 2\pi_t(q_{\mathrm{at}})\pi_t(X)\right)\left(\beta\sqrt{\epsilon\gamma}\right)^{-1}d\nu_1'(t),\end{aligned}$$

with the modified innovations process $\nu_1'(t)$ given by,

$$\nu_1'(t) = y_{\mathrm{m}}(t) - 2\beta\epsilon\sqrt{2\kappa\gamma}\int_0^t \pi_s(q_{\mathrm{at}})ds.$$

In the above, $H' = H_a - \hbar|\alpha|^2 q_{\mathrm{at}}$ and $\kappa = 2k_0^2 g_0^4|\alpha|^2/(\gamma\Delta^2)$. If we assume that H_a is quadratic, then the filtering equation above becomes a linear quantum Kalman filtering equation. This will be the case for a harmonically trapped atom as considered in [5], where $H_a = \frac{1}{2m}p_{\mathrm{at}}^2 + \frac{m_{\mathrm{at}}\omega^2}{2}q_{\mathrm{at}}^2$. Here, m_{at} is the mass of the atom and ω its oscillation frequency.

Given the reduced approximate atomic quantum filtering equation, let $x = (q_{\mathrm{at}}, p_{\mathrm{at}})^\top$, where p_{at} is the atomic momentum operator and we have the commutation relation $[q_{\mathrm{at}}, p_{\mathrm{at}}] = \imath\hbar$. Consider a control Hamiltonian of the form $H_c(t) = x^\top \mathbb{J} u_1(t)$, with u_1 having two components, and set $u_2(t) = 0$. That is, the system will only be controlled through $H_c(t)$ alone. This corresponds to $B_{c,1} = I_2$ (and of course, $B_{c,2} = 0$). One can consider the LQG problem of minimizing the cost function (5.10) with

$$C_x = \begin{bmatrix} \sqrt{m_{\mathrm{at}}}\omega & 0 \\ 0 & 1/\sqrt{m_{\mathrm{at}}} \\ 0 & 0 \\ 0 & 0 \end{bmatrix}, \quad D_u = r\begin{bmatrix} 0 & 0 \\ 0 & 0 \\ \sqrt{m_{\mathrm{at}}}\omega & 0 \\ 0 & 1/\sqrt{m_{\mathrm{at}}} \end{bmatrix},$$

where r is a positive weighting constant. In the harmonically trapped scenario considered here, the symmetrized covariance matrix satisfies the matrix Riccati differential equation,

$$\dot{P} = \begin{bmatrix} \dot{P}_q & \dot{P}_{qp} \\ \dot{P}_{qp} & \dot{P}_p \end{bmatrix}$$
$$= \begin{bmatrix} 2P_{qp}/m_{\mathrm{at}} - 8\kappa\epsilon P_q^2 & P_p/m_{\mathrm{at}} - m_{\mathrm{at}}\omega^2 P_q - 8\kappa\epsilon P_{qp} P_q \\ P_p/m_{\mathrm{at}} - m_{\mathrm{at}}\omega^2 P_q - 8\kappa\epsilon P_{qp} P_q & -2m_{\mathrm{at}}\omega^2 P_{qp} - 8\kappa\epsilon P_{qp}^2 + 2\kappa\hbar^2 \end{bmatrix},$$

with a steady-state solution that can be analytically computed by setting the left-hand side of the above equation to zero and solving for constants P_q, P_{qp} and P_p (explicit expressions for these can be found in [5]). In this case, it turns out that the infinite horizon measurement-based LQG problem has an analytical solution minimizing the cost (5.10) given by $u_{\mathrm{opt},1}(t) = K_\infty \pi_t(x)$, with K_∞ a constant diagonal matrix, $K_\infty = (1/r) I_2$.

5.2 Coherent Feedback Quantum LQG Control

An alternative approach to quantum feedback control is to dispense of the measurement process and taking the controller to be another quantum system that is attached to the plant in a feedback interconnection, as illustrated in Fig. 5.2. This control paradigm where a quantum system is controlled by another quantum system is often referred to as *coherent (quantum) feedback control*. The interconnection is achieved by passing some output fields of the plant to the controller and vice versa. All the information that is being exchanged in the loop are quantum information. There are several reasons for considering coherent feedback controllers, among them:

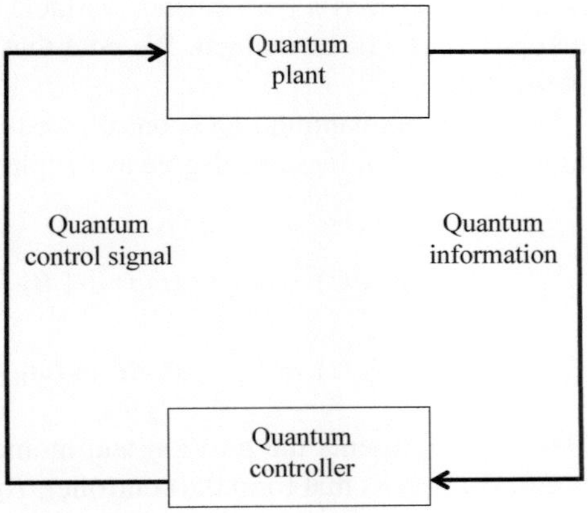

Fig. 5.2 Coherent feedback control loop

- Being a quantum system, such a controller can match the speed of the response of the quantum plant.
- A coherent feedback controller could potentially be fabricated on the same physical platform as the quantum plant.
- Avoids issues of hardware mismatch between a quantum plant and the measurement apparatus. The former is typically "cold," operating at very low temperatures, while the latter is typically "hot," operating at room temperature.
- The quantum nature of the controller has the potential to offer a performance improvement over classical measurement-based controllers by exploiting quantum effects.

A coherent feedback version of the quantum LQG control problem was first formulated in [12] and referred to as coherent quantum LQG control. This type of control is the main content of this section.

We consider *plants* that are linear quantum systems described by the linear QSDE,

$$\begin{aligned} dx(t) &= Ax(t)dt + [\, B \;\, B_w \,] \begin{bmatrix} du(t) \\ dw(t) \end{bmatrix}, \\ dy(t) &= Cx(t)dt + D_w dw(t), \\ z(t) &= C_z x(t) + D_z \beta_u(t). \end{aligned} \tag{5.11}$$

Here, $w(t)$ is a quantum Wiener disturbance vector, $\beta_u(t)$ is an adapted, self-adjoint process commuting with $x(t)$ (i.e., $[\beta_u(t), x(t)^\top] = 0$), and $u(t)$ is a *control* input of the form

$$du(t) = \beta_u(t)dt + d\tilde{u}(t). \tag{5.12}$$

In the decomposition of $u(t)$, $\beta_u(t)$ is identified as the "signal" part while $\tilde{u}(t)$ is the noise part. The vectors $w(t)$ and $\tilde{u}(t)$ are assumed to be distinct vacuum quantum noises, though this assumption is straightforward to relax. The dimensions of $x(t)$, $u(t)$, $w(t)$, and $y(t)$, are $2n$, $2n_u$, $2n_w$, and $2n_y$, respectively. Also, $z(t)$ has been introduced as a performance variable through which a cost function that will be optimized will be defined. We note that $z(t)$ need not correspond to a physical observable.

In coherent quantum LQG control, we take as the controller another linear quantum system with the same degree as the plant, described by the linear QSDE:

$$\begin{aligned} dx_K(t) &= A_K x_K(t)dt + [\, B_{K1} \;\, B_{K2} \;\, B_{K3} \,] \begin{bmatrix} dw_{K1}(t) \\ dw_{K2}(t) \\ dw_{K3}(t) \end{bmatrix}, \\ du(t) &= C_K x_K(t)dt + dw_{K1}(t). \end{aligned} \tag{5.13}$$

Here, x_K represents the position and momentum quadratures of the quantum harmonic oscillators that form the controller, B_{K2} is a square matrix of the same dimension as A_K, and B_{K1} has the same number of columns as there are rows of C_K. The vectors w_{Ki}, $i = 1, 2$, are noise vectors of non-commutative Wiener processes (in vacuum states) with dimensions $2n_{w_{Ki}}$, and y is an input to the controller coming from the plant (5.11). The plant and controller are interconnected in a fully quantum feedback loop by the identification $w_{K3}(t) \equiv y(t)$, $\beta_u(t) \equiv C_K\xi(t)$, and $\tilde{u}(t) \equiv w_{K1}(t)$, resulting in a closed-loop system that is another linear quantum system with internal dynamics given by the QSDE,

$$\begin{aligned} d\tilde{x}(t) &= \tilde{A}\tilde{x}(t)dt + \tilde{B}dw_{cl}(t), \\ z(t) &= \tilde{C}\tilde{x}(t), \end{aligned} \tag{5.14}$$

where $\tilde{x}(t) = [x(t)^\top \; x_K(t)^\top]^\top$, and

$$w_{cl}(t) = \begin{bmatrix} w(t) \\ w_{K1}(t) \\ w_{K2}(t) \end{bmatrix}; \; \tilde{A} = \begin{bmatrix} A & BC_K \\ B_{K3}C & A_K \end{bmatrix};$$
$$\tilde{B} = \begin{bmatrix} B_w & B & 0_{2\times 2} \\ B_{K3}D_w & B_{K1} & B_{K2} \end{bmatrix}; \quad \tilde{C} = \begin{bmatrix} C_z & D_zC_K \end{bmatrix}.$$

We can associate with the closed-loop system (5.14) a quadratic performance index

$$J_T(u) = \int_0^T \langle z(t)^\top z(t)\rangle dt. \tag{5.15}$$

Here, $\langle\cdot\rangle$ refers to quantum expectation with respect to the initial state of the composite system consisting of the plant, controller, and all quantum noises. Let us define the symmetrized covariance matrix $P(t)$ by

$$P(t) = 1/2\langle \tilde{x}(t)\tilde{x}(t)^\top + (\tilde{x}(t)\tilde{x}(t)^\top)^\top\rangle. \tag{5.16}$$

We will now express the performance index in terms of $P(t)$ and its dynamical equation. Applying the quantum Itō rule, we have

$$\begin{aligned} dP(t) &= 1/2(\langle d\tilde{x}(t)\,\tilde{x}(t)^\top\rangle + \langle (d\tilde{x}(t)\,\tilde{x}(t)^\top)^\top\rangle + \\ &\qquad \langle \tilde{x}(t)\,d\tilde{x}(t)^\top\rangle + \langle (\tilde{x}(t)\,d\tilde{x}(t)^\top)^\top\rangle + (\tilde{B}\tilde{B}^\top + (\tilde{B}\tilde{B}^\top)^\top)dt) \\ &= (\tilde{A}P(t) + P(t)\tilde{A}^\top + \tilde{B}\tilde{B}^\top)dt. \end{aligned}$$

Therefore, $P(\cdot)$ satisfies the differential equation

$$\dot{P}(t) = \tilde{A}P(t) + P(t)\tilde{A}^\top + \tilde{B}\tilde{B}^\top; \quad P(0) = P_0. \tag{5.17}$$

From the symmetry of $\tilde{C}^\top \tilde{C}$ and P, it follows that

$$\begin{aligned}\langle z^\top z\rangle &= \langle \tilde{x}^\top \tilde{C}^\top \tilde{C}\tilde{x}\rangle \\ &= \langle \mathrm{Tr}(\tilde{x}^\top \tilde{C}^\top \tilde{C}\tilde{x})\rangle \\ &= \tfrac{1}{2}\langle \mathrm{Tr}(\tilde{C}^\top \tilde{C}[\tilde{x}\tilde{x}^\top + (\tilde{x}\tilde{x}^\top)^\top])\rangle \\ &= \mathrm{Tr}(\tilde{C}^\top \tilde{C}P).\end{aligned}$$

Hence, we may express (5.15) as

$$J_T(u) = \int_0^T \mathrm{Tr}(\tilde{C}^\top \tilde{C}P(t))dt, \tag{5.18}$$

where $P(t)$ solves (5.17). We now focus on the infinite horizon case by passing to the limit $T \uparrow \infty$. If the matrix $\tilde{A}$ is Hurwitz, standard results on Lyapunov equations give us $\lim_{t\to\infty} P(t) = P$, where P is the unique symmetric positive definite solution to the Lyapunov equation:

$$\tilde{A}P + P\tilde{A}^\top + \tilde{B}\tilde{B}^\top = 0. \tag{5.19}$$

Following [12], we consider the infinite horizon coherent quantum LQG problem with an infinite horizon cost function given by

$$J_\infty(u) = \limsup_{T\to\infty} \frac{1}{T}\int_0^T \langle z(s)^\top z(s)\rangle ds. \tag{5.20}$$

From (5.19) and standard methods of analysis, we have

$$J_\infty(u) = \mathrm{Tr}(\tilde{C}^\top \tilde{C}P) = \mathrm{Tr}(\tilde{C}P\tilde{C}^\top).$$

To represent a linear quantum system, the controller is required to be physically realizable; thus, the matrices A_K, B_{K1}, B_{K2}, B_{K3}, and C_K must satisfy the physical realizability constraints (2.28)–(2.30). Since in this case (2.30) is automatically satisfied, this gives the following two constraints,

$$A_K\mathbb{J}_n + \mathbb{J}_n A_K^\top + \sum_{j=1} B_{Kj}\mathbb{J}_{m_j} B_{Kj}^\top = 0, \tag{5.21}$$

$$\mathbb{J}_n C_K^\top + B_{K1}\mathbb{J}_{m_1} = 0, \tag{5.22}$$

where $m_j = n_{w_{Kj}}$. The cost-bounded coherent quantum LQG problem can now be formulated.

Problem 5.1 Given a cost-bound parameter $\gamma > 0$, find system matrices A_K, B_{K1}, B_{K2}, B_{K3}, and C_K for the controller (5.13) such that the following requirements are satisfied.

F1. There exists a symmetric matrix $P > 0$ satisfying (5.19).
F2. $J_\infty = \mathrm{Tr}(\tilde{C} P \tilde{C}^\top) < \gamma$.
F3. The physical realizability constraints (5.21) and (5.22) are satisfied.

Problem 5.1 is a formidable non-convex optimization problem in the variables A_K, B_{K1}, B_{K2}, B_{K3}, and C_K since the physical realizability constraints are already non-convex in these variables. Unlike the measurement-based LQG control problem, there is currently no known analytical solution to the coherent quantum LQG problem. The first treatment of this problem in [12] reformulated it into a rank-constrained linear matrix inequality (LMI) problem and proposed numerical methods to solve the reformulation using specialized optimization algorithms. Several other approaches to solve this problem have since been proposed in the literature, e.g., [13–15]. In particular, [13] also considers the possibility of allowing a direct quadratic Hamiltonian coupling between the plant and the controller besides the interconnection via the quantum fields that is considered here. We will now describe the original approach developed in [12] to solving the cost-bounded coherent quantum LQG control problem.

5.2.1 Reformulating the Quantum LQG Problem into a Rank-Constrained LMI Problem

The purpose of this section is to show how Problem 5.1 can be transformed into a rank-constrained LMI problem. The latter problem can then be treated with appropriate numerical methods. The key tool will be a matrix lifting and linearization technique for matrix-valued polynomials in matrix variables. To keep the exposition simple, we shall focus, for the purpose of illustration, on a plant and controller of degree n (recall that we seek a controller which is of the same order as the plant) with $n_y = n_u = n$ and B_{K1}, B_{K2}, B_{K3}, C_K all of dimension $2n \times 2n$. The matrix-lifting technique is too complicated to describe in a general form. Besides this, the choice of matrix-lifting variables is not unique and, for efficiency, the choice needs to be considered on a case-by-case basis and tailored to the particular plant at hand, in order take advantage of existing structures in a particular problem.

Consider now a plant (5.11) of degree n with $n_y = n_u = n$ and a linear quantum controller (5.13) of the same degree with $n_{w_{K1}} = n_{w_{K2}} = n$ (hence $B_{K1}, B_{K2} \in \mathbb{R}^{2n \times 2n}$). In this case, we have that P will be a symmetric $4n \times 4n$ matrix. We now transform the constraints (5.19) and $J_\infty < \gamma$ into an LMI constraint. We do this by exploiting a nonlinear change of variables that was proposed in [16, Eq. (35)]. However, this requires a suitable redefinition of the plant and controller without altering the closed-loop equations. To this end, we redefine our plant equations as,

$$\begin{aligned} dx(t) &= Ax(t)dt + B\beta_u(t)dt + B'_{w'}dw'(t); \quad x(0) = x, \\ dy'(t) &= C'x(t)dt + D'_{w'}dw'(t), \\ z(t) &= C_z x(t) + D_z \beta_u(t), \end{aligned} \tag{5.23}$$

with $w' = [\, w^\top \; w_{K1}^\top \; w_{K2}^\top \,]^\top$, $B'_{w'} = [\, B_w \; B \; 0_{2n\times 2n} \,]$, $C' = [\, 0_{2n\times 2n} \; 0_{2n\times 2n} \; C^\top \,]^\top$, and

$$D'_{w'} = \begin{bmatrix} 0_{2n\times 2n_w} & I_{2n} & 0_{2n\times 2n} \\ 0_{2n\times 2n_w} & 0_{2n\times 2n} & I_{2n} \\ D_w & 0_{2n\times 2n} & 0_{2n\times 2n} \end{bmatrix}.$$

Here, y' is the output equation for the modified plant. It contains the quantum noise w_{K2} which entered in the original controller equations but not in the original plant equations. This redefinition means that all noises can be viewed as coming from the modified plant, as in standard classical LQG problems. We now also redefine our controller equations as,

$$\begin{aligned} dx_K(t) &= A_K x_K(t)dt + B_K dy'(t), \\ \beta_u(t) &= C_K x_K(t), \end{aligned} \tag{5.24}$$

with $B_K = [\, B_{K1} \; B_{K2} \; B_{K3} \,]$. A simple inspection shows that interconnecting (5.23) and (5.24) gives the same closed-loop equation (5.14). We are now in the setup of [16], with $D_K = 0$ in [16, Eq. (2)].

Introduce the auxiliary variables N, M, $\mathbf{X}$, $\mathbf{Y}$, $Q \in \mathbb{R}^{2n\times 2n}$, with $\mathbf{X}$, $\mathbf{Y}$, Q symmetric. Application of the nonlinear change of variables from [16, Sect. IV-B] with $\hat{\mathbf{D}} = D_K = 0$) yields:

$$\mathbf{A} = NA_K M^\top + NB_K C'\mathbf{X} + \mathbf{Y}BC_K M^\top + YA\mathbf{X}, \tag{5.25}$$

$$\mathbf{B} = NB_K, \tag{5.26}$$

$$\mathbf{C} = C_K M^\top. \tag{5.27}$$

In terms of the newly defined variables above, the constraints (5.19) and $J_\infty < \gamma$ can be recast as the following LMI constraint [16, Eq. (14)]:

$$\begin{bmatrix} A\mathbf{X} + \mathbf{X}A^\top + B\mathbf{C} + (B\mathbf{C})^\top & \mathbf{A}^\top + A & B'_{w'} \\ \mathbf{A} + A^\top & A^\top\mathbf{Y} + \mathbf{Y}A + \mathbf{B}C' + (\mathbf{B}C')^\top & \mathbf{Y}B'_{w'} + \mathbf{B}D'_{w'} \\ (\mathbf{Y}B'_{w'} + \mathbf{B}D'_{w'})^\top & (B'_{w'})^\top & -I \end{bmatrix} < 0, \tag{5.28}$$

$$\begin{bmatrix} \mathbf{X} & I & (C_z\mathbf{X} + D_z\mathbf{C})^\top \\ I & \mathbf{Y} & C_z^\top \\ C_z\mathbf{X} + D_z\mathbf{C} & C_z & Q \end{bmatrix} > 0. \tag{5.29}$$

$$\mathrm{Tr}(Q) < \gamma. \tag{5.30}$$

Since the controller is of the same degree as the plant, the matrices N and M in the above may be freely chosen to be any pair of (invertible) square matrices satisfying $MN^\top = I - \mathbf{XY}$.

If there exists matrices $\mathbf{A}$, $\mathbf{B}$, $\mathbf{C}$, $\mathbf{X}$, $\mathbf{Y}$, Q verifying the LMIs (5.28)–(5.30) as well as matrices N and M satisfying $MN^\top = I - \mathbf{XY}$, the controller matrices A_K, B_K, C_K can be recovered from the transformed variables (5.25)–(5.27) as [16, Eq. (40)]:

$$C_K = \mathbf{C}M^{-\top}, \tag{5.31}$$

$$B_K = N^{-1}\mathbf{B}, \tag{5.32}$$

$$A_K = N^{-1}(\mathbf{A} - NB_K C'\mathbf{X} - \mathbf{Y}BC_K M^\top - \mathbf{Y}A\mathbf{X})M^{-\top}. \tag{5.33}$$

Let us now multiply the left- and right-hand sides of (5.21) with N and $N^\top$, respectively, and define the new variables $\breve{N} = N\mathbb{J}_n$, $\breve{A}_K = NA_K$, and $\breve{B}_{Ki} = NB_{Ki}$, $i = 1, 2, 3$. In terms of the new variables, the physical realizability constraints (5.21) and (5.22) can be cast as,

$$\begin{aligned}(-\mathbf{A}M^{-\top} + (\breve{B}_{K3}C + \mathbf{Y}A)\mathbf{X}M^{-\top} + \mathbf{Y}BC_K)\breve{N}^\top \\ +\breve{N}(\mathbf{A}M^{-\top} - (\breve{B}_{K3}C + \mathbf{Y}A)\mathbf{X}M^{-\top} - \mathbf{Y}BC_K)^\top \\ +\sum_{i=1}^{3} \breve{B}_{Ki}\,\mathbb{J}_n\,\breve{B}_{Ki}^\top = 0,\end{aligned} \tag{5.34}$$

$$\breve{B}_{K1} = \breve{N}C_K^\top\,\mathbb{J}_n. \tag{5.35}$$

Conversely, if A_K, B_{K1}, B_{K2}, B_{K3}, C_K solve Problem 5.1 and P solves (5.19), then (5.28)–(5.30) are satisfied for some pairs M and N of square matrices satisfying $MN^\top = I - \mathbf{XY}$; see [16, Sect. IV-B]. Furthermore, since the associated controller solving Problem 5.1 is physically realizable by hypothesis, the constraints (5.34) and (5.35) are thus automatically satisfied. Therefore, we may state following result:

Theorem 5.1 *Under the assumptions of this section, Problem 5.1 has a solution for a given* $\gamma > 0$ *if and only if there exist matrices* $\mathbf{A}$, $\breve{B}_{K1}$, $\breve{B}_{K2}$, $\breve{B}_{K3}$, $\mathbf{C}$, $\mathbf{X}$, $\mathbf{Y}$, $\breve{N}$, M, N, C_K *satisfying the LMIs (5.28)–(5.30) (with* $\mathbf{B} = [\,\breve{B}_{K1}\ \breve{B}_{K2}\ \breve{B}_{K3}\,]$*) and the constraints (5.34) and (5.35),* $\breve{N} = N\mathbb{J}_n$, $NM^\top = I - \mathbf{YX}$ *and* $\mathbf{C} = C_K M^\top$.

It can be seen that the constraints (5.34) and (5.35) are essentially polynomial matrix equality constraints in the (non-commuting) matrix variables $\mathbf{A}$ $\breve{B}_{K1}$,$\breve{B}_{K2}$,$\breve{B}_{K3}$,$\mathbf{C}$,$\mathbf{X}$, $\mathbf{Y}$,$\breve{N}$, and $M^{-\top}$. That is, (5.34) and (5.35) are equality constraints involving matrix-valued multivariate polynomials with non-commuting matrix-valued variables. One approach to deal with these constraints is to turn them into a collection of scalar multivariate polynomial equality constraints with scalar decision variables, by taking the decision variables to be the elements of the matrix variables (we need only take the upper triangular elements of symmetric variables such as $\mathbf{X}$). It is known that a collection of scalar multivariate polynomial equality and/or inequality constraints can be "linearized" and converted into a set of linear equality and inequality constraints in some symmetric positive definite matrix variable Z together with the rank constraint $\text{rank}(Z) = 1$. This can be done by introducing auxiliary variables known as lifting variables together with auxiliary equality constraints on these lifting variables [17, 18]. However, this conversion into scalar variables may not be desirable since converting polynomial matrix constraints into a collection of scalar polynomial constraints can result in scalar multivariate polynomials of orders much higher than the order of the original matrix polynomial. This is impractical for the case that we are considering here since there will be many scalar decision variables involved, leading to a scalar polynomial program that may become too large to handle numerically. Indeed, if we are to *scalarize* (5.34) we would end up with 34 decision variables and constraints in polynomials of order 4, a substantially large problem. Therefore, we would like to keep the matrix structure of our problem by pursuing the idea of defining suitable matrix-lifting variables instead. This will proceed in a very similar fashion to the scalar situation but being mindful of the fact that matrices, unlike scalars, do not in general commute with one another.

We will now look at linearizing (5.34) and (5.35) by defining suitable matrix-lifting variables and related equality constraints, and subsequently transforming the coherent quantum LQG problem into an LMI problem with a rank $2n$ constraint. To this end, let us now set $M = I_{2n}$ and $N = I - \mathbf{YX}$. This removes one free matrix variable, namely $M^{-\top}$, to reduce the complexity of the problem. We will use 14 matrix-lifting variables $W_1, W_2, \ldots, W_{14} \in \mathbb{R}^{2n\times 2n}$ defined as follows: $W_i = \breve{B}_{Ki}\mathbb{J}$, $i = 1, 2, 3$, $W_4 = \mathbf{Y}B$, $W_5 = \breve{B}_{K3}C + \mathbf{Y}A$, $W_6 = \breve{N}\mathbf{C}^\top$, $W_7 = \breve{N}\mathbf{X}$, $W_8 = \mathbf{A}\breve{N}^\top$, $W_9 = \mathbf{YX}$, $W_{10} = W_4 W_6^\top$, $W_{11} = W_5 W_7^\top$, $W_{12} = W_1 \breve{B}_{K1}^\top$, $W_{13} = W_2 \breve{B}_{K2}^\top$ and $W_{14} = W_3 \breve{B}_{K3}^\top$. Now, let Z be a $46n \times 46n$ symmetric matrix,

$$\mathbf{Z}_{i,j} = [Z_{kl}]_{k=2in+1,(i+1)2n,l=2jn+1,(j+1)2n},$$

$$x = (x_1, \ldots, x_8) = (1, 2, \ldots, 8),$$

and

$$v = (v_1, \ldots, v_{14}) = (9, 10, \ldots, 22).$$

We impose that Z satisfies the constraints,

$$\left.\begin{array}{cc}
Z \geq 0 & \mathbf{Z}_{v_6,1} - \mathbf{Z}_{x_8,x_5} = 0 \\
\mathbf{Z}_{0,0} - I_{2n} = 0 & \mathbf{Z}_{v_7,1} - \mathbf{Z}_{x_8,x_6} = 0 \\
\mathbf{Z}_{1,x_6} - \mathbf{Z}_{x_6,1} = 0 & \mathbf{Z}_{v_8,1} - \mathbf{Z}_{x_1,x_8} = 0 \\
\mathbf{Z}_{1,x_7} - \mathbf{Z}_{x_7,1} = 0 & \mathbf{Z}_{v_9,1} - \mathbf{Z}_{x_7,x_6} = 0 \\
\mathbf{Z}_{v_1,1} - \mathbf{Z}_{x_2,1}\,\mathbb{J}_n = 0 & \mathbf{Z}_{v_{10},1} - \mathbf{Z}_{v_4,v_6} = 0 \\
\mathbf{Z}_{v_2,1} - \mathbf{Z}_{x_3,1}\,\mathbb{J}_n = 0 & \mathbf{Z}_{v_{11},1} - \mathbf{Z}_{v_5,v_7} = 0 \\
\mathbf{Z}_{v_3,1} - \mathbf{Z}_{x_4,1}\,\mathbb{J}_n = 0 & \mathbf{Z}_{v_{12},1} - \mathbf{Z}_{v_1,x_2} = 0 \\
\mathbf{Z}_{v_4,1} - \mathbf{Z}_{x_7,1}B = 0 & \mathbf{Z}_{v_{13},1} - \mathbf{Z}_{v_2,x_3} = 0 \\
\mathbf{Z}_{v_5,1} - \mathbf{Z}_{x_4,1}C - \mathbf{Z}_{x_7,1}A = 0 & \mathbf{Z}_{v_{14},1} - \mathbf{Z}_{v_3,x_4} = 0 \\
\mathbf{Z}_{x_8,1} - \mathbb{J}_n + \mathbf{Z}_{v_9,1}\mathbb{J}_n = 0. &
\end{array}\right\} \tag{5.36}$$

Terms of the form $\mathbf{Z}_{a,b}$ with $a, b \in \{x_1, \ldots, x_8\} \cup \{v_1, \ldots, v_{14}\}$ in the above should be identified with $\mathbf{Z}_{a,1}(\mathbf{Z}_{b,1})^\top$. We can then express the LMI constraints (5.28)–(5.30) in terms of Z by replacing $\mathbf{A}, \mathbf{B}, \mathbf{C}, \mathbf{X}, \mathbf{Y}$ with $\mathbf{Z}_{x_1,1}$, $\left[\mathbf{Z}_{x_2,1}\ \mathbf{Z}_{x_3,1}\ \mathbf{Z}_{x_4,1}\right]$, $\mathbf{Z}_{x_5,1}$, $\mathbf{Z}_{x_6,1}$, $\mathbf{Z}_{x_7,1}$, respectively. Meanwhile, the physical realizability constraints (5.34) and (5.35) become the following linear equality constraints:

$$\begin{aligned} &-\mathbf{Z}_{v_8,1} + \mathbf{Z}_{v_8,1}^\top + \mathbf{Z}_{v_{11},1} - \mathbf{Z}_{v_{11},1}^\top + \mathbf{Z}_{v_{10},1} - \mathbf{Z}_{v_{10},1}^\top \\ &\quad + \mathbf{Z}_{v_{12},1} + \mathbf{Z}_{v_{13},1} + \mathbf{Z}_{v_{14},1} = 0, \\ &\mathbf{Z}_{x_2,1} - \mathbf{Z}_{v_6,1}\mathbb{J}_n = 0. \end{aligned} \tag{5.37}$$

Finally, we also demand that Z satisfies a rank $2n$ constraint:

$$\text{rank}(Z) \leq 2n. \tag{5.38}$$

Now, to understand the relationship of the rank-constrained LMI to our original constraints, suppose that we can find Z satisfying (5.36)–(5.38) as well as the LMI constraints (expressed in terms of block elements of Z) and the rank constraint. Due to the rank constraint, we have the factorization $Z = VV^\top$, where $V \in \mathbb{R}^{46n\times 2n}$ and satisfy $[V_{ij}]_{i,j=1,\ldots,n} = I_{2n}$. Using (5.36), we can recover $\mathbf{A}$, $\breve{B}_{Ki}$ $(i = 1, 2, 3)$, $\mathbf{C}, \mathbf{X}, \mathbf{Y}, \breve{N}$ as, respectively, $\mathbf{Z}_{x_1,1}, \ldots, \mathbf{Z}_{x_8,1}$, and $W_i = \mathbf{Z}_{v_i,1}$, $i = 1, \ldots, 14$. It then follows that $N = \breve{N}\mathbb{J}_n$ and $\mathbf{B}_{Ki} = N^{-1}\breve{B}_{Ki}$ $(i = 1, 2, 3)$. The controller matrices A_K, B_K, C_K can then be determined using (5.31)–(5.33) and by construction they will satisfy (5.28)–(5.30), (5.21) and (5.22). Therefore, we have obtained a solution to Problem 5.1.

We now make the remark that due to the simplifying assumptions $M = I$ and $N = I - \mathbf{YX}$ introduced earlier, the solvability of the rank-constrained LMI problem above is only *sufficient* for solvability of the coherent quantum LQG problem. However, it is not difficult to see that we can remove these simplifying assumptions to produce a rank-constrained LMI problem that is necessary and sufficient for the solvability of Problem 5.1. It can be inspected that this can be done by (i) introducing additional variables M, N, $C_K \equiv \mathbf{C}M^{-\top}$, $\breve{\mathbf{A}} \equiv \mathbf{A}M^{-\top}$, $\breve{\mathbf{X}} \equiv \mathbf{X}M^{-\top}$, (ii) additional lifting variables $W_{15} = NM^\top$, $W_{16} = \breve{\mathbf{A}}M^\top$, $W_{17} = \breve{\mathbf{X}}M^\top$, $W_{18} = C_K M^\top$ and associated constraints $W_{15} - I + W_9 = 0$, $\mathbf{A} - W_{16} = 0$, $\mathbf{X} - W_{17} = 0$, $\breve{N} - N\mathbb{J}_n = 0$, $\mathbf{C} - W_{18} = 0$, (iii) redefining $W_6 = \breve{N}C_K^\top$, $W_7 = \breve{N}\breve{\mathbf{X}}^\top$ and $W_8 = \breve{\mathbf{A}}\breve{N}^\top$, and (iv) enlarging and redefining Z as well as the set of constraints (5.36) and (5.37) accordingly. However, this of course comes at the expense of solving a larger problem.

5.2.2 *Numerically Solving the Rank-Constrained LMI Problem*

The rank-constrained LMI problem formulated above can be treated numerically using iterative algorithms that attempt to directly search for a solution to the problem. Most of these algorithms are based on alternating projections (see [19] and the references therein for details). Compared to converting the problem to a scalar polynomial programming problem and employing LMI relaxations techniques based on the theory of moments and the dual theory of sum of squares (SOS) polynomials [20–23], these algorithms are much more efficient for the problem at hand in terms of size, i.e., the number of variables and constraints that have to be considered. However, alternating projection algorithms have a drawback that they do not necessarily converge to a solution from arbitrary starting points, even if a solution to the rank-constrained LMI problem exists.

Following [12], we will employ an algorithm that was developed in [19]. This algorithm has been implemented in the freely available LMIRank MATLAB toolbox [24]. This toolbox can be called and executed from within the Yalmip optimization prototyping environment [25]. The algorithm belongs to the family of alternating projection algorithms but is equipped with a built-in Newton step that could potentially accelerate convergence. We will later employ this toolbox to numerically solve some example coherent quantum LQG control problems.

The LMIRank algorithm needs an initial starting point. For a given $\gamma > 0$, as a heuristic choice of starting point for the LMIRank solver, one can first solve (5.28)–(5.30) to obtain $\mathbf{A}$, $\mathbf{B}$, $\mathbf{C}$, $\mathbf{X}$, $\mathbf{Y}$, Q. Then set $M = I_{2n}$ and $N = I - \mathbf{YX}$ and calculate $\breve{B}_{K1}$, $\breve{B}_{K2}$, $\breve{B}_{K3}$, $\breve{N}$ as well as the matrix-lifting variables $W_1, \ldots, W_{14}$ following the definitions given in Sect. 5.2.1. Let

$$V_0 = [\, I_{2n}\ \mathbf{A}^\top\ \breve{B}_{K1}^\top\ \breve{B}_{K2}^\top\ \breve{B}_{K3}^\top\ \mathbf{C}^\top\ \mathbf{X}^\top\ \mathbf{Y}^\top\ \breve{N}^\top\ W_1^\top \\ \ldots\ W_{14}^\top\,]^\top.$$

Then, a heuristic starting point for the LMIRank algorithm is $Z = V_0 V_0^\top$.

5.2.3 An Extension of the Numerical Procedure

We will now develop an extension of the rank-constrained LMI approach by formulating a problem that is more natural for the coherent quantum LQG problem but will involve the optimization of additional matrix variables. The starting point is that for the LQG cost, it is the transfer function that matters rather than the system matrices (A, B, C, D) (recall the discussion in Sect. 3.3). Application of the symplectic similarity transformation $(A, B, C, D) \mapsto (VAV^{-1}, VB, CV^{-1}, D)$ for any invertible matrix V does not change the LQG cost. However, with this transformation the new internal variable $\tilde{x}_K = Vx_K$ satisfies the commutation relation $[\tilde{x}_K(t), \tilde{x}_K(t)^\top] = 2\imath\Theta_K$, with $\Theta_K = V\mathbb{J}_n V^\top$. Therefore, we can formulate a relaxed optimization problem by introducing Θ_K as a new matrix variable with the constraint that it is skew symmetric, invertible, and can be expressed as $\Theta_K = V\mathbb{J}_n V^\top$ for some invertible matrix V. We thus arrive at the following generalized formulation of the coherent quantum LQG problem:

Problem 5.2 Given a cost-bound parameter $\gamma > 0$, find real matrices A_K, B_{K1}, B_{K2}, B_{K3}, C_K, and an invertible skew symmetric matrix Θ_K such that:

1. There exists a symmetric matrix $P > 0$, satisfying (5.19).
2. $J_\infty = \mathrm{Tr}(\tilde{C} P \tilde{C}^\top) < \gamma$.
3. The resulting controller is physically realizable. That is, A_K, B_{K1}, B_{K2}, B_{K3}, C_K, and Θ_K satisfy (5.21) and (5.22) with $\mathbb{J}_n$ replaced by Θ_K.

The lemma below is straightforward and shows that solving Problem 5.2 also solves Problem 5.1.

Lemma 5.1 *Suppose that the matrices $\hat{A}_K$, $\hat{B}_{K1}$, $\hat{B}_{K2}$, $\hat{B}_{K3}$, $\hat{C}_K$, and $\hat{\Theta}_K$ solve Problem 5.2, and $\hat{\Theta}_K = V \mathbb{J}_n V^\top$ for some real invertible matrix V. Then, the matrices*

$$A_K = V^{-1} \hat{A}_K V; \;\; B_{Ki} = V^{-1} \hat{B}_{Ki} \; i = 1, 2, 3; \;\; C_K = \hat{C}_K V, \tag{5.39}$$

solve Problem 5.1.

Proof Since $\hat{\Theta}_K$ is real skew symmetric and invertible, we can find an invertible matrix V such that $\hat{\Theta}_K = V \mathbb{J}_n V^\top$. Now, we have that

$$\hat{A}_K \hat{\Theta}_K + \hat{\Theta}_K \hat{A}_K^\top + \sum_{j=1}^{3} \hat{B}_{Kj} \, \mathbb{J}_n \, \hat{B}_{Kj}^\top = 0, \tag{5.40}$$

$$\hat{B}_{K1} = \hat{\Theta}_K \hat{C}_K^\top \mathbb{J}_{n_u}. \tag{5.41}$$

After substitution of (5.39) and $\hat{\Theta}_K = V \mathbb{J}_n V^\top$ into (5.40) and (5.41), and some algebraic manipulations, it is easily obtained that A_K, B_K, and C_K satisfy

$$A_K \mathbb{J}_n + \mathbb{J}_n A_K^\top + \sum_{j=1}^{3} B_{Kj} \, \mathbb{J}_n \, B_{Kj}^\top = 0, \tag{5.42}$$

$$B_{K1} = \mathbb{J}_n C_K^\top \mathbb{J}_{n_u}. \tag{5.43}$$

Since the LQG cost is invariant under a similarity transformation of the controller state-space matrices, these matrices solve Problem 5.1, as claimed. □

The following corollary is then immediate.

Corollary 5.1 *Suppose that $A_K, B_{K1}, B_{K2}, B_{K3}, C_K$ solve the standard cost-bounded LQG problem for a given $\gamma > 0$ (i.e., (5.19) and $J_\infty < \gamma$ are satisfied), and there exists an invertible real skew symmetric $2n \times 2n$ matrix Z satisfying*

$$A_K Z + Z A_K^\top + \sum_{j=1}^{3} B_{Kj} \, \mathbb{J}_n \, B_{Kj}^\top = 0, \tag{5.44}$$

$$B_{K1} = Z C_K^\top \mathbb{J}_{n_u}. \tag{5.45}$$

Moreover, let V *be a real invertible matrix such that* $VZV^\top = \mathbb{J}_n$. *Then, the matrices* $\hat{A}_K$, $\hat{B}_{K1}$, $\hat{B}_{K2}$, $\hat{B}_{K3}$, $\hat{C}_K$ *given by:*

$$\begin{aligned}&\hat{A}_K = VA_KV^{-1}; \; \hat{B}_{Ki} = SB_{Ki}\, i = 1, 2, 3;\\ &\hat{C}_K = C_KV^{-1},\end{aligned} \tag{5.46}$$

solve Problem 5.1.

The corollary states that a solution A_K, B_K, C_K to the standard LQG problem is also a solution to the coherent quantum LQG problem (Problem 5.1) *if and only if* there exists a matrix Z satisfying the equalities stated in the corollary.

Problem 5.2 can also be treated using the rank-constrained LMI procedure developed in Sect. 5.2.1. Let us do this again under the simplifying assumptions $M = I$ and $N = I - \mathbf{YX}$ (that can easily be removed if desired). Introduce the additional variable Θ_K and substitute A_K, B_{Ki} ($i = 1, 2, 3$), C_K with, respectively, $\hat{A}_K$, $\hat{B}_{Ki}$ ($i = 1, 2, 3$), $\hat{C}_K$ (see Lemma 5.1). We also redefine $\breve{N} = N\Theta_K$, $x = (x_1, \ldots, x_{10})$, and Z to be a real symmetric matrix of dimension $50n \times 50n$, and replace (5.36) with the following constraints:

$$\left.\begin{array}{c}
Z \geq 0 \qquad \mathbf{Z}_{x_8,1} + \mathbf{Z}_{x_9,x_{10}} = 0\\
\mathbf{Z}_{0,0} - I_{2n} = 0 \quad \mathbf{Z}_{v_4,1} - \mathbf{Z}_{x_7,1}B = 0\\
\mathbf{Z}_{1,x_6} - \mathbf{Z}_{x_6,1} = 0 \quad \mathbf{Z}_{v_6,1} - \mathbf{Z}_{x_8,x_5} = 0\\
\mathbf{Z}_{1,x_7} - \mathbf{Z}_{x_7,1} = 0 \quad \mathbf{Z}_{v_7,1} - \mathbf{Z}_{x_8,x_6} = 0\\
\mathbf{Z}_{v_9,1} - \mathbf{Z}_{x_7,x_6} = 0 \quad \mathbf{Z}_{v_8,1} - \mathbf{Z}_{x_1,x_8} = 0\\
\mathbf{Z}_{v_{10},1} - \mathbf{Z}_{v_4,v_6} = 0 \quad \mathbf{Z}_{x_{10},1} + \mathbf{Z}_{1,x_{10}} = 0\\
\mathbf{Z}_{v_{11},1} - \mathbf{Z}_{v_5,v_7} = 0 \quad \mathbf{Z}_{v_{13},1} - \mathbf{Z}_{v_2,x_3} = 0\\
\mathbf{Z}_{v_{12},1} - \mathbf{Z}_{v_1,x_2} = 0 \quad \mathbf{Z}_{v_{14},1} - \mathbf{Z}_{v_3,x_4} = 0\\
\mathbf{Z}_{v_1,1} - \mathbf{Z}_{x_2,1}\,\mathbb{J}_n = 0\\
\mathbf{Z}_{v_2,1} - \mathbf{Z}_{x_3,1}\,\mathbb{J}_n = 0\\
\mathbf{Z}_{x_9,1} - I_{2n} + \mathbf{Z}_{v_9,1} = 0\\
\mathbf{Z}_{v_3,1} - \mathbf{Z}_{x_4,1}\,\mathbb{J}_n = 0\\
\mathbf{Z}_{v_5,1} - \mathbf{Z}_{x_4,1}C - \mathbf{Z}_{x_7,1}A = 0
\end{array}\right\} \tag{5.47}$$

In the above, $\mathbf{Z}_{x_{10},1} + \mathbf{Z}_{1,x_{10}} = 0$ corresponds to the constraint that $\Theta_K = -\Theta_K^\top$, while $\mathbf{Z}_{x_8,1} + \mathbf{Z}_{x_9,x_{10}} = 0$ corresponds to the constraint $\breve{N} = N\Theta_K$. The rest of the constraints from Sect. 5.2.1, (5.37), and $\mathrm{rank}(Z) \leq 2n$ are unchanged. Notice that now the variable N is an independent variable, whereas in Sect. 5.2.1 N was linearly related to $\breve{N}$ by the identity $N = \breve{N}\mathbb{J}_n$. As the last step, we formulate a heuristic for the initial guess Θ_K^0 for Θ_K. Possible choices for Θ_K^0 include $\Theta_K^0 = 0$ or $\Theta_K^0 = \mathbb{J}_n$. When Θ_K^0 has been selected, set $N = I - \mathbf{YX}$, $\breve{N}^0 = N\Theta_K^0$, and

$$\begin{aligned}V_0 = [\, &I \;\mathbf{A}^\top \;\breve{B}_{K1}^\top \;\breve{B}_{K2}^\top \;\breve{B}_{K3}^\top \;\mathbf{C}^\top \;\mathbf{X}^\top \;\mathbf{Y}^\top \;(\breve{N}^0)^\top \;N^\top\\ &(\Theta_K^0)^\top \;W_1^\top \ldots W_{14}^\top\,]^\top.\end{aligned}$$

The LMIRank routine can then be executed with $Z = V_0 V_0^\top$ as the starting point.

We conclude the discussion by noting that the requirement for Θ_K to be invertible has not been built into the above numerical procedure. Therefore, the procedure could, in principle, potentially return a solution with a singular Θ_K (up to errors due to finite machine precision). Although such a solution can be interpreted as a controller with mixed quantum-classical dynamics (see the discussion in Sect. 5.2.6), an additional constraint can in principle be added to enforce the invertibility condition on Θ_K. Next, we present some numerical examples of coherent quantum LQG design using the methods presented here.

5.2.4 Quantum LQG Control Design Examples

We will now apply the approach Sect. 5.2.1 to find a cost-bounded quantum LQG controller that asymptotically stabilizes a marginally stable quantum plant. The numerical implementation was carried out in MATLAB using the Yalmip prototyping environment and LMIRank. The semidefinite program solver used for LMIRank was SeDuMi Version 1.1 Release 3; see [26]. For comparison, we also compute a classical LQG controller that controls the plant by driving it with a modulated optical beam.

The quantum plant under consideration is a one degree of freedom linear quantum system with Hamiltonian matrix R and coupling matrix K given by

$$R = 1/2 \begin{bmatrix} \Delta & 0 \\ 0 & \Delta \end{bmatrix}, \quad K = \begin{bmatrix} \sqrt{\kappa_1} & 0 \\ \sqrt{\kappa_2} & 0 \\ \sqrt{\kappa_3} & 0 \end{bmatrix},$$

with $\Delta = 0.1$ and $k_1 = k_2 = k_3 = 10^{-2}$. Its evolution is governed by the linear QSDE

$$\begin{aligned} dx &= \begin{bmatrix} 0 & \Delta \\ -\Delta & 0 \end{bmatrix} x dt + \begin{bmatrix} 0 & 0 \\ 0 & -2\sqrt{k_1} \end{bmatrix} du + \\ &\quad \begin{bmatrix} 0 & 0 & 0 & 0 \\ 0 & -2\sqrt{k_2} & 0 & -2\sqrt{k_3} \end{bmatrix} \begin{bmatrix} dw_1 \\ dw_2 \end{bmatrix}, \\ dy &= \begin{bmatrix} 2\sqrt{k_2} & 0 \\ 0 & 0 \end{bmatrix} x dt + dw_1. \end{aligned} \tag{5.48}$$

Notice that for the given coupling matrix K, the quantum noise inputs couple only with the position operator of the system. Such a coupling is typically sought in schemes for quantum non-demolition continuous measurement of position. This yields a marginally stable plant having two mutually conjugate eigenvalues on the imaginary axis.

5.2.4.1 Quantum LQG Controller Design Example I

Let us seek another linear quantum system as a quantum LQG controller to asymptotically stabilize the given plant. We choose the performance variable to be $z = x + \beta_u$; i.e., $C_z = I_{2\times 2} = D_z$. Selecting $\gamma = 5.75$ and solving Problem 5.1 numerically following Sects. 5.2.1 and 5.2.2, LMIRank yields the physically realizable controller (numerical results are presented only up to four digits behind the decimal point),

$$\begin{aligned} dx_K &= \begin{bmatrix} -2.3907 & 0.8420 \\ -5.5518 & 1.9380 \end{bmatrix} x_K dt + \begin{bmatrix} -0.3029 & 0.5042 \\ -0.6603 & 1.0819 \end{bmatrix} dw_{K1} \\ &\quad + 10^{-10} \begin{bmatrix} 0.0241 & -0.0471 \\ 0.0576 & -0.1136 \end{bmatrix} dw_{K2} + \begin{bmatrix} 3.3626 & 2.1470 \\ 7.6699 & 5.0302 \end{bmatrix} dy, \\ du(t) &= \begin{bmatrix} -1.0819 & 0.5042 \\ -0.6603 & 0.3029 \end{bmatrix} xdt + dw_{K1}. \end{aligned} \tag{5.49}$$

This controller asymptotically stabilizes the closed-loop system, and the closed-loop LQG cost achieved is $J_\infty = 5.7382$.

We can see that elements of B_{K2} are very small, of the order 10^{-10}. We note that B_{K2} is the coefficient for the quantum noise w_{K2}, which only enters in the controller and did not originate from the plant. This additional controller noise contributes to the LQG cost, but since the control goal is to bound this cost the algorithm finds a controller with a small B_{K2} term in order to minimize the impact of the variance of w_{K2} on the LQG cost. Up to the numerical precision of MATLAB, the numerical results give

$$\begin{aligned} & A_K \mathbb{J} + \mathbb{J} A_K^\top + \sum_{k=1}^{3} B_{Ki} \mathbb{J} B_{Ki}^\top \\ &= 10^{-13} \begin{bmatrix} 0 & 0.2896 \\ -0.2896 & 0 \end{bmatrix}, \end{aligned} \tag{5.50}$$

while removing the term $B_{K2} \mathbb{J} B_{K2}^\top$ (i.e., setting $B_{K2} = 0$) returns an identical numerical result on the right-hand side of (5.50). This indicates that the contribution of B_{K2} to (5.50) is less than the numerical precision of MATLAB, and to obtain a simpler controller we may simply set $B_{K2} = 0$.

5.2.4.2 Classical LQG Controller Design

Now that we have obtained a fully quantum controller, we can pose a natural question: Does this controller offer any improvement over a classical measurement-based feedback controller driven by continuous measurements of a quadrature of the plant output y (e.g., by homodyne detection as discussed in Sect. 4.4.1)? To address this question, we consider performing continuous measurements of one quadrature of

y (in this case, its first element). This entails replacing the output y in (5.48) with another output y' (another classical signal) defined as:

$$y' = [2\sqrt{k_2}\ 0]x dt + [1\ 0]dw_1.$$

We now wish to seek a classical controller of the form:

$$\begin{aligned} dx_K &= A_K x_K dt + B_K dy'; \\ \beta_u &= C_K x_K \end{aligned} \tag{5.51}$$

whose output will modulate an optical beam and produce the control signal u:

$$\begin{aligned} du &= \beta_u dt + dw_{K1} \\ &= C_K x_K dt + dw_{K1}. \end{aligned} \tag{5.52}$$

To do this, we can apply the standard LQG machinery (with $C_z = D_z = I_2$ as before) to the following modified plant (to account for the presence of the noise w_{K1} in the controller output u, recall the discussion in Sect. 5.2.1) with β_u being viewed as the "control signal":

$$\begin{aligned} dx &= \begin{bmatrix} 0 & \Delta \\ -\Delta & 0 \end{bmatrix} x dt + \begin{bmatrix} 0 & 0 \\ 0 & -2\sqrt{k_1} \end{bmatrix} \beta_u dt \\ &\quad + \begin{bmatrix} 0 & 0 & 0 & 0 & 0 & 0 \\ 0 & -2\sqrt{k_1} & 0 & -2\sqrt{k_2} & 0 & -2\sqrt{k_3} \end{bmatrix} \begin{bmatrix} dw_{K1} \\ dw_1 \\ dw_2 \end{bmatrix}, \\ dy' &= [2\sqrt{k_2}\ 0]x dt + [1\ 0]dw_1. \end{aligned}$$

The optimal classical controller was then found to be:

$$\begin{aligned} dx_K &= \begin{bmatrix} -0.0658 & 0.1 \\ -0.1217 & -0.2 \end{bmatrix} x_K dt + \begin{bmatrix} 0.3291 \\ 0.1083 \end{bmatrix} dy'; \\ du &= \begin{bmatrix} -1 & 0 \\ 0 & 1 \end{bmatrix} x_K dt + dw_{K1}, \end{aligned}$$

and the optimal LQG cost achieved is $J_\infty = 4.8468$.

We find that the cost obtained by this classical controller is actually lower than the one achieved by the quantum LQG controller, compare $J_\infty = 4.8468$ for the former with $J_\infty = 5.7382$ for the latter. Moreover, this may not yet be the best performance that can be achieved by a classical linear controller because there is still the freedom to select arbitrary rotated quadratures of y and performing heterodyne detection on both quadratures. This can be implemented by passing y through a phase shifter and mixing the phase shifter output with an additional vacuum noise at a beam splitter [27, 28], followed by homodyne detection of the amplitude quadrature of one beam splitter output and phase quadrature of the other output. This supplies *noisy* information about *both* quadratures of y the controller. To describe this measurement process, denote the vacuum noise going into the beamsplitter by $w_0 = (w_{0,1}, w_{0,2})$ (in quadrature notation). The beam splitter divides power in its incoming signals between its two output ports according to the ratio $\epsilon^2 : \delta^2$, where $\epsilon^2 + \delta^2 = 1$, $0 \le \epsilon, \delta \le 1$. The classical measurement signal y'' at the output of this measurement scheme is given by,

$$dy'' = \begin{bmatrix} \epsilon & 0 \\ 0 & -\delta \end{bmatrix} \begin{bmatrix} \cos(\theta) & -\sin(\theta) \\ \sin(\theta) & \cos(\theta) \end{bmatrix} dy + \begin{bmatrix} \delta & 0 \\ 0 & \epsilon \end{bmatrix} dw_0,$$

where θ is the phase shift introduced by the phase shifter.

We now compute classical LQG controllers of the form (5.51) and (5.52) with y' replaced by y'' for θ taking values in $(-\pi, \pi]$ and ϵ in $[0, 1]$. Notice that measurement of y' is a special case of measurement of y'' corresponding to setting $\theta = 0$ and $\epsilon = 1$. A surface plot of the LQG cost J_∞ achieved by these controllers versus the value of θ and ϵ is shown in Fig. 5.3. The minimum value of J_∞ over θ and ϵ is about 4.8468 achieved at $\theta \approx 0$ and $\epsilon \approx 1$, so the general measurement scheme with y'' for this example offers no noticeable advantage over just measuring y'. In fact, it is reasonable to conjecture that measuring y' is already optimal for this example.[1] Thus, it is now natural to ask, does there exist at all a fully quantum LQG controller that can beat the LQG cost achievable by a classical controller for this example? It turns out that the answer to the question is affirmative and a fully quantum controller that beats all classical LQG controllers can be found using the extended numerical procedure developed in Sect. 5.2.3. This will be shown next.

5.2.5 *Quantum LQG Controller Design Example II*

We now turn our attention to numerically solving Problem 5.2 to find a fully quantum controller that achieves a cost $J_\infty < 4.8468$. We will do this by applying the extended numerical scheme in Sect. 5.2.3 with $\gamma = 5$ and $\Theta_K^0 = 0_{2\times 2}$. The LMIRank routine returns the following solution:

$$\Theta_K = \begin{bmatrix} 0 & -0.1820 \\ 0.1820 & 0 \end{bmatrix}; \; \hat{A}_K = \begin{bmatrix} -0.2125 & 0.0666 \\ -0.3789 & 0.0257 \end{bmatrix};$$

$$\hat{B}_{K1} = \begin{bmatrix} 0.0642 & -0.0547 \\ 0.0480 & -0.2556 \end{bmatrix};$$

[1] We note that [12] suggested that a lower LQG cost of 4.444 can be attained by a classical linear controller by measuring y'' with the setting $\theta = 0$ and $\epsilon \approx 0.715$. This is incorrect and the discrepancy arose out of a coding error therein. However, this error does not negate the findings of [12]. In fact, it only strengthens the conclusion that there can exist a coherent quantum LQG controller which outperforms all classical LQG controllers for the same cost function and actuation structure.

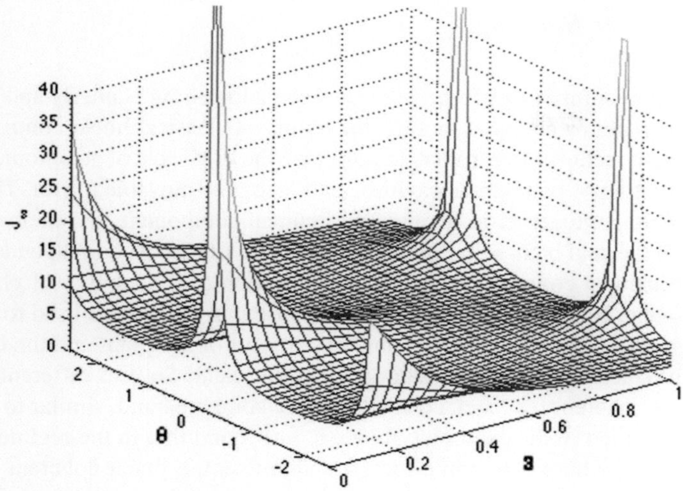

Fig. 5.3 Plot of J_∞ versus θ and α

$$\hat{B}_{K2} = 10^{-10}\begin{bmatrix} -0.0468 & 0.0255 \\ -0.1160 & 0.0114 \end{bmatrix};$$

$$\hat{B}_{K3} = \begin{bmatrix} 0.3522 & -0.0215 \\ 0.4393 & -0.0842 \end{bmatrix};\ \hat{C}_K = \begin{bmatrix} -1.4044 & 0.3008 \\ -0.2639 & 0.3526 \end{bmatrix},$$

with the cost $J_\infty = 4.1793$. Since $\Theta_K = VJV^\top$ with $V = \sqrt{0.182}\begin{bmatrix} 0 & 1 \\ 1 & 0 \end{bmatrix}$, from Lemma 5.1 we find the matrices A_K, B_{Ki} $(i = 1, 2, 3)$, and C_K of the controller solving Problem 5.1 are

$$\begin{aligned} A_K &= \begin{bmatrix} 0.0257 & -0.3789 \\ 0.0666 & -0.2125 \end{bmatrix};\ B_{K1} = \begin{bmatrix} 0.1126 & -0.5992 \\ 0.1504 & -0.1283 \end{bmatrix}; \\ B_{K2} &= 10^{-10}\begin{bmatrix} -0.2721 & 0.0272 \\ -0.1096 & 0.0601 \end{bmatrix}; \\ B_{K3} &= \begin{bmatrix} 1.0297 & -0.1974 \\ 0.8255 & -0.0503 \end{bmatrix};\ C_K = \begin{bmatrix} 0.1283 & -0.5992 \\ 0.1504 & -0.1126 \end{bmatrix}. \end{aligned} \tag{5.53}$$

As with design example I, we again find that B_{K2} has negligibly small entries. So we may similarly set $B_{K2} = 0$ to obtain a simpler controller. That we could find a fully quantum controller that outperforms all classical LQG controllers, compare $J_\infty = 4.1793$ in the former against $J_\infty \approx 4.8468$ in the latter, indicates the intriguing potential for a coherent feedback controller to offer performance advantage over classical controllers, to be discussed further below.

5.2.6 Further Reading

The coherent quantum LQG formalism has been adopted by Hamerly and Mabuchi in [6] to compare, on an equal footing, the ability of classical linear controllers and linear coherent feedback controllers to minimize the LQG cost of an optomechanical system consisting of two optical cavities, each with one movable mirror. The moveable mirrors move together as a single mechanical component, and their motion is coupled to a thermal bath with noise power k_n. The LQG cost corresponds to cooling the mechanical component, that is, minimization of the number of phonons in this component. Equal footing here is in the sense that the signals used to drive the controllers were obtained from the same output port of the plant and the controller in turn drives/actuates the same input port of the plant. Several different types of classical and coherent feedback controllers were compared and, similar to what has been demonstrated in the examples above, it was found that in the regime of small noise power k_n, where quantum effects are significant, a linear coherent feedback controller can be found that outperformed the linear classical controllers. The work [6] attributes this advantage, for the system considered therein, to the ability of the controller to simultaneously process non-commuting quadratures of the output field. A deeper analysis of why a linear coherent feedback controller can beat a linear measurement feedback controller was given more recently in [29].

The coherent feedback LQG control theory was originally developed in [12] in a more general setting that allowed controllers to possibly have mixed quantum-classical dynamics. The treatment here has been restricted to controllers that are linear quantum systems with no classical dynamics. The numerical approach described herein can be straightforwardly adapted to this more general setting.

5.3 Coherent Feedback H^∞ Control

Control laws are usually designed based on a mathematical model for the dynamics of the plant to be controlled. These models are necessarily imperfect. There can be many sources of imperfections, for instance, uncertainties about the model parameter values, and unmodelled dynamics. Such uncertainties or errors in the assumed model may lead to instability in the closed feedback loop when the feedback controller is designed based on the imperfect model. A natural question that can then be asked is, how can controllers be designed so that they can maintain closed-loop stability and acceptable control performance in the presence of uncertainties and disturbances? One framework for achieving such a robust design of controllers is H^∞ control synthesis. This method was initially developed for classical linear time-invariant systems based on minimizing the H^∞ norm of the closed-loop transfer function and was solved using the state-space formulation in the landmark paper [30]. A first principles approach to the solution of the so-called non-singular H^∞ control problem was presented in [31], and the LMI approach can be found in [16, 32].

The H^∞ methodology has subsequently been extended to nonlinear systems, see, e.g., [33].

The adaptation of the H^∞ control method to the quantum setting was presented in [34] as a framework for designing robust linear quantum controllers that can tolerate unwanted disturbances/noise and modeling uncertainties. The approach builds on a formulation of H^∞ control based upon powerful concepts from modern control theory, including dissipation inequalities for open classical dynamical systems that originated in the work of Willems [35], and generalizations of this idea to the stochastic setting, classical [36] and quantum [37]. In this section, we will give an exposition of the H^∞ control approach to coherent feedback quantum control, following [34]. We begin by formulating the dissipation properties of linear quantum systems.

5.3.1 Dissipation Properties

We now recall some important dissipation properties that are frequently exploited in control engineering, suitably adapted to the quantum setting. These properties quantify the effect of disturbance inputs on energy transfers and stability. In particular, a quantum adaptation of the Strict Bounded Real Lemma (Corollary 5.3) will be presented and used for quantum H^∞ controller synthesis. We will be working with linear quantum systems of the form

$$\begin{aligned} dx(t) &= Ax(t)dt + [\, B \;\, G\,][\, dw(t)^\top \;\, dv(t)^\top \,]^\top, \\ dz(t) &= Cx(t)dt + [\, D \;\, H\,][\, dw(t)^\top \;\, dv(t)^\top \,]^\top. \end{aligned} \tag{5.54}$$

The input to the quantum system has two components $w(t)$ and $v(t)$; $w(t)$ represents disturbance signals, while v represents additional noise sources. The input $w(t)$ is composed of two components,

$$dw = \beta_w dt + d\tilde{w}. \tag{5.55}$$

Here β_w is an adapted process that is a component of dw which could, say, be passed to the system as input from another quantum system, while $\tilde{w}$ is a vector of non-commutative quantum Wiener noise processes. The vector v is also a vector of non-commutative quantum Wiener noise processes and is independent of $\tilde{w}(t)$. In the treatment of H^∞ control in this chapter, the noises are not necessarily vacuum noises but may be arbitrary zero-mean Gaussian noise as elaborated in Chap. 2. In (5.54), we do not write down the output equation, as it is not yet required in this section, and $z(t)$ is a new vector that is a linear combination of $x(s)$, $w(s)$, and $v(s)$, $0 \leq s \leq t$. As in the coherent feedback LQG problem, the vector $z(t)$ is not necessarily a physical observable but it will be relevant for formulating the control objective in the ensuing H^∞ control design.

In the following, we will identify a Gaussian state of oscillators or bosonic fields with its density operator ρ. We will also use the notation $\langle \rho, X \rangle = \mathrm{Tr}(\rho X) = \langle X \rangle$.

Definition 5.1 Given an operator-valued quadratic form

$$r(x, \beta_w) = [x^\top \beta_w^\top] \mathsf{R} \begin{bmatrix} x \\ \beta_w \end{bmatrix}$$

where

$$\mathsf{R} = \begin{bmatrix} \mathsf{R}_{11} & \mathsf{R}_{12} \\ \mathsf{R}_{12}^\top & \mathsf{R}_{22} \end{bmatrix}$$

is a given real symmetric matrix, the system (5.54) is said to be *dissipative with supply rate* $r(x, \beta_w)$ if there exists a positive operator-valued quadratic form $V(x) = x^\top X x$ (where X is a real positive definite symmetric matrix) and a constant $\lambda > 0$ such that

$$\langle V(x(t)) \rangle + \int_0^t \langle r(x(s), \beta_w(s)) \rangle ds \leq \langle V(x(0)) \rangle + \lambda t, \ \forall t > 0, \qquad (5.56)$$

for all initial Gaussian states ρ of the system.

The system (5.54) is said to be *strictly dissipative* if there exists a constant $\epsilon > 0$ such that inequality (5.56) holds with the matrix R replaced by the matrix $\mathsf{R} + \epsilon I$.

In the definition above, the term $\langle V(x(t)) \rangle$ can be interpreted as an abstract notion of internal "energy" for the linear quantum system at time t. Meanwhile, the term $\langle r(x(t), \beta_w(t)) \rangle$ generalizes to quantum systems the notion of abstract energy flow in and out of a system at time t. These notions have been extensively used in the stability analysis of classical deterministic dynamical systems [35, 38]. The dissipation inequality (5.56) was introduced for classical stochastic systems in [36] and subsequently adapted to linear quantum networks in [37].

The following result relates the dissipation properties to certain LMIs, with the proof given in the appendix to this chapter.

Theorem 5.2 *Given a quadratic form $r(x, \beta_w)$ defined as above, then the quantum stochastic system (5.54) is dissipative with supply rate $r(x, \beta_w)$ if and only if there exists a real positive definite symmetric matrix X such that the following LMI is satisfied:*

$$\begin{pmatrix} A^\top X + XA + \mathsf{R}_{11} & \mathsf{R}_{12} + XB \\ B^\top X + \mathsf{R}_{12}^\top & \mathsf{R}_{22} \end{pmatrix} \leq 0. \qquad (5.57)$$

Furthermore, the system is strictly dissipative if and only if there exists a real positive definite symmetric matrix X such that the following LMI is satisfied:

$$\begin{pmatrix} A^\top X + XA + \mathsf{R}_{11} & \mathsf{R}_{12} + XB \\ B^\top X + \mathsf{R}_{12}^\top & \mathsf{R}_{22} \end{pmatrix} < 0. \qquad (5.58)$$

Moreover, if either of (5.57) or (5.58) holds, then the required constant $\lambda \geq 0$ can be chosen as $\lambda = \lambda_0$, where

$$\lambda_0 = \mathrm{Tr}\left[\begin{bmatrix} B^\top \\ G^\top \end{bmatrix} X \begin{bmatrix} B & G \end{bmatrix} F\right]. \tag{5.59}$$

Here, the matrix F is defined by via the relation:

$$F dt = \begin{bmatrix} dw \\ dv \end{bmatrix} \begin{bmatrix} dw^\top & dv^\top \end{bmatrix}. \tag{5.60}$$

Remark 5.1 We refer to F defined in (5.60) as an Itō matrix. It can be defined more generally for any $2m \times 1$ vector $\nu(t)$ of adapted quantum processes such that $d\nu(t) d\nu(t)^\top = F_\nu dt$, where F_ν would be the Itō matrix for $\nu(t)$. When $F_\nu = I_{2m} + \imath \mathbb{J}_m$, then we say that F is canonical. This means that $d\nu(t) = \beta_\nu(t) dt + S_\nu(t) dw_\nu(t)$ with β_ν, S_ν adapted and the latter satisfying $S_\nu(t) S_\nu(t)^\top = I_{2m}$ for each t, and w_ν a vacuum quantum Wiener noise vector. The notion of canonical and non-canonical noises was introduced in [34] to treat rather general linear non-commutative systems that can have mixed classical and quantum linear dynamics, driven by possibly non-vacuum Gaussian input fields. Here, we do not consider systems with mixed quantum-classical dynamics, see Sect. 5.4 for an additional discussion.

The corollaries below state the application of the theorem above to a special case of the matrix R, defined in terms of error output operator $\beta_z(t) = Cx(t) + D\beta_w(t)$. This operator is the adapted dt component of the performance variable $z(t)$.

Definition 5.2 The quantum stochastic system (5.54) is said to be *Bounded Real with disturbance attenuation g* if the system (5.54) is dissipative with supply rate

$$\begin{aligned} r(x, \beta_w) &= \beta_z^\top \beta_z - g^2 \beta_w^\top \beta_w \\ &= [x^\top \beta_w^\top] \begin{bmatrix} C^\top C & C^\top D \\ D^\top C & D^\top D - g^2 I \end{bmatrix} \begin{bmatrix} x \\ \beta_w \end{bmatrix}. \end{aligned}$$

Also, the quantum stochastic system (5.54) is said to be *Strictly Bounded Real with disturbance attenuation g* if the system (5.54) is strictly dissipative with this supply rate.

With the above definition in hand, we can state following corollary of Theorem 5.2 (see [39] for the corresponding classical result).

Corollary 5.2 *The quantum stochastic system (5.54) is bounded real with disturbance attenuation g if and only if there exists a positive definite symmetric matrix $X \in \mathbb{R}^{2n \times 2n}$ (n denoting the degrees of freedom of (5.54)) such that the following matrix inequality is satisfied:*

$$\begin{pmatrix} A^\top X + XA + C^\top C & C^\top D + XB \\ B^\top X + D^\top C & D^\top D - g^2 I \end{pmatrix} \leq 0.$$

Furthermore, the quantum stochastic system is strictly bounded real with disturbance attenuation g *if and only if there exists a positive definite symmetric matrix* $X \in \mathbb{R}^{2n\times 2n}$ *such that the following matrix inequality is satisfied:*

$$\begin{pmatrix} A^\top X + XA + C^\top C & C^\top D + XB \\ B^\top X + D^\top C & D^\top D - g^2 I \end{pmatrix} < 0.$$

Moreover, in both cases the required constant $\lambda \geq 0$ *can be chosen as* $\lambda = \lambda_0$*, where* λ_0 *is defined by (5.59).*

We can now combine the corollary with the standard Strict Bounded Real Lemma from modern control theory (e.g., see [31, 40]) to arrive at the next corollary. Note the terminology used that a square matrix A is said to be *stable* or is a *stability matrix* if it is Hurwitz (all its eigenvalues have a negative real part).

Corollary 5.3 *The following statements are equivalent*

(i) The quantum stochastic system (5.54) is strictly bounded real with disturbance attenuation g.
(ii) A *is a stable matrix, and* $\|C(sI - A)^{-1}B + D\|_\infty < g$.
(iii) $g^2 I - D^\top D > 0$*, and there exists a positive definite matrix* $\tilde{X} > 0$ *such that*

$$A^\top \tilde{X} + \tilde{X}A + C^\top C + (\tilde{X}B + C^\top D)(g^2 I - D^\top D)^{-1}(B^\top \tilde{X} + D^\top C) < 0.$$

(iv) $g^2 I - D^\top D > 0$ *and the algebraic Riccati equation*

$$A^\top X + XA + C^\top C + (XB + C^\top D)(g^2 I - D^\top D)^{-1}(B^\top X + D^\top C) = 0$$

has a stabilizing solution $X \geq 0$.

Furthermore, if these statements hold then $X < \tilde{X}$.

5.3.2 H^∞ Controller Synthesis

We will now consider the problem of H^∞ controller design for linear quantum systems. We begin by defining the controller and closed-loop plant–controller system in Sect. 5.3.2.1. This is followed in Sect. 5.3.2.3 by the application of the Strict Bounded Real Lemma to the closed-loop system to obtain the main results. Physical realizability of the resulting linear quantum controllers is addressed in Sect. 5.3.2.4.

5.3.2.1 The Closed-Loop Plant–Controller System

We shall consider linear quantum systems in the real quadrature form (2.5). Systems of this form will be the prototype for the interconnection of components in the quantum control system. Let us now describe the plant and controller models and the closed-loop system formed by the plant and controller.

Plants are taken to be linear quantum systems of the following form,

$$\begin{aligned} dx(t) &= Ax(t)dt + [\, B_0 \; B_1 \; B_2 \,][\, dv(t)^\top \; dw(t)^\top \; du(t)^\top \,]^\top; \\ &\qquad x(0) = x_0, \\ dz(t) &= C_1x(t)dt + D_{12}du(t), \\ dy(t) &= C_2x(t)dt + [\, D_{20} \; D_{21} \; 0_{2n_y \times 2n_u} \,] \\ &\qquad \times [\, dv(t)^\top \; dw(t)^\top \; du(t)^\top \,]^\top. \end{aligned} \tag{5.61}$$

Here, x_0 is a vector of the position and momentum operator pairs of the single-mode oscillators forming the plant. The input $w(t)$ represents a *disturbance* signal of the form (5.55). The signal $u(t)$ is a *control* input of the form

$$du(t) = \beta_u(t)dt + d\tilde{u}(t), \tag{5.62}$$

where $\tilde{u}(t)$ is the noise part of $u(t)$, and $\beta_u(t)$ is a vector of adapted, self-adjoint processes. Also, $dv(t)$ represents any additional quantum noise inputs that may enter the plant. The vectors $v(t)$, $\tilde{w}(t)$, and $\tilde{u}(t)$ are quantum noises with Itō matrices F_v, $F_{\tilde{w}}$, and $F_{\tilde{u}}$ which are all non-negative Hermitian (see Remark 5.1).

Controllers are taken to be linear quantum systems of the form

$$\begin{aligned} dx_K(t) &= A_Kx_K(t)dt + [\, B_{K1} \; B_K \,][\, dv_K(t)^\top \; dy(t)^\top \,]^\top; \\ &\qquad x_K(0) = x_{K,0}, \\ du(t) &= C_Kx_K(t)dt + [\, B_{K0} \; 0_{2n_u \times 2n_y} \,][\, dv_K(t)^\top \; dy(t)^\top \,]^\top, \end{aligned} \tag{5.63}$$

where $x_K(t) = [\, x_{K1}(t) \; \dots \; x_{Kn_K}(t) \,]^\top$ is a vector of self-adjoint controller variables. The vector $v_K(t) = [\, v_{K1}(t) \; \dots \; v_{KK_v}(t) \,]^\top$ is a quantum noise vector consisting of non-commutative Wiener processes in the vacuum state, and $x_{K,0}$ is a vector of the position and momentum operator pairs of the single-mode oscillators forming the controller.

The *closed-loop system* can be formed by making the identification $\beta_u(t) = C_Kx_K(t)$ and interconnecting (5.61) and (5.63) to give

$$
\begin{aligned}
d\tilde{x}(t) &= \begin{bmatrix} A & B_2C_K \\ B_KC_2 & A_K \end{bmatrix}\tilde{x}(t)dt + \\
&\quad \begin{bmatrix} B_0 & B_2B_{K0} \\ B_KD_{20} & B_{K1} \end{bmatrix}\begin{bmatrix} dv(t) \\ dv_K(t) \end{bmatrix} + \begin{bmatrix} B_1 \\ B_KD_{21} \end{bmatrix} dw(t), \\
dz(t) &= \begin{bmatrix} C_1 & D_{12}C_K \end{bmatrix}\tilde{x}(t)dt + \begin{bmatrix} 0 & D_{12}B_{K0} \end{bmatrix}\begin{bmatrix} dv(t) \\ dv_K(t) \end{bmatrix},
\end{aligned} \tag{5.64}
$$

where $\tilde{x}(t) = [x(t)^\top \; x_K(t)^\top]^\top$. We can write the closed-loop system in the form

$$
\begin{aligned}
d\tilde{x}(t) &= \tilde{A}\tilde{x}(t)dt + \tilde{B}dw(t) + \tilde{G}d\tilde{v}(t) \\
&= \tilde{A}\tilde{x}(t)dt + \begin{bmatrix} \tilde{B} & \tilde{G} \end{bmatrix}\begin{bmatrix} dw(t) \\ d\tilde{v}(t) \end{bmatrix}, \\
dz(t) &= \tilde{C}\tilde{x}(t)dt + \tilde{H}d\tilde{v}(t) \\
&= \tilde{C}\tilde{x}(t)dt + \begin{bmatrix} 0 & \tilde{H} \end{bmatrix}\begin{bmatrix} dw(t) \\ d\tilde{v}(t) \end{bmatrix},
\end{aligned} \tag{5.65}
$$

where

$$
\begin{aligned}
\tilde{v}(t) &= \begin{bmatrix} v(t) \\ v_K(t) \end{bmatrix}; \; \tilde{A} = \begin{bmatrix} A & B_2C_K \\ B_KC_2 & A_K \end{bmatrix}; \; \tilde{B} = \begin{bmatrix} B_1 \\ B_KD_{21} \end{bmatrix}; \\
\tilde{G} &= \begin{bmatrix} B_0 & B_2B_{K0} \\ B_KD_{20} & B_{K1} \end{bmatrix}; \; \tilde{C} = \begin{bmatrix} C_1 & D_{12}C_K \end{bmatrix}; \; \tilde{H} = \begin{bmatrix} 0 & D_{12}B_{K0} \end{bmatrix}.
\end{aligned}
$$

The closed-loop system (5.65) is therefore a linear quantum system of the quadrature form (2.5).

5.3.2.2 H^∞ Control Objective

In the H^∞ controller synthesis problem, we have a disturbance attenuation parameter $g > 0$ as a design parameter that represents the minimum accepted attenuation of the disturbance signal β_w in the performance variable $z(t)$, in a sense that will be defined below. Typically, it is desired that the parameter g be as small as possible. Given this parameter, the goal of H^∞ controller synthesis is to find a controller (5.63) such that the closed-loop system (5.65) satisfies:

$$
\begin{aligned}
&\int_0^t \langle \beta_z(s)^\top \beta_z(s) + \epsilon \tilde{x}(s)^\top \tilde{x}(s) \rangle ds \\
&\quad \leq (g^2 - \epsilon) \int_0^t \langle \beta_w(s)^\top \beta_w(s) \rangle ds + \mu_1 + \mu_2 t, \; \forall t > 0
\end{aligned} \tag{5.66}
$$

for some real constants $\epsilon, \mu_1, \mu_2 > 0$. Here $\beta_z(t) = \tilde{C}\eta(t)$ is the dt component of the performance variable $z(t)$ in the closed-loop system (5.65). The design criteria can be interpreted as the controller bounding the "energy" transfer from the disturbance signal $\beta_w(t)$ to the error signal $\beta_z(t)$ as a component of $z(t)$.

The closed-loop system (5.65) will meet the objective (5.66) if it is designed to be strictly bounded real with disturbance attenuation g. This follows directly from Definition 5.2. According that definition, (5.65) will be strictly bounded real with disturbance attenuation g if and only if there exists a real positive definite symmetric matrix X and a constant $\lambda > 0$ such that

$$\begin{aligned}\langle \tilde{x}(t)^\top X \tilde{x}(t)\rangle + \int_0^t \langle \beta_z(s)^\top \beta_z(s) - g^2 \beta_w(s)^\top \beta_w(s) + \epsilon \tilde{x}(s)^\top \tilde{x}(s) \\ + \epsilon \beta_w(s)^\top \beta_w(s)\rangle ds \leq \langle \tilde{x}(0)^\top X \tilde{x}(0)\rangle + \lambda t, \quad \forall t > 0. \qquad (5.67)\end{aligned}$$

Inequality (5.66) now follows with $\mu_1 = \langle \tilde{x}(0)^\top X \tilde{x}(0)\rangle$ and $\mu_2 = \lambda$.

In the following, we will give necessary and sufficient conditions for the existence of a controller achieving the H^∞ control objective (5.66) for a given g. Explicit formulas for the controller matrices A_K, B_K, and C_K will be obtained. These results mirror the corresponding well-known results for classical linear systems as given in, e.g., [30, 31].

5.3.2.3 Necessary and Sufficient Conditions

In order to establish necessary and sufficient conditions for the existence of an H^∞ controller, we need to impose some assumptions on the plant (5.61). They are the following.

Assumption 5.1

1. $D_{12}^\top D_{12} = E_1 > 0$.
2. $D_{21} D_{21}^\top = E_2 > 0$.
3. The matrix $\begin{bmatrix} A - j\omega I & B_2 \\ C_1 & D_{12} \end{bmatrix}$ is full rank for all $\omega \geq 0$.
4. The matrix $\begin{bmatrix} A - j\omega I & B_1 \\ C_2 & D_{21} \end{bmatrix}$ is full rank for all $\omega \geq 0$.

The results will be stated in terms of the following pair of algebraic Riccati equations:

$$\begin{aligned}&(A - B_2 E_1^{-1} D_{12}^\top C_1)^\top X + X(A - B_2 E_1^{-1} D_{12}^\top C_1) \\ &+X(B_1 B_1^\top - g^2 B_2 E_1^{-1} B_2')X \\ &+g^{-2} C_1^\top (I - D_{12} E_1^{-1} D_{12}^\top) C_1 = 0; \qquad (5.68)\end{aligned}$$

$$\begin{aligned}&(A - B_1 D_{21}^\top E_2^{-1} C_2)Y + Y(A - B_1 D_{21}^\top E_2^{-1} C_2) \\ &+Y(g^{-2} C_1^\top C_1 - C_2^\top E_2^{-1} C_2)Y \\ &+B_1(I - D_{21}^\top E_2^{-1} D_{21}) B_1^\top = 0. \qquad (5.69)\end{aligned}$$

We assume that the solutions to these Riccati equations satisfy the following.

Assumption 5.2

(i) $A - B_2 E_1^{-1} D_{12}^\top C_1 + (B_1 B_1^\top - g^2 B_2 E_1^{-1} B_2')X$ is a stability matrix.
(ii) $A - B_1 D_{21}^\top E_2^{-1} C_2 + Y(g^{-2} C_1^\top C_1 - C_2^\top E_2^{-1} C_2)$ is a stability matrix.
(iii) The matrix XY has a spectral radius strictly less than one.

We are now ready to state the main result on H^∞ controller synthesis.

Theorem 5.3 Necessity. *Consider the system (5.61) and suppose that Assumption 5.1 is satisfied. If there exists a controller of the form (5.63) such that the resulting closed-loop system (5.65) is strictly bounded real with disturbance attenuation g, then the Riccati equations (5.68) and (5.69) will have stabilizing solutions $X \geq 0$ and $Y \geq 0$ satisfying Assumption 5.2.*

Sufficiency. *Suppose the Riccati equations (5.68) and (5.69) have stabilizing solutions $X \geq 0$ and $Y \geq 0$ satisfying Assumption 5.2. If the controller (6.53) is such that the matrices A_K, B_K, C_K are constructed as*

$$\begin{aligned} A_K &= A + B_2 C_K - B_K C_2 + (B_1 - B_K D_{21}) B_1^\top X, \\ B_K &= (I - YX)^{-1}(Y C_2^\top + B_1 D_{21}^\top) E_2^{-1}, \\ C_K &= -E_1^{-1}(g^2 B_2^\top X + D_{12}^\top C_1), \end{aligned} \tag{5.70}$$

then the resulting closed-loop system (5.65) will be strictly bounded real with disturbance attenuation g. Also the constant $\lambda \geq 0$ in Definition 5.1 can be chosen as in (5.59), $\lambda = \lambda_0$ with B, G, F replaced by the corresponding matrices of the closed-loop system.

Proof Using the Strict Bounded Real Lemma Corollary 5.3, the theorem follows directly from the corresponding classical H^∞ result; e.g., see [10, 31, 41]. □

Notice that Theorem 5.3 does not specify the remaining controller matrices B_{K0}, B_{K1}, and the noise vector v_K. This is because they are free as far as the H^∞ objective is concerned. However, they will play a role when considering physical realizability of the resulting controllers.

5.3.2.4 Physical Realization of Controllers

The matrices A_K, B_K, and C_K resulting from the H^∞ controller synthesis of Theorem 5.3 do not in general define or completely specify a physically realizable linear quantum system. Fortunately, due to the freedom of choosing B_{K0}, B_{K1}, and adding the noise v_K without affecting the H^∞ control objective, it will be shown in the next theorem that we can always choose these remaining matrices to yield a physically realizable H^∞ controller.

Theorem 5.4 *Assume $F_y = D_{20}F_v D_{20}^\top + D_{21}F_w D_{21}^\top$ is canonical (Remark 5.1). Let $\{A_K, B_K, C_K\}$ be an arbitrary triple (such as given by (5.70)). Then, there exists controller parameters B_{K0}, B_{K1}, and the controller noise v_K such that the controller (5.63) is physically realizable.*

The theorem follows directly from the next lemma.

Lemma 5.2 *Let F_y be canonical and $\{A_K, B_K, C_K\}$ be such that $A_K \in \mathbb{R}^{2n_K\times 2n_K}$, $B_K \in \mathbb{R}^{2n_K\times 2n_y}$, and $C_K \in \mathbb{R}^{2n_u\times 2n_K}$. Then, there exist an integer $n_{v_K} \geq n_u$ and $B_{K1} \in \mathbb{R}^{2n_K\times 2n_{v_K}}$ such that the system (5.63) is physically realizable with (taking K to be the coupling matrix in the coupling operator L)*

$$B_{K0} = [\, I_{2n_u} \; 0_{2n_u\times 2(n_{v_K}-n_u)} \,],$$

$$R = 1/2(Z + Z^\top), \tag{5.71}$$

$$B_{K1} = \left[\, B_{K1,1} \; B_{K1,2} \,\right], \tag{5.72}$$

$$\mathsf{K} = \left[\, \tfrac{1}{2} C_K^\top P_{n_u}^\top \begin{bmatrix} I \\ \imath I \end{bmatrix} \mathsf{K}_{b1}^\top \; \mathsf{K}_{b2}^\top \,\right]^\top, \tag{5.73}$$

$$B_{K1,1} = \mathbb{J}_{n_K} C_K^\top \mathbb{J}_{n_u}, \tag{5.74}$$

$$\mathsf{K}_{b2} = -\imath \left[\, I_{n_y\times n_y} \; 0_{n_y\times n_y} \,\right] P_{n_y}(I_{2n_y} \otimes M) B_K^\top \mathbb{J}_{n_K}, \tag{5.75}$$

$$B_{K1,2} = 2\imath \mathbb{J}_{n_K} \left[-\mathsf{K}_{b1}^* \; \mathsf{K}_{b1}^\top \right] P_{n_{v_K}-n_u} \mathrm{diag}_{n_{v_K}-n_u}(M), \tag{5.76}$$

where $Z = -\frac{1}{2}\mathbb{J}_{n_K} A_K$ and $n_{v_K} \geq n_u + 1$, P_m is a permutation matrix as defined in Sect. 2.3.2, and

$$M = 1/2 \begin{bmatrix} 1 & \imath \\ 1 & -\imath \end{bmatrix}.$$

Here, K_{b1} is any complex $(n_{v_K} - n_u) \times 2n_K$ matrix such that

$$\mathsf{K}_{b1}^* \mathsf{K}_{b1} = W + \imath \Big(1/2(Z - Z^\top) - 1/4 C_K^\top P_{N_u}^\top \begin{bmatrix} 0 & I \\ -I & 0 \end{bmatrix} P_{n_u} C_K - \Im(\mathsf{K}_{b2}^* \mathsf{K}_{b2}) \Big), \tag{5.77}$$

where W is any real symmetric $2n_K \times 2n_K$ matrix such that the right-hand side of (5.77) is non-negative definite.

The proof of Lemma 5.2 is given in the appendix, Sect. 5.4.

Remark 5.2 Note that for (5.63) and (5.61) to be compatible, it is required that the Itō matrices of u and v_K satisfy the condition

$$F_u = B_{K0} F_{v_K} B_{K0}^\top. \tag{5.78}$$

5.3.2.5 Robust Stability

We will now develop further results that show how the H^∞ synthesis results of Sect. 5.3.2 can be used to guarantee stability robustness in the presence of real parameter uncertainties. We assume that the true closed-loop quantum system corresponding to the system (5.65) is of the form

$$d\tilde{x}(t) = \bar{A}\tilde{x}(t)dt + \tilde{G}d\tilde{v}(t), \tag{5.79}$$

where $\bar{A} = \tilde{A} + \tilde{B}\Delta\tilde{C}$, and Δ is a constant but unknown uncertainty matrix satisfying

$$\Delta^\top \Delta \leq \frac{1}{g^2}I. \tag{5.80}$$

The stability robustness will be a consequence of the fact that the H^∞ control design leads to a closed-loop quantum system which is strictly bounded real with disturbance attenuation g. We start with the following definition.

Definition 5.3 The closed-loop quantum system (5.79) is said to be *mean square stable* if there exists a real positive definite matrix $X > 0$ and a constant $\lambda > 0$ such that

$$\langle \tilde{x}(t)^\top X\tilde{x}(t)\rangle + \int_0^t \langle \tilde{x}(s)^\top \tilde{x}(s)\rangle ds \leq \langle \tilde{x}(0)^\top X\tilde{x}(0)\rangle + \lambda t \quad \forall t > 0,$$

for all Gaussian initial states ρ of the plant and controller.

The next lemma and theorem connect the robust stability of the above system to its H^∞ properties. The proofs for these results can be found in the appendix.

Lemma 5.3 *The quantum system (5.79) is mean square stable if and only if the matrix $\bar{A}$ is a stable matrix.*

Proof We first observe that the system (5.79) is mean square stable if and only if it is dissipative with a supply rate defined by the matrix $\mathsf{R} = \mathrm{diag}(I, 0)$. It follows from Theorem 5.2 that the system (5.79) is mean square stable if and only if there exists a real positive definite symmetric matrix X such that $\bar{A}^\top X + X\bar{A} + I \leq 0$. Hence, using a standard Lyapunov result (e.g., see [10]), it follows that the system (5.79) is mean square stable if and only if the matrix $\bar{A}$ is asymptotically stable. □

Theorem 5.5 *If the closed-loop quantum system (5.65) is strictly bounded real with disturbance attenuation g, then the true closed loop system (5.79) is mean square stable for all Δ satisfying (5.80).*

Proof It follows from Corollary 5.3 that the closed-loop quantum system (5.65) is strictly bounded real with disturbance attenuation g, then $\tilde{A}$ is a stable matrix and $\|\tilde{C}(sI - \tilde{A})^{-1}\tilde{B} + \tilde{D}\|_\infty < g$. From this, it follows using the standard small gain theorem (e.g., see Theorem 9.1 on page 212 of [10]) that the matrix $\bar{A} = \tilde{A} + \tilde{B}\Delta\tilde{C}$ is stable for all Δ satisfying (5.80). Hence using Lemma 5.3, it follows that the true closed-loop system (5.79) is mean square stable for all Δ satisfying (5.80). □

5.3.3 H^{∞} Synthesis in Quantum Optics

We will now illustrate some H^{∞} controller designs for a quantum optical plant in the form of an optical cavity. The type of control discussed in the first example has been experimentally demonstrated in [42] in bulk quantum optics and in [43] in silicon photonics, in the context of broadband attenuation of a coherent light that is injected into one port of an optical cavity at the output of another port of the cavity.

5.3.3.1 Quantum Controller Synthesis

The plant that will be considered is an optical cavity which is coupled to three optical inputs v, w, and u, as shown in Fig. 5.4. As our control objective, we wish to attenuate the effect of the signal w as a disturbance input to the system on the output z. In physical terms, this means to dim the light emerging from z resulting from light shone in at w across a broad range of frequencies. Thus, most of the optical power in w will be redirected to output fields other than z; again, see [42, 43] for the details of some physical realizations of this example.

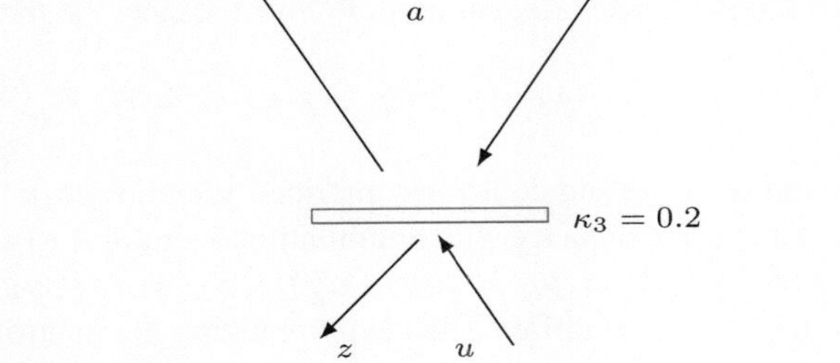

Fig. 5.4 An optical cavity (plant). Reprinted, with permission, from [34] ©2008 IEEE

The dynamics of this optical cavity system is given in the complex mode form by:

$$\begin{aligned}
da(t) &= -\frac{\gamma}{2}a(t)dt - \sqrt{\kappa_1}d\mathcal{A}_1(t) - \sqrt{\kappa_2}d\mathcal{A}_2(t)\sqrt{\kappa_3}d\mathcal{A}_3(t);\\
&\quad a(0) = a,\\
da^*(t) &= -\frac{\gamma}{2}a^*(t)dt - \sqrt{\kappa_1}d\mathcal{A}_1^*(t) - \sqrt{\kappa_2}d\mathcal{A}_2^*(t) - \sqrt{\kappa_3}d\mathcal{A}_3^*(t);\\
&\quad a^*(0) = a^*,\ (\gamma = \kappa_1 + \kappa_2 + \kappa_3),\\
d\mathcal{Y}_3(t) &= \sqrt{\kappa_3}a(t)dt + d\mathcal{A}_3(t),\\
d\mathcal{Y}_2(t) &= \sqrt{\kappa_2}a(t)dt + d\mathcal{A}_2(t),
\end{aligned} \tag{5.81}$$

where $\mathcal{A}_1(t)$, $\mathcal{A}_2(t)$, and $\mathcal{A}_3(t)$ are, respectively, the annihilation processes of the input fields in channels v, w, and u (with v and u in the vacuum state, and w in a coherent state), while $\mathcal{Y}_3(t)$ and $\mathcal{Y}_2(t)$ are, respectively, the output fields of channels u and w.

Let us introduce the following quadrature components of (5.81), $x_1(t) = q(t) = a(t) + a^*(t)$, $x_2(t) = p(t) = (a(t) - a^*(t)/\imath$, $v(t) = (v_1(t) = \mathcal{A}_1(t) + \mathcal{A}_1^*(t), v_2(t) = (\mathcal{A}_1(t) - \mathcal{A}_1^*(t))/\imath)^\top$, $w(t) = (w_1(t) = \mathcal{A}_2(t) + \mathcal{A}_2^*(t), w_2(t) = (\mathcal{A}_2(t) - \mathcal{A}_2^*(t))/\imath)^\top$, $u(t) = (u_1(t) = \mathcal{A}_3(t) + \mathcal{A}_3^*(t), u_2(t) = (\mathcal{A}_3(t) - \mathcal{A}_3^*(t))/\imath)^\top$, $z(t) = (z_1(t) = \mathcal{Y}_3(t) + \mathcal{Y}_3^*(t), z_2(t) = (\mathcal{Y}_3(t) - \mathcal{Y}_3^*(t))/\imath)^\top$ and $y(t) = (y_1(t) = \mathcal{Y}_2(t) + \mathcal{Y}_2^*(t), y_2(t) = (\mathcal{Y}_2(t) - \mathcal{Y}_2^*(t))/\imath)^\top$. This leads to the quadrature form (5.61) with the following system matrices:

$$\begin{aligned}
&A = -\frac{\gamma}{2}I;\ B_0 = -\sqrt{\kappa_1}I;\ B_1 = -\sqrt{\kappa_2}I;\ B_2 = -\sqrt{\kappa_3}I;\\
&C_1 = \sqrt{\kappa_3}I;\ D_{12} = I;\ C_2 = \sqrt{\kappa_2}I;\ D_{21} = I.
\end{aligned}$$

In this example, suppose that the coupling coefficients of the cavity are $\kappa_1 = 2.6$, $\kappa_2 = \kappa_3 = 0.2$, so that the total cavity decay rate is $\kappa = \kappa_1 + \kappa_2 + \kappa_3 = 3$. Setting $g = 0.1$, the Riccati equations (5.68) and (5.69) were found to have stabilizing solutions satisfying Assumption 5.2 given by $X = Y = 0_{2\times 2}$. If a controller of the form (6.53) is employed with matrices A_K, B_K, C_K determined as in (5.70), then following Theorem 5.2 the resulting closed-loop system will be strictly bounded real with disturbance attenuation g. In this instance, the matrices are given by

$$A_K = -1.1I,\ B_K = -0.447I,\ C_K = -0.447I.$$

From the resulting controller matrices, it can be seen that (6.53) can be realized as another optical cavity with annihilation operator a_K (with quadratures $x_{K1} = q_K = a_K + a_K^*$, $x_{K2} = p_K = (a_K - a_K^*)/\imath$, $x_K = (q_K, p_K)^\top$), connected at the output to a 180° phase shifter. The cavity realizing the controller has coupling coefficients $\kappa_{K1} = 0.2$, $\kappa_{K2} = 1.8$, $\kappa_{K3} = 0.2$, and $\kappa_K = 2.2$. For suitable choices of the matrices B_{K0} and B_{K1}, the H^∞ controller takes the quadrature form:

$$\begin{aligned}
dx_K(t) &= A_K x_K(t)dt + [\,B_{K1}\ B_K\,][\,dv_K^\top\ dy^\top\,]^\top,\\
d\tilde{u}(t) &= -C_K\xi(t)dt + [\,I_2\ 0_{2\times 4}\,][\,dv_K^\top\ dy^\top\,]^\top,
\end{aligned}$$

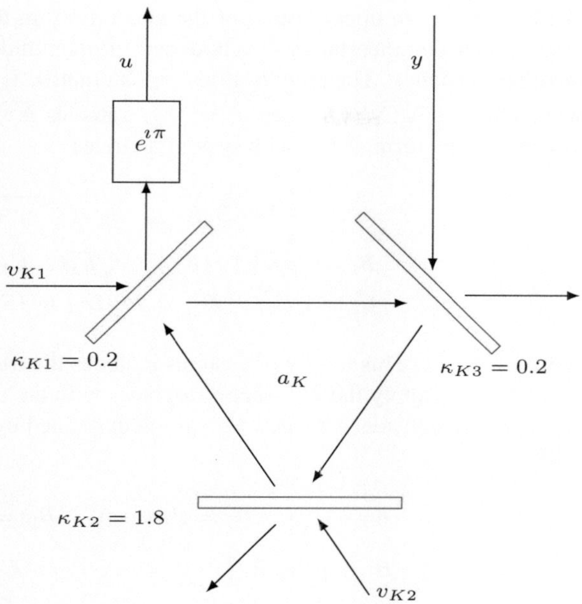

Fig. 5.5 An optical cavity quantum realization of the controller for the plant shown in Fig. 5.4. Reprinted, with permission, from [34] © 2008 IEEE

where $v_K(t) = (v_{K11}(t), v_{K12}(t), v_{K21}(t), v_{K22}(t))^\top$ are the quadratures of two distinct quantum noises, $B_{K1} = [-0.447I \; -1.342I]$, and $\tilde{u}(t)$ is the output of the cavity. Note that B_{K1} (whose choice does not affect the H^∞ performance) has been specifically chosen so that this controller can be realized as another optical cavity. The output of the controller is $u(t)$, given by $u(t) = K_s\tilde{u}(t)$, where $K_s = -I_2$. Here, K_s models the 180° phase shift at the cavity output. Thus, the controller is of the form (5.63) with $B_{K0} = [-I \; 0]$ and B_{K1} as given above. This controller is illustrated in Fig. 5.5.

5.3.3.2 Robust Stability in Quantum Optics

To illustrate the application of H^∞ controller synthesis for stability robustness, we modify the previous example by allowing for uncertainty in the value of the coupling coefficient to the input v. That is, we assume the actual value of the coupling coefficient is $\kappa_1 + \delta$, where δ is an unknown parameter modeling the uncertainty. For this modified example, the equations (5.61) describing the optical cavity now have matrices

$$\begin{aligned} A &= -\frac{\gamma+\delta}{2}I; \quad B_0 = \sqrt{\kappa_1+\delta}I; \quad B_1 = -\sqrt{\kappa_2}I; \\ B_2 &= -\sqrt{\kappa_3}I; \quad C_1 = \sqrt{\kappa_3}I; \quad D_{12} = I; \\ C_2 &= \sqrt{\kappa_2}I; \quad D_{21} = I. \end{aligned} \tag{5.82}$$

We first derive an upper bound of the uncertainty in the matrix A. To do this, we assume that the uncertainly δ is bounded in magnitude, $|\delta| < \mu$ for some known positive constant μ. Then for any non-singular matrix S, we have that $-\frac{\delta}{2}I = \tilde{B}_1\Delta\tilde{C}_1$ where $\tilde{B}_1 = \frac{\mu}{2}S$, $\tilde{C}_1 = S^{-1}$ and $\Delta = -\frac{\delta}{\mu}I$ satisfies $\Delta^\top\Delta \leq I$. Consider a family of systems of the form (5.61) with system matrices

$$\begin{aligned}
A &= -\frac{\gamma}{2}I + \tilde{B}_1\Delta\tilde{C}_1; \quad B_0 = \sqrt{\kappa_1 + \delta I}; \\
B_1 &= -\sqrt{\kappa_2}I; \quad B_2 = -\sqrt{\kappa_3}I; \quad C_1 = \sqrt{\kappa_3}I; \\
C_2 &= \sqrt{\kappa_2}I; \quad D_{12} = I; \quad D_{21} = I,
\end{aligned} \tag{5.83}$$

with $\Delta^\top\Delta \leq I$. This family of systems includes the true system as a member.

In order to apply the H^∞ theory together with the results of Sect. 5.3.2.5 to this example, we consider a related H^∞ problem defined by a system of the form (5.61) with

$$\begin{aligned}
A &= -\frac{\gamma}{2}I; \quad B_0 = \sqrt{\kappa_1 + \delta I}; \quad B_{10} = -\sqrt{\kappa_2}I; \\
B_1 &= \begin{bmatrix} B_{10} & \tilde{B}_1 \end{bmatrix}; \quad B_2 = -\sqrt{\kappa_3}I; \quad C_{10} = \sqrt{\kappa_3}I; \\
C_1 &= \begin{bmatrix} C_{10} \\ g\tilde{C}_1 \end{bmatrix}; \quad D_{120} = I; \quad D_{12} = \begin{bmatrix} D_{120} \\ 0 \end{bmatrix}; \\
C_2 &= \sqrt{\kappa_2}I; \quad D_{210} = I; \quad D_{21} = \begin{bmatrix} D_{210} & 0 \end{bmatrix}.
\end{aligned} \tag{5.84}$$

Here, g is the disturbance attenuation parameter in the H^∞ control problem to be considered. We note that B_0 depends on the unknown parameter δ, but this quantity will not be involved in the calculation of the H^∞ controller.

We set the parameter values for the *nominal* plant cavity to be the same as in the previous example, $\kappa_1 = 2.6$, $\kappa_2 = \kappa_3 = 0.2$. Also, suppose that $\mu = 0.1$. With a disturbance attenuation constant of $g = 0.1$ and $S = 1.5I$, the Riccati equations (5.68) and (5.69) were found to have stabilizing solutions satisfying Assumption 5.2 given by $X = 0.1733I$, $Y = 0.0022I$. Thus, the corresponding controller matrices are as follows:

$$A_K = -1.0997I, \; B_K = -0.4464I, \; C_K = -0.4464I. \tag{5.85}$$

As in the previous example, the controller defined by the matrices (5.85) can be realized by another optical cavity. In this case, the controller parameters are $\kappa_{K1} = 0.1993$, $\kappa_{K2} = 1.8008$, $\kappa_{K3} = 0.1993$, and $\gamma_K = 2.1993$. The physical realization of the controller is analogous to the one shown in Fig. 5.5.

The resulting closed-loop system is of the form (5.65), with

$$\tilde{A} = \begin{bmatrix} A & B_2 C_K \\ B_K C_2 & A_K \end{bmatrix};$$
$$\tilde{B} = \begin{bmatrix} B_1 \\ B_K D_{21} \end{bmatrix} = \begin{bmatrix} B_{10} & \tilde{B}_1 \\ B_K D_{210} & 0 \end{bmatrix};$$
$$\tilde{G} = \begin{bmatrix} B_0 & B_2 B_{K0} \\ B_K D_{20} & B_{K1} \end{bmatrix};$$
$$\tilde{C} = \begin{bmatrix} C_1 & D_{12} C_K \end{bmatrix} = \begin{bmatrix} C_{10} & D_{120} C_K \\ g\tilde{C}_1 & 0 \end{bmatrix};$$
$$\tilde{D} = \begin{bmatrix} 0 & D_{12} B_{K0} \end{bmatrix},$$

and, by Theorem 5.2, satisfies the strict bounded real condition with disturbance attenuation g. As a consequence of Corollary 5.3, we have that $\|\tilde{C}(sI - \tilde{A})^{-1}\tilde{B}\|_\infty < g$, and find that

$$\left\| \begin{bmatrix} C_{10} & D_{120} C_K \end{bmatrix} (sI - \tilde{A})^{-1} \begin{bmatrix} B_{10} \\ B_K D_{210} \end{bmatrix} \right\|_\infty < g \tag{5.86}$$

and

$$\left\| \begin{bmatrix} \tilde{C}_1 & 0 \end{bmatrix} (sI - \tilde{A})^{-1} \begin{bmatrix} \tilde{B}_1 \\ 0 \end{bmatrix} \right\|_\infty < 1. \tag{5.87}$$

By Corollary 5.3, (5.86) implies that the nominal closed-loop system is strictly bounded real with disturbance attenuation g, while (5.87) implies that the closed-loop system

$$\begin{aligned} d\tilde{x}(t) &= \begin{bmatrix} A & B_2 C_K \\ B_K C_2 & A_K \end{bmatrix} \tilde{x}(t)dt + \\ &\quad \begin{bmatrix} B_0 & B_2 B_{K0} \\ B_K D_{20} & B_{K1} \end{bmatrix} \begin{bmatrix} dv(t) \\ dv_K(t) \end{bmatrix} + \begin{bmatrix} \tilde{B}_1 \\ 0 \end{bmatrix} dw(t), \\ dz(t) &= \begin{bmatrix} \tilde{C}_1 & 0 \end{bmatrix} \tilde{x}(t)dt, \end{aligned}$$

is strictly bounded real with unity disturbance attenuation. Therefore, by Theorem 5.5, we conclude that the closed-loop uncertain system

$$\begin{aligned} d\tilde{x}(t) &= \begin{bmatrix} A + \tilde{B}_1 \Delta \tilde{C}_1 & B_2 C_K \\ B_K C_2 & A_K \end{bmatrix} \tilde{x}(t)dt + \\ &\quad \begin{bmatrix} B_0 & B_2 B_{K0} \\ B_K D_{20} & B_{K1} \end{bmatrix} \begin{bmatrix} dv(t) \\ dv_K(t) \end{bmatrix}, \end{aligned}$$

is mean square stable for all matrices Δ such that $\Delta^\top \Delta \leq I$. Hence, the true closed-loop system must also be mean square stable. Furthermore, for this example, it is also possible to verify that the true closed-loop system must in fact also be strictly bounded real with disturbance attenuation g.

5.4 Further Reading

The exposition of H^∞ control for linear quantum systems in this chapter has been limited to coherent feedback controllers in the form of linear quantum systems. However, the original treatment in [34], on which the exposition here is based, was quite general allowing for linear classical and mixed quantum-classical H^∞ controllers. The notion of physical realizability for systems with mixed quantum-classical dynamics and the associated conditions were introduced in [34, 44], based on the notion of a fully quantum augmentation of mixed quantum-classical linear systems, and developed further in [45–47]. Application of the theory to the construction of quantum optical realizations of classical linear stochastic systems was developed in [48].

The H^∞ controller design allowed for some free parameters that can be used to yield physically realizable controllers. However, this may entail adding additional noise channels to satisfy the physical realizability constraints, which is undesirable and needs to be minimized. The minimal number of additional quantum noise channels that are required to make a linear quantum controller physically realizable for any triplet (A_K, B_K, C_K) has been investigated in [49, 50], where the minimal number can be characterized in terms of rank of a certain matrix that depends upon (A_K, B_K, C_K).

Appendices

Appendix A: Proof of Theorem 5.2

The proof of Theorem 5.2 will use the following lemma.

Lemma 5.4 *Consider a real symmetric matrix X and corresponding operator-valued quadratic form $x^\top X x$ for the system (5.54). Then the following statements are equivalent:*

(i) There exists a constant $\lambda \geq 0$ such that $\langle \rho, x^\top X x \rangle \leq \lambda$ for all Gaussian states ρ.
(ii) The matrix X is negative semidefinite.

Proof $(i) \Rightarrow (ii)$. To establish this part of the lemma, consider a Gaussian state ρ which has mean $\bar{x}$ and covariance matrix $Y = [\mathrm{Tr}(\rho x_i x_j)]$, with symmetrized covariance $1/2(Y + Y^\top) \geq 0$ and $Y - Y^\top = 2\imath \mathbb{J}_n$. Then, we can write

$$\begin{aligned}\langle\rho, x^\top X x\rangle &= \sum_{i=1}^{n}\sum_{j=1}^{n} X_{ij}\langle\rho, x_i x_j\rangle \\ &= \sum_{i=1}^{n}\sum_{j=1}^{n} X_{ij}[Y_{ij} + \bar{x}_i\bar{x}_j] \\ &= \bar{x}^T X\bar{x} + \mathrm{Tr}[XY]. \end{aligned} \tag{5.88}$$

Now for any constant $\alpha > 0$, consider the inequality of part (i) where ρ is a Gaussian state with mean $\alpha\bar{x}$ and covariance matrix Y. Then, it follows from the bound $\langle\rho, x^\top X x\rangle \leq \lambda$ and (5.88) that $\alpha^2\bar{x}^\top X\bar{x} + \mathrm{Tr}[XY] \leq \lambda$ for all $\alpha > 0$. From this, it immediately follows that $\bar{x}^\top X\bar{x} \leq 0$. However, $\bar{x}$ can be chosen arbitrarily. Hence, we can conclude that condition (ii) of the lemma is satisfied.

$(ii) \Rightarrow (i)$. Suppose that the matrix X is negative semidefinite and let ρ be any Gaussian state and suppose that ρ has mean $\bar{x}$ and covariance matrix $Y \geq 0$. Then, it follows from (5.88) that $\langle\rho, x^\top X x\rangle = \bar{x}^\top X\bar{x} + \mathrm{Tr}[XY]$. However, $X \leq 0$, $Y + Y^\top \geq 0$, and $Y - Y^\top = 2\imath\mathbb{J}_n$ implies $\bar{x}^\top X\bar{x} \leq 0$, and $\mathrm{Tr}[XY] = 1/2\mathrm{Tr}[X(Y + Y^\top)] \leq 0$. Hence, $\langle\rho, x^\top X x\rangle \leq 0$ and condition (ii) is satisfied with $\lambda = 0$. □

Proof of Theorem 5.2. Let the system be dissipative with $V(x) = x^\top X x$. By Itō's rule, the product table (5.60), and the quantum stochastic differential equation (5.54), we have

$$\begin{aligned}& d\langle V(x(t))\rangle \\ &= \langle dx^\top(t)Xx(t) + x^\top(t)Xdx(t) + dx^\top(t)Xdx(t)\rangle \\ &= \langle x^\top(t)(A^\top X + XA)x(t) + \beta_w^\top(t)B^\top Xx(t) + \\ &\quad x^\top(t)XB\beta_w(t) + \lambda_0\rangle dt, \end{aligned} \tag{5.89}$$

where λ_0 is given by (5.59). We now note that $\langle V(x(t))\rangle = \langle\rho, E_0[V(x(t))]\rangle$, where E_t $(t \geq 0)$ denotes the conditional expectation map[2] with respect to the vacuum state $|\Omega\rangle$ of the field (e.g., see [51, p. 215]), and ρ is an initial Gaussian state of the system. Combining this with the integral of (5.89) and (5.56) we find that

$$\begin{aligned}&\Big\langle\rho, \int_0^t E_0[x^\top(s)(A^\top X + XA)x(s) + \beta_w^\top(s)B^\top Xx(s) + \\ &\quad x^\top(s)XB\beta_w(s) + \lambda_0 + r(x(s), \beta_w(s)]ds\Big\rangle \leq \lambda t.\end{aligned}$$

Let $t \to 0$ to obtain

$$\begin{aligned}&\Big\langle\rho, x^\top(A^\top X + XA)x + \beta_w^\top B^\top X + x^\top XB\beta_w + \lambda_0 + \\ &\quad [x^\top\beta_w^\top]R\begin{bmatrix} x \\ \beta_w\end{bmatrix}\Big\rangle \leq \lambda.\end{aligned}$$

[2]This map is distinct from the quantum conditional expectation used in quantum filtering theory.

Here, x and β_w denote the initial conditions. An application of Lemma 5.4 implies (5.57). Also, (5.58) is a straightforward consequence of this inequality when R is replaced by $\mathsf{R} + \epsilon I$ where $\epsilon > 0$.

To establish the converse part of the theorem, we first assume that (5.57) is satisfied. Then with $V(x) = x^\top X x$, it follows from (5.89) that

$$\langle V(x(t))\rangle - \langle V(x(0))\rangle + \int_o^t \langle r(x(s), \beta_w(s))\rangle ds \leq \lambda_0 t$$

for all $t > 0$ and all $\beta_w(t)$. Hence, inequality (5.56) is satisfied with λ given by (5.59).

If matrix inequality (5.58) is satisfied, then it follows by similar reasoning that there exists an $\epsilon > 0$ such that

$$\begin{aligned}\langle V(x(t))\rangle - \langle V(x(0))\rangle + \int_0^t \langle r(x(s), \beta_w(s)) + \\ \epsilon(x(s)^\top x(s) + \beta_w(s)^\top \beta_w(s))\rangle ds \leq \lambda_0 t.\end{aligned}$$

Hence, inequality (5.56) is satisfied with $\lambda = \lambda_0$ given by (5.59) and with R replaced by $\mathsf{R} + \epsilon I$. □

Appendix B: Proof of Lemma 5.2

The proof of Lemma 5.2 will use the following lemma.

Lemma 5.5 *If S is a Hermitian matrix, then there is a real constant α_0 such that $\alpha I + S \geq 0$ for all $\alpha \geq \alpha_0$.*

Proof Since S is Hermitian, it has real eigenvalues and is diagonalizable. Hence, $S = V^* E V$ for some real diagonal matrix E and orthogonal matrix V. Now let $\alpha_0 = -\lambda$, where λ is the smallest eigenvalue of S. The result follows since $\alpha I + S = V^*(\alpha I + E)V$ while $\alpha I + E \geq 0$ for all $\alpha \geq \alpha_0$. □

Proof of Lemma 5.2. The main idea is to explicitly construct matrices $R \in \mathbb{R}^{2n_K \times 2n_K}$, $\mathsf{K} \in \mathbb{C}^{(n_{v_K}+n_y)\times 2n_K}$, $B_{K1} \in \mathbb{R}^{2n_K \times 2n_{v_K}}$, and $B_{K0} \in \mathbb{R}^{2n_u \times 2n_{v_K}}$, with $n_{v_K} \geq n_u$, such that (see Chap. 2)

$$A_K = 2\mathbb{J}_{n_K}(R + \Im\{\mathsf{K}^*\mathsf{K}\}), \tag{5.90}$$

$$\begin{bmatrix} B_{K1} & B_K \end{bmatrix} = 2\imath\mathbb{J}_{n_K}[-\mathsf{K}^* \; \mathsf{K}^\top]\Gamma_{n_{v_K}+n_y}, \tag{5.91}$$

$$\begin{aligned} C_K &= [\, I_{2n_u\times 2n_u} \; 0_{2n_u\times 2(n_{v_K}+n_y-n_u)}\,]P^\top_{n_{v_K}+n_y} \\ &\quad \times \begin{bmatrix} \mathsf{K} + \mathsf{K}^\# \\ -\imath\mathsf{K} + \imath\mathsf{K}^\# \end{bmatrix}, \end{aligned} \tag{5.92}$$

$$\begin{bmatrix} B_{K0} & 0_{2n_u\times 2(n_{v_K}+n_y)} \end{bmatrix} = [\, I_{2n_u\times 2n_u} \; 0_{2n_u\times 2(n_{v_K}+n_y-n_u)}\,], \tag{5.93}$$

are satisfied. To this end, let $Z = \frac{1}{2}\mathbb{J}_{n_K}^{-1}A = -\frac{1}{2}\mathbb{J}_{n_K}A$. We first construct matrices K_{b2}, K_{b1}, $B_{K1,1}$, and $B_{K1,2}$ according to the following procedure:

1. Construct the matrix K_{b2} according to (5.75).
2. Construct a real symmetric $2n_K \times 2n_K$ matrix W_1 such that the matrix

$$\begin{aligned} W_2 = {} & W_1 \\ & + \imath\left(\frac{Z - Z^\top}{2} - \frac{1}{4}C_K^\top P_{n_u}^\top \begin{bmatrix} 0 & I \\ -I & 0 \end{bmatrix} P_{n_u} C_K - \Im(\mathsf{K}_{b2}^* \mathsf{K}_{b2})\right), \end{aligned}$$

   is non-negative definite. It follows from Lemma 5.5 that such a matrix W_1 always exists.
3. Construct a matrix K_{b1} such that $\mathsf{K}_{b1}^*\mathsf{K}_{b1} = W_2$, where K_{b1} has *at least* 1 row. This can be done, for example, using the singular value decomposition of W_2 (in this case, K_{b1} will have $2n_K$ rows).
4. Construct the matrices $B_{K1,1}$ and $B_{K1,2}$ according to Eqs. (5.74) and (5.76), respectively.

Let $R = \frac{1}{2}(Z + Z^\top)$. We now show that there exists an integer $n_{v_K} \geq n_u$ such conditions (5.90)–(5.93) are satisfied with the matrix R as defined and with $B_{K1} = [\, B_{K1,1} \;\; B_{K1,2}\,]$ and

$$\mathsf{K} = \begin{bmatrix} 1/2\,[\, I \;\; \imath I\,]\, P_{n_u} C_k \\ \mathsf{K}_{b1} \\ \mathsf{K}_{b2} \end{bmatrix}. \tag{5.94}$$

First note that necessarily $n_{v_K} \geq n_u + 1 > n_u$ since B_{K1} has at least $2n_u + 2$ columns. Also, by virtue of our choice of K_{b1}, we have

$$\begin{aligned} \Im(\mathsf{K}_{b1}^*\mathsf{K}_{b1}) &= \Im\{W_2\} \\ &= 1/2(Z - Z^\top) - 1/4 C_K^\top P_{n_u}^\top \begin{bmatrix} 0 & I \\ -I & 0 \end{bmatrix} P_{n_u} C_K - \Im\{\mathsf{K}_{b2}^*\mathsf{K}_{b2}\}, \end{aligned}$$

and hence

$$\begin{aligned} \Im(\mathsf{K}^*\mathsf{K}) &= \Im(\mathsf{K}_{b1}^*\mathsf{K}_{b1}) + \Im(\mathsf{K}_{b2}^*\mathsf{K}_{b2}) + 1/4 C_K^\top P_{n_u}^\top \begin{bmatrix} 0 & I \\ -I & 0 \end{bmatrix} P_{n_u} C_K \\ &= 1/2(Z - Z^\top). \end{aligned}$$

Since $R = \frac{Z+Z^\top}{2}$, we have $R + \Im(\mathsf{K}^*\mathsf{K}) = Z$. Therefore, (5.90) is satisfied.

Now, we observe that $\imath\mathbb{J}_{n_K} B_K \mathrm{diag}_{n_y}(M^*) P_{n_y}^\top = [\, T \;\; -T^\# \,]$ for some $2n_K \times n_y$ complex matrix T. But by taking the conjugate transpose of both sides of (5.75) which defined K_{b2}, we conclude that $T = -\mathsf{K}_{b2}^*$. Hence,

$$B_K = 2\imath\mathbb{J}_{n_K}[\, -\mathsf{K}_{b2}^* \;\; \mathsf{K}_{b2}^\top \,] P_{n_y} \mathrm{diag}_{n_y}(M). \tag{5.95}$$

From (5.74) which defined $B_{K1,1}$, we obtain

$$\begin{aligned} B_{K1,1} &= \mathbb{J}_{n_K} C_K^\top \mathrm{diag}_{n_u}(\mathbb{J}) \\ &= \mathbb{J}_{n_K} C_K^\top \mathrm{diag}_{n_u}(\mathbb{J})(2\mathrm{diag}_{n_u}(M^*))\mathrm{diag}_{n_u}(M) \\ &= \imath \mathbb{J}_{n_K} C_K^\top \mathrm{diag}_{n_u}\left(\begin{bmatrix} -1 & 1 \\ \imath & \imath \end{bmatrix}\right) \mathrm{diag}_{n_u}(M) \\ &= \imath \mathbb{J}_{n_K} C_K^\top P_{n_u}^\top \begin{bmatrix} -I & I \\ \imath I & \imath I \end{bmatrix} P_{n_u} \mathrm{diag}_{n_u}(M). \end{aligned} \tag{5.96}$$

Combining (5.76), (5.95) and (5.96) gives us

$$\begin{aligned} &[\, B_{K1,1} \; B_{K1,2} \; B_K \,] \\ &= 2\imath \mathbb{J}_{n_K} \left[{}^1\!/\!_2 C_K^\top P_{n_u}^\top \begin{bmatrix} -I & I \\ \imath I & \imath I \end{bmatrix} P_{n_u} \left[-\mathsf{K}_{b1}^* \; \mathsf{K}_{b1}^\top \right] P_{n_{v_K}-n_u} \right. \\ &\qquad \left. \left[-\mathsf{K}_{b2}^* \; \mathsf{K}_{b2}^\top \right] P_{n_y} \right] P_{n_{w_K}}^\top P_{n_{w_K}} \mathrm{diag}_{n_{w_K}}(M) \\ &= 2\imath \mathbb{J}_{n_K} \left[-{}^1\!/\!_2 C_K^\top P_{n_u}^\top \begin{bmatrix} I \\ -\imath I \end{bmatrix} -\mathsf{K}_{b1}^* \; -\mathsf{K}_{b2}^* \right. \\ &\qquad \left. {}^1\!/\!_2 C_K^\top P_{n_u}^\top \begin{bmatrix} I \\ \imath I \end{bmatrix} \mathsf{K}_{b1}^\top \; \mathsf{K}_{b2}^\top \right] P_{n_{w_K}} \mathrm{diag}_{n_{w_K}}(M) \\ &= 2\imath \mathbb{J}_{n_K} \left[-{}^1\!/\!_2 C_K^\top P_{n_u}^\top \begin{bmatrix} I \\ -\imath I \end{bmatrix} -\mathsf{K}_{b1}^* \; -\mathsf{K}_{b2}^* \right. \\ &\qquad \left. {}^1\!/\!_2 C_K^\top P_{n_u}^\top \begin{bmatrix} I \\ \imath I \end{bmatrix} \mathsf{K}_{b1}^\top \; \mathsf{K}_{b2}^\top \right] \Gamma \\ &= 2\imath \mathbb{J}_{n_K} \left[-\mathsf{K}^* \; \mathsf{K}^\top \right] \Gamma. \end{aligned}$$

Therefore, (5.91) is also satisfied. Moreover, it is straightforward to verify (5.92) by substituting K as defined by (5.94) into the right-hand side of (5.92). Finally, since $n_{v_K} \geq n_u$, it follows that $[\, B_{K0} \; 0_{2n_u \times 2(n_{v_K}+n_y-n_u)} \,]$ is precisely the right-hand side of (5.93). This completes the proof of Theorem 5.4. □

References

1. L. Bouten, Filtering and control in quantum optics, Ph.D. dissertation, Catholic University of Nijmegen (2004)
2. L. Bouten, R. van Handel, On the separation principle of quantum control, in *Quantum Stochastics and Information: Statistics, Filtering and Control (University of Nottingham, UK, 15–22 July 2006)*, ed. by V.P. Belavkin, M. Guta (World Scientific, Singapore, 2008), pp. 206–238
3. J. Gough, M.R. James, The series product and its application to quantum feedforward and feedback networks. IEEE Trans. Autom. Control **54**(11), 2530–2544 (2009)

4. D.A. Steck, K. Jacobs, H. Mabuchi, S. Habib, T. Bhattacharya, Feedback cooling of atomic motion in cavity QED. Phys. Rev. A **74**, 012322 (2006)
5. A.C. Doherty, K. Jacobs, Feedback-control of quantum systems using continuous state-estimation. Phys. Rev. A **60**, 2700 (1999)
6. R. Hamerly, H. Mabuchi, Advantages of coherent feedback for cooling quantum oscillators. Phys. Rev. Lett. **109**, 173602 (2012)
7. A.H. Jazwinski, *Stochastic Processes and Filtering Theory* (Academic Press, New York, 1970)
8. A. Bagchi, *Optimal Control of Stochastic Systems*. International Series in Systems and Control Engineering (Prentice Hall, Upper Saddle River, 1993)
9. W.L. Brogan, *Modern Control Theory*, 3rd edn. (Prentice-Hall, Upper Saddle River, 1991)
10. K. Zhou, J.C. Doyle, K. Glover, *Robust and Optimal Control* (Prentice-Hall, Upper Saddle River, 1995)
11. V.P. Belavkin, S.C. Edwards, Quantum filtering and optimal control, in *Quantum Stochastics and Information: Statistics, Filtering and Control (University of Nottingham, UK, 15–22 July 2006)*, ed. by V.P. Belavkin, M. Guta (World Scientific, Singapore, 2008), pp. 143–205
12. H.I. Nurdin, M.R. James, I.R. Petersen, Coherent quantum LQG control. Automatica **45**, 1837–1846 (2009) © 2009 Reprinted, with permission from Elsevier
13. G. Zhang, M.R. James, Direct and indirect couplings in coherent feedback control of linear quantum systems. IEEE Trans. Autom. Control **56**(7), 1535–1550 (2011)
14. I.G. Vladimirov, I.R. Petersen, A gradient descent approach to optimal coherent quantum LQG controller design, in *Proceedings of the 2015 American Control Conference (ACC)* (2015), pp. 1487–1492
15. A.K. Sichani, I.G. Vladimirov, I.R. Petersen, A gradient descent approach to optimal coherent quantum LQG controller design, in *Proceedings of the 2015 American Control Conference (ACC)* (2015), pp. 1487–1492
16. C.W. Scherer, P. Gahinet, M. Chilali, Multiobjective output-feedback control via LMI optimization. IEEE Trans. Autom. Control **42**(7), 896–911 (1997)
17. Y. Nesterov, A. Nemirovskii, *Interior-Point Polynomial Algorithms in Convex Programming* (SIAM, Philadelphia, 1994)
18. S. Boyd, L. Vanderberghe, Semidefinite programming relaxations of non-convex problems in control and combinatorial optimization, in *Communications, Computation, Control and Signal Processing - A Tribute to Thomas Kailath*, ed. by A. Paulraj, V. Roywhcowdhury, C. Schaper (Kluwer Academic Publishers, New York, 1997)
19. R. Orsi, U. Helmke, J.B. Moore, A Newton-like method for solving rank constrained linear matrix inequalities. Automatica **42**(11), 1875–1882 (2006)
20. J.B. Lasserre, Global optimization with polynomials and the problem of moments. SIAM J. Optim. **11**(3), 796–817 (2001)
21. M. Kojima, Sums of squares relaxations of polynomial semidefinite programs. Department of Mathematics and Computer Science, Tokyo Institute of Technology, Tokyo, Japan. Technical report B-397 (2003)
22. C.W.J. Hol, C. Scherer, Sum of squares relaxations for polynomial semidefinite programming, in *Proceedings of the 16th International Symposium on Mathematical Theory and Networks (MTNS) 2004 (Leuven, Belgium, July 5–9, 2004)* (2004)
23. D. Henrion, J.B. Lasserre, Convergent relaxations of polynomial matrix inequalities and static output feedback. IEEE Trans. Autom. Control **51**(2), 192–202 (2006)
24. R. Orsi, LMIRank: Software for Rank Constrained LMI Problems (2005), http://rsise.anu.edu.au/~robert/lmirank/
25. J. Löfberg, YALMIP: a toolbox for modeling and optimization in MATLAB, in *Proceedings of the CACSD Conference, Taipei, Taiwan* (2004), http://control.ee.ethz.ch/~joloef/yalmip.php
26. Advanced Optimization Lab, McMaster University: SeDuMi v1.1R3 (2006), http://sedumi.mcmaster.ca/
27. H. Bachor, T. Ralph, *A Guide to Experiments in Quantum Optics*, 2nd edn. (Wiley-VCH, Weinheim, 2004)

28. C.W. Gardiner, P. Zoller, *Quantum Noise: A Handbook of Markovian and Non-Markovian Quantum Stochastic Methods with Applications to Quantum Optics*, 3rd edn. (Springer, Berlin, 2004)
29. K. Jacobs, H.I. Nurdin, F.W. Strauch, M.R. James, Comparing resolved-sideband cooling and measurement-based feedback cooling on an equal footing: analytical results in the regime of ground-state cooling. Phys. Rev. A **91**, 043812 (2015)
30. J.C. Doyle, K. Glover, P.P. Khargonekar, B. Francis, State-space solutions to the standard H_2 and H_∞ control problems. IEEE Trans. Autom. Control **34**(8), 831–847 (1989)
31. I. Petersen, B. Anderson, E. Jonckheere, A first principles solution to the non-singular H^∞ control problem. Int. J. Robust Nonlinear Control **1**(3), 171–185 (1991)
32. C.W. Scherer, Multiobjective H_2/H_∞ control. IEEE Trans. Autom. Control **40**(6), 1054–1062 (1995)
33. J. Helton, M. James, *Extending H^∞ Control to Nonlinear Systems: Control of Nonlinear Systems to Achieve Performance Objectives*, Advances in Design and Control, vol. 1 (SIAM, Philadelphia, 1999)
34. M.R. James, H.I. Nurdin, I.R. Petersen, H^∞ control of linear quantum stochastic systems. IEEE Trans. Autom. Control **53**(8), 1787–1803 (2008) Reprinted, with permission, © 2008 IEEE
35. J. Willems, Dissipative dynamical systems - Part I: general theory. Arch. Rational Mech. Anal. **45**, 321–351 (1972)
36. P. Dupuis, M. James, I. Petersen, Robust properties of risk-sensitive control. Math. Control Syst. Signals **13**, 318–332 (2000)
37. C. D'Helon, M.R. James, Stability, gain, and robustness in quantum feedback networks. Phys. Rev. A **73**, 053803 (2006)
38. A. van der Schaft, *L_2-Gain and Passivity Techniques in Nonlinear Control* (Springer, New York, 1996)
39. S. Boyd, L. El Ghaoui, E. Feron, V. Balakrishnan, *Linear Matrix Inequalities in System and Control Theory* (SIAM, Philadelphia, 1994)
40. K. Zhou, P. Khargonekar, An algebraic Riccati equation approach to H_∞ optimization. Syst. Control Lett. **11**, 85–91 (1988)
41. M. Green, D. Limebeer, *Linear Robust Cotrol* (Prentice-Hall, Englewood Cliffs, 1995)
42. H. Mabuchi, Coherent-feedback quantum control with a dynamic compensator. Phys. Rev. A **78**, 032323 (2008)
43. M. Sarovar, D.B.S. Soh, J. Cox, C. Brif, C.T. DeRose, R. Camacho, P. Davids, Silicon nanophotonics for scalable quantum coherent feedback networks. EPJ Quantum Technol. **3**(14), 1–18 (2016)
44. H.I. Nurdin, Topics in classical and quantum linear stochastic systems, Ph.D. dissertation, The Australian National University (2007)
45. H.I. Nurdin, Network synthesis of mixed quantum-classical linear stochastic systems, in *Proceedings of the 2011 Australian Control Conference (AUCC), Engineers Australia* (2011), pp. 68–75
46. S. Wang, H.I. Nurdin, G. Zhang, M.R. James, Synthesis and structure of mixed quantum-classical linear systems, in *Proceedings of the 51st IEEE Conference on Decision and Control (Maui, Hawaii, Dec. 10–13, 2012)* (2012), pp. 1093–1098
47. S. Wang, H.I. Nurdin, G. Zhang, M.R. James, Network synthesis for a class of mixed quantum-classical linear stochastic systems (2014), arXiv:1403.6928 (arXiv preprint)
48. S. Wang, H.I. Nurdin, G. Zhang, M.R. James, Quantum optical realization of classical linear stochastic systems. Automatica **49**(10), 3090–3096 (2013)
49. S. Vuglar, I.R. Petersen, How many quantum noises need to be added to make an LTI system physically realizable? in *Proceedings of the 2011 Australian Control Conference (AUCC)* (2011), pp. 363–367
50. S. Vuglar, I.R. Petersen, Quantum implemention of an LTI system with the minimal number of additional quantum noise inputs, in *Proceedings of the 2013 European Control Conference (ECC)* (2013), pp. 2724–2727
51. K.R. Parthasarathy, *An Introduction to Quantum Stochastic Calculus* (Birkhauser, Berlin, 1992)

Chapter 6
Linear Systems and Control Theory for Quantum Information

Abstract This chapter illustrates several example applications of the theory of linear quantum systems to the analysis of problems of interest in quantum information processing and discusses two experimental demonstrations of real-time coherent feedback and measurement-based feedback control from the literature. The problems covered are dissipative generation of Gaussian states of single-mode oscillators, efficient enhancement of entanglement between traveling Gaussian fields, back-action evasion, perfect state transfer in a linear quantum network, and robust quantum amplification. The two experiments are demonstrations of enhancement of optical squeezing via static coherent feedback and generation of a spin-squeezed state in an atomic ensemble via measurement-based feedback control.

In this chapter, we discuss several applications of linear systems and control theory for *continuous-variable* (CV) quantum information processing. Here, the meaning of CV is that the spectrum of the observables q or p is continuous in $\mathbb{R}$, in contrast to the discrete variable case where, for instance, the spectrum of $\sigma_z = \mathrm{diag}(1, -1)$ is discrete. The topics discussed in this chapter are as follows:

- Section 6.1: Dissipative generation of arbitrary pure Gaussian states,
- Section 6.2: Entanglement enhancement via coherent feedback networks,
- Section 6.3: Back-action evasion via feedback for force sensing,
- Section 6.4: Quantum memory with decoherence-free subsystem,
- Section 6.5: Robust quantum amplification via coherent feedback control, and
- Section 6.6: Quantum feedback experiments for squeezing enhancement.

The topics in Sects. 6.1 and 6.4 do not deal with feedback control problems but discuss problems of system synthesis so that the engineered system achieves state generation in the former case and state transfer in the latter. Both cases fully utilize system theoretic notions/tools such as controllability and zero dynamics. On the

---

Section 6.1 contains reprinted excerpt with permission from [10]. Copyright (2012) by the American Physical Society.
Section 6.2 contains some materials reprinted from [22] with permission of Springer.
Section 6.5 contains reprinted excerpt with permission from [65]. Copyright (2016) by the American Physical Society.
Section 6.6 contains some materials reprinted, with permission, from [21] © 2012 IEEE.

H.I. Nurdin and N. Yamamoto, *Linear Dynamical Quantum Systems*,
Communications and Control Engineering, DOI 10.1007/978-3-319-55201-9_6

other hand, in Sects. 6.2, 6.3, and 6.5, we elaborate upon coherent feedback networks for enhanced CV entanglement generation, back-action evasion, and robust quantum amplification, respectively. Lastly, Sect. 6.6 is devoted to showing two experimental demonstrations of quantum feedback for optical and spin-squeezing enhancement. It is hoped that these examples will illustrate to the readers that systems and control theory provides powerful and pertinent tools for synthesizing and controlling quantum systems as quantum information processors.

6.1 Dissipative Generation of Pure Gaussian States

Our first topic deals with the problem of generating a target Gaussian state in a linear quantum system. In general, a Gaussian state is useful for quantum information, such as for quantum teleportation and secure communication [1, 2]. In particular, it is known that a *Gaussian cluster state* [3, 4] serves as an essential resource for measurement-based quantum computation. The method for generating a Gaussian state presented here belongs to the so-called *environment engineering* (or *reservoir engineering*) approach [5] described as follows. Recall now that for a closed system, generation of a target final state $|\psi_f\rangle$ is implemented by applying a unitary operation to a given initial state $|\psi_i\rangle$ so that $|\psi_i\rangle \to |\psi_f\rangle = U|\psi_i\rangle$; but in such an implementation, the final state $|\psi_f\rangle$ is very sensitive to the initial state $|\psi_i\rangle$ and the parameters of the unitary operator U, which includes the duration time. The environment engineering approach overcomes these issues. The point of this approach is to exploit open systems, rather than closed systems. The dynamics of the statistics of an open dissipative system is modeled by the master equation (4.20):

$$\frac{d\rho}{dt} = -i[H, \rho] + \sum_{k=1}^{m} \Big(L_k \rho L_k^* - \frac{1}{2} L_k^* L_k \rho - \frac{1}{2} \rho L_k^* L_k \Big), \tag{6.1}$$

where ρ is the (unconditional) quantum state, H is the Hamiltonian, and L_k is the Lindblad operator representing the kth system–bath coupling. Note again that the second Lindblad term on the right-hand side of (6.1) represents the dissipation. In the environment engineering approach, we aim to design H and L_k so that ρ evolves toward a target pure state: $\rho(t) \to |\psi_f\rangle\langle\psi_f|$ as $t \to \infty$. The notable advantage of this method is that the dissipation-induced final state $|\psi_f\rangle$ is independent of the initial state $\rho(0)$, and also, we do not need to stop the evolution at a certain precise time. That is, this method can realize robust generation of $|\psi_f\rangle$. Thus, what we need to use the environment engineering method is the condition for this desired time evolution to occur. In the finite-dimensional case, this has been given in [6, 7], and its application to quantum information processing has been proposed [8, 9].

In this section, we discuss the infinite-dimensional Gaussian version of the environment engineering method [10]. The discussion begins with a necessary and sufficient condition for the master equation (6.1) to have a unique pure Gaussian steady state, followed by a complete parameterization of the linear quantum system having this property. Then, a case study is given to demonstrate a pure Gaussian cluster state generation. Finally, a QSDE formulation [11] is provided, showing the general passivity property of the desired dissipative dynamics and a specific trade-off appearing in the problem of pure Gaussian cluster state generation.

6.1.1 General Condition for Pure Gaussian State Generation

In this section, we define x as $x = (q^\top, p^\top)^\top = (q_1, \dots, q_n, p_1, \dots, p_n)^\top$, which satisfies

$$xx^\top - (xx^\top)^\top = \imath\bar{\mathbb{J}}_n, \;\; \bar{\mathbb{J}}_n = \mathbb{J} \otimes I_n = \begin{bmatrix} 0 & I_n \\ -I_n & 0 \end{bmatrix}.$$

As described in Sect. 2.7.1, a quantum Gaussian state can be completely characterized by only the mean vector $\langle x \rangle$ and the symmetrized covariance matrix $V = \langle \Delta x \Delta x^\top + (\Delta x \Delta x^\top)^\top \rangle/2,\ \Delta x = x - \langle x \rangle$. For an open linear system with a Gaussian state, the time evolution equations of $\langle x \rangle$ and V can be obtained in a very simple form as follows. That is, for a linear quantum system with Hamiltonian $H = x^\top R x/2$ and coupling operator $L = Kx$, where $R = R^\top \in \mathbb{R}^{2n\times 2n}$ and $K \in \mathbb{C}^{m\times 2n}$, the time evolution of $\langle x \rangle$ and V is given by

$$\frac{d\langle x \rangle}{dt} = A\langle x \rangle, \;\; \frac{dV}{dt} = AV + VA^\top + \mathbb{D}, \tag{6.2}$$

where $A = \bar{\mathbb{J}}_n(R + \Im\{K^*K\})$ and $\mathbb{D} = \bar{\mathbb{J}}_n \Re\{K^*K\}\bar{\mathbb{J}}_n^\top$. The state $\rho(t)$ is Gaussian with mean $\langle x(t) \rangle$ and covariance matrix $V(t)$ for all t.

The environment engineering problem for Gaussian state generation can be formulated as follows. Now, when A is Hurwitz, the steady state of the above linear quantum system is given by a Gaussian state with mean $\langle x(\infty) \rangle = 0$ and covariance matrix V_s, which is the unique solution to the algebraic Lyapunov equation

$$AV_s + V_s A^\top + \mathbb{D} = 0. \tag{6.3}$$

Therefore, our goal is to synthesize the system matrices R and K so that V_s coincides with the covariance matrix of a target pure Gaussian state. To attack this problem, the following result is useful:

Theorem 6.1 *Let V_s be a covariance matrix corresponding to a pure Gaussian state. Then, this is the unique steady state of a linear quantum system characterized by (R, K), if and only if*

$$\ker\left(V_s + \frac{\imath}{2}\bar{\mathbb{J}}_n\right) = \mathrm{range}(\mathbb{K}^\top), \tag{6.4}$$

where

$$\mathbb{K} = [K^\top \; R\bar{\mathbb{J}}_n^\top K^\top \; \dots \; (R\bar{\mathbb{J}}_n^\top)^{2n-1}K^\top]^\top.$$

For the proof, see [10]. Note that this result does not state whether a given system has a unique pure steady state; for this purpose, Theorem 6.1 in [10] can be used.

6.1.2 Synthesizing a Dissipative Gaussian System

Here, we show that Theorem 6.1 leads to a complete parametrization of (R, K), which as a consequence provides us with an explicit procedure for synthesizing a desired open Gaussian system. Let us start from the fact that in general, the covariance matrix V_s of a pure Gaussian state can be represented as follows [3, 4, 12]:

$$V_s = \frac{1}{2}SS^\top, \quad S = \begin{bmatrix} Y^{-1/2} & 0 \\ XY^{-1/2} & Y^{1/2} \end{bmatrix}, \tag{6.5}$$

where X and Y are $n \times n$ real symmetric and real positive definite matrices (i.e., $Y = Y^\top > 0$), respectively. Note that S is $\bar{\mathbb{J}}_n$-symplectic, satisfying $S\bar{\mathbb{J}}_n S^\top = \bar{\mathbb{J}}_n$. Then, defining $Z = X + \imath Y$ leads to

$$V_s + \frac{\imath}{2}\bar{\mathbb{J}}_n = \frac{1}{2}\begin{bmatrix} I \\ Z^* \end{bmatrix} Y^{-1}[I \ Z].$$

Now, because the rank of $[I \ Z]$ is n, we have

$$\ker\left(V_s + \frac{\imath}{2}\bar{\mathbb{J}}_n\right) = \text{range}\left(\begin{bmatrix} -Z \\ I \end{bmatrix}\right).$$

Hence, (6.4) is equivalent to

$$\text{range}\left(\begin{bmatrix} -Z \\ I \end{bmatrix}\right) = \text{range}\left(\begin{bmatrix} K^\top & R\bar{\mathbb{J}}_n^\top K^\top & \dots & (R\bar{\mathbb{J}}_n^\top)^{2n-1}K^\top \end{bmatrix}\right).$$

To satisfy this condition, it is required that $\text{range}(K^\top)$ is contained in $\text{range}\left([-Z \ I]^\top\right)$ and $\text{range}\left([-Z \ I]^\top\right)$ is invariant under $R\bar{\mathbb{J}}_n^\top$. These conditions are expressed as

$$K^\top = \begin{bmatrix} -Z \\ I \end{bmatrix} P, \quad R\bar{\mathbb{J}}_n^\top \begin{bmatrix} -Z \\ I \end{bmatrix} = \begin{bmatrix} -Z \\ I \end{bmatrix} Q, \tag{6.6}$$

where P and Q are some $n \times m$ and $n \times n$ complex matrices. From these equations, $\mathbb{K}^\top$ can be expressed as

$$\mathbb{K}^\top = \begin{bmatrix} -Z \\ I \end{bmatrix} [P \ QP \ \dots \ Q^{2n-1}P].$$

Consequently, the necessary and sufficient condition for $\text{range}(\mathbb{K}^\top)$ to coincide with $\text{range}\left([-Z \ I]^\top\right)$ is that there exist P and Q satisfying (6.6) and the rank condition

$$\mathrm{rank}\left([P \;\; QP \;\; \dots \;\; Q^{n-1}P]\right) = n. \tag{6.7}$$

Now, let us decompose R into the following form:

$$R = \begin{bmatrix} R_1 & R_2 \\ R_2^\top & R_3 \end{bmatrix},$$

where R_1 and R_3 are real $n \times n$ symmetric matrices and R_2 a real $n \times n$ matrix. Then, the latter equation in (6.6) leads to

$$R_1 + R_2 X + X R_2^\top + X R_3 X - Y R_3 Y = 0,$$
$$(R_2 + X R_3)Y + Y(R_2 + X R_3)^\top = 0.$$

The second equation equivalently means that there exists a real skew symmetric matrix Γ (i.e., $\Gamma + \Gamma^\top = 0$) satisfying $(R_2 + X R_3)Y = \Gamma$. Hence, by writing $P' = R_3$, we find that R_2 can be expressed as $R_2 = -XP' + \Gamma Y^{-1}$. Then, it follows from (6.6) that Q is of the form $Q = -\imath P'Y - Y^{-1}\Gamma^\top$. As a consequence, the complete parameterizations of K and R are obtained as follows:

$$K = P^\top[-Z \;\; I],$$
$$R = \begin{bmatrix} XP'X + YP'Y - \Gamma Y^{-1}X - XY^{-1}\Gamma^\top & -XP' + \Gamma Y^{-1} \\ -P'X + Y^{-1}\Gamma^\top & P' \end{bmatrix}. \tag{6.8}$$

Again, P (complex), P' (real symmetric), and Γ (real skew) are the parameterizing matrices. Thus, a procedure to synthesize a linear quantum system generating a target pure Gaussian state has now been constructed; given a target pure Gaussian state with covariance matrix (6.5), which determines X and Y, choose the free matrices P, P', and Γ so that both the system–field coupling $L = Kx$ and the Hamiltonian $H = x^\top Rx/2$ with the system matrices (6.8) would be physically implementable, while P and $Q = -\imath P'Y - Y^{-1}\Gamma^\top$ satisfy the rank condition (6.7).

Lastly, note that there always exists a triplet (P, P', Γ) satisfying the rank condition (6.7), if no additional constraint is required for those matrices; that is, for an arbitrary given pure Gaussian state, we can always find a Gaussian dissipative system that produces that state as its unique steady state.

6.1.3 Gaussian Cluster State Generation via Dissipation

Menicucci et al. [4] developed a unified graphical calculus applicable to all pure Gaussian states, characterized by the matrix $Z = X + \imath Y$. A particularly important result is that the canonical CV cluster state can be represented in a general form as

$$Z = X + \imath e^{-2r} I, \tag{6.9}$$

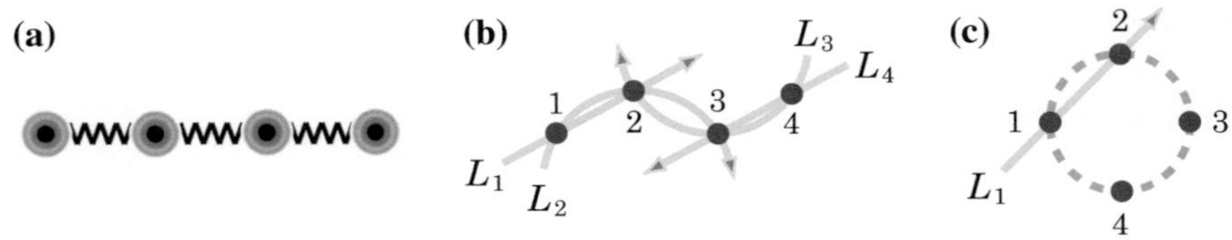

Fig. 6.1 **a** Shape of a one-dimensional harmonic chain with 4 oscillators. **b** Dissipative channels (indicated by the *arrows*) for generating a harmonic chain state. **c** A quasi-local dissipative system generating the same harmonic chain state in **b**; the *dotted circle* represents the system Hamiltonian. Reprinted with permission from [10] © (2012) by the American Physical Society

where X is the symmetric adjacent matrix representing the structure of the cluster state. Also, $Y = e^{-2r}I$ $(r \in \mathbb{R})$ represents the approximation error of the actual state with covariance matrix (6.5) to the ideal cluster state corresponding to X; note that the ideal state corresponds to the case with $r \to \infty$ (usually interpreted as infinite squeezing) and is thus not physically realizable.

Let us now apply the procedure obtained in the previous section to synthesize a linear quantum system generating a typical cluster state, a one-dimensional *harmonic chain state* with equal weight: This is the state of a set of quantum oscillators connected in a chain form as depicted in Fig. 6.1a. In particular, here we consider the case $n = 4$. Now, the adjacent matrix X and the resultant graph matrix (6.9) are, respectively, given by

$$X = \begin{bmatrix} 0 & 1 & & \\ 1 & 0 & 1 & \\ & 1 & 0 & 1 \\ & & 1 & 0 \end{bmatrix}, \quad Z = \begin{bmatrix} \imath e^{-2r} & 1 & & \\ 1 & \imath e^{-2r} & 1 & \\ & 1 & \imath e^{-2r} & 1 \\ & & 1 & \imath e^{-2r} \end{bmatrix},$$

The above X represents that for instance, the first and the second subsystems are coupled, with weight 1. A simple choice of the matrices in (6.8) is $P = I_4$, $P' = 0$, and $\Gamma = 0$, leading to $K = [-Z\ I]$ and $R = 0$. In this case, the system is driven only by the dissipation term characterized by the following Lindblad operator $L = Kx$:

$$\begin{aligned} L_1 &= (-\imath e^{-2r}q_1 + p_1) - q_2, \quad L_2 = -q_1 + (-\imath e^{-2r}q_2 + p_2) - q_3, \\ L_3 &= -q_2 + (-\imath e^{-2r}q_3 + p_3) - q_4, \quad L_4 = -q_3 + (-\imath e^{-2r}q_4 + p_4). \end{aligned} \tag{6.10}$$

Figure 6.1b depicts how the system couples with the dissipation channels. It is observed from this figure that each dissipative channel acts on at most three subsystems; in general, if the channel L acts only on a few subsystems, then it is called *quasi-local* [7].

Of course, the way to design a desired system is not unique. For instance, let us try to find a system that generates the same chain state by a quasi-local dissipative process acting on at most *two* adjacent subsystems. This goal is simply motivated by the fact that engineering such a dissipative environment might be easier to do than the previous case, apart from that we may need an additional Hamiltonian. Here, we take $P = (1, 0, 0, 0)^\top$, implying that the system has one dissipative channel L_1 in (6.10), which indeed acts on only two subsystems. To determine the Hamiltonian matrix R in (6.8), we have some freedom, but let us choose

$$P' = X^{-1} = \begin{bmatrix} 0 & 1 & 0 & -1 \\ 1 & 0 & 0 & 0 \\ 0 & 0 & 0 & 1 \\ -1 & 0 & 1 & 0 \end{bmatrix}, \quad \Gamma = 0.$$

This leads to $Q = -\imath P'Y - Y^{-1}\Gamma^\top = -\imath e^{-2r}X^{-1}$, and it is readily verified that the matrix $[P \ QP \ Q^2P \ Q^3P]$ is of full rank. Then, R is given by

$$R = \begin{bmatrix} X + e^{-2r}X^{-1} & -I \\ -I & X^{-1} \end{bmatrix}.$$

The loci of nonzero entries of this matrix show that the Hamiltonian $H = x^\top Rx/2$ has a ring-type structure where the 1–2, 2–3, 3–4, and 4–1 subsystems are connected. Therefore, we now see that the four-mode harmonic chain state can also be generated in the dissipative system whose system–field couplings and the Hamiltonian are both quasi-local, as shown in Fig. 6.1c.

6.1.4 A QSDE Formalism

Let us begin with the fact that a pure Gaussian state $|\psi_Z\rangle$ having the covariance matrix (6.5) always satisfies

$$\mathcal{N}|\psi_Z\rangle = 0, \quad \mathcal{N} := [-Z \ I]x = p - Zq = \begin{bmatrix} p_1 \\ \vdots \\ p_n \end{bmatrix} - Z \begin{bmatrix} q_1 \\ \vdots \\ q_n \end{bmatrix}. \tag{6.11}$$

That is, for $Z = (z_{ij})$, this equation means that $(p_i - \sum_j z_{ij}q_j)|\psi_Z\rangle = 0$ for all i. Conversely, if a pure Gaussian state $|\psi\rangle$ satisfies $\mathcal{N}|\psi\rangle = 0$, it follows that $|\psi\rangle = |\psi_Z\rangle$. The vector of operators, $\mathcal{N}$, is called the *nullifier* associated with the pure Gaussian state $|\psi_Z\rangle$.

Clearly, it should be worth examining the time evolution of $\mathcal{N}$ in the Heisenberg picture, when the conditions given in Theorem 6.1 are imposed. For this purpose, let us recall the following QSDE with system variable $x = (q^\top, p^\top)^\top$:

$$\frac{dx(t)}{dt} = Ax(t) - \imath\bar{\mathbb{J}}_n K^* \xi(t) + \imath\bar{\mathbb{J}}_n K^\top \xi(t)^\#, \tag{6.12}$$

where $A = \bar{\mathbb{J}}_n(R + \Im\{K^*K\})$. Recall that the output equation of the system is given by

$$\eta(t) = Kx(t) + \xi(t). \tag{6.13}$$

Now, from (6.6), we have

$$\begin{aligned}
[-Z\ I]A &= [-Z\ I]\bar{\mathbb{J}}_n R + \frac{1}{2\imath}[-Z\ I]\bar{\mathbb{J}}_n(K^*K - K^\top K^\#) \\
&= Q^\top[-Z\ I] - \frac{1}{2\imath}[I\ Z]\Big\{\begin{bmatrix} -Z^\# \\ I \end{bmatrix} P^\# P^\top[-Z\ I] - \begin{bmatrix} -Z \\ I \end{bmatrix} PP^*[-Z^\#\ I]\Big\} \\
&= Q^\top[-Z\ I] + \frac{Z^\# - Z}{2\imath} P^\# P^\top[-Z\ I] = (Q^\top - YP^\# P^\top)[-Z\ I],
\end{aligned}$$

$$[-Z\ I](-\imath\bar{\mathbb{J}}_n K^*) = -\imath[-Z\ I]\bar{\mathbb{J}}_n \begin{bmatrix} -Z^\# \\ I \end{bmatrix} P^\# = \imath(Z - Z^\#)P^\# = -2YP^\#,$$

$$[-Z\ I](\imath\bar{\mathbb{J}}_n K^\top) = \imath[-Z\ I]\bar{\mathbb{J}}_n \begin{bmatrix} -Z \\ I \end{bmatrix} P = 0.$$

Thus, multiplying both sides of (6.12) by $[-Z\ I]$ from the left, we have

$$\frac{d\mathcal{N}(t)}{dt} = (Q^\top - YP^\# P^\top)\mathcal{N}(t) - 2YP^\#\xi(t). \tag{6.14}$$

Regarding the output process (6.13), because $Kx = P^\top[-Z\ I]x = P^\top\mathcal{N}$, it can be represented by

$$\eta(t) = P^\top\mathcal{N}(t) + \xi(t). \tag{6.15}$$

The above two Eqs. (6.14) and (6.15) reveal several important properties of the system that uniquely generates a pure Gaussian state in a dissipative way. First of all, it must be a completely passive linear system as defined in Sect. 2.4.1; in fact, as a property of completely passivity, the system equations contain only the annihilation-driving noise ξ. Recall that the system equations of an optical empty cavity, which is a very simple completely passive system, are given by (1.25) and (1.26). Hence, we now find that the linear quantum system focused in this section is a non-trivial extension of those simple passive optical systems, in the sense that they have a non-vacuum pure steady state.

The above result also clarifies the meaning of rank condition (6.7), which is indeed hard to see without the QSDE formalism discussed here. That is, we have the following fact (the proof is given in [11]): *The matrix* $Q^\top - YP^\# P^\top$ *is Hurwitz if and only if the rank condition* (6.7) *is satisfied.* That is, the rank condition is necessary for the target pure Gaussian state $|\psi_Z\rangle$ to be a unique stable steady state of the system. This fact can be seen by multiplying all the entries of (6.14) by the composite state vector $|\Psi\rangle = |\psi\rangle \otimes |\Omega\rangle$ from the right ($|\Omega\rangle$ is the vacuum state on the field); that is, due to the relation $\xi|\Omega\rangle = 0$, we have

$$\frac{d}{dt}\mathcal{N}(t)|\Psi\rangle = (Q^\top - YP^\# P^\top)\mathcal{N}(t)|\Psi\rangle.$$

Therefore, if the rank condition (6.7) is satisfied, or equivalently $Q^\top - YP^\# P^\top$ is Hurwitz, the nullifier vector $\mathcal{N}(t)|\Psi\rangle = U(t)^*\mathcal{N}U(t)|\Psi\rangle$ converges to zero. Hence, in the Schrödinger picture, $U(\infty)|\Psi\rangle$ is the common zero eigenstate of all the elements of $\mathcal{N}$, implying that from the reason given at the beginning of this section, the system state becomes the target $|\psi_Z\rangle$ as $t \to \infty$. As mentioned above, our system is an extension of simple passive optical systems; but note again that while the steady state generated in such an optical system is merely a vacuum state, the steady state of the system (6.14) and (6.15) can be assigned to the highly non-trivial state $|\psi_Z\rangle$.

Based on the above discussion, we here provide a remark. That is, the system (6.14) and (6.15) with the Hurwitz property is *all-pass*, meaning that the output equation at steady state does not have any information about the system, which is indeed the reason why a pure state is generated in the end. This property can be seen immediately in the Fourier domain; the linear transformation from the input $\xi[\imath\omega]$ to the output $\eta[\imath\omega]$ is given by

$$\eta[\imath\omega] = F[\imath\omega]\xi[\imath\omega], \quad F[\imath\omega] := I - 2P^\top(\imath\omega - Q^\top + YP^\# P^\top)^{-1}YP^\#.$$

Then, we can see that the transfer function matrix $F[\imath\omega]$ is unitary, i.e., $F[\imath\omega]$ $F[\imath\omega]^* = I$, for all ω. This means that the output is a white noise process with flat power spectrum density, i.e., $\langle\eta[\imath\omega]^*\eta[\imath\omega]\rangle = \langle\xi[\imath\omega]^*\xi[\imath\omega]\rangle = I$. Thus, as expected, the output at steady state does not contain any information about the system state.

Lastly, let us examine the case where the target state is assigned to the canonical CV cluster sate (6.9). In this case, the nullifier dynamics (6.14) is of the form

$$\frac{d\mathcal{N}}{dt} = (Q^\top - e^{-2r}P^\# P^\top)\mathcal{N} - 2e^{-2r}P^\#\xi.$$

The real part of the eigenvalue of $Q^\top - e^{-2r}P^\# P^\top$ is calculated as $\Re\{\lambda\} = -e^{-2r}\|P^\top x\|$, where x is the corresponding eigenvector. Thus, if we make r bigger, or equivalently if we assign the final state closer to the ideal cluster state, then the eigenvalue moves toward the imaginary axis, and thus, the stability of the system becomes worse (i.e., the system is closed to a marginally stable system). We can make this interesting observation more concrete, by defining the convergence time to the target as $T = 1/\min\Re(\lambda)$ and the approximation error of the state to the ideal cluster state as $\epsilon = e^{-2r}$. It is then straightforward to obtain the relation $T\epsilon \geq c$ with c a constant. Therefore, there is a clear trade-off between the convergence time and the approximation error; in other words, to obtain a Gaussian state that is very close to an ideal cluster state by dissipation, we need a lot of time.

6.1.5 Remarks and Further Reading

As remarked at the end of Sect. 6.1.2, if there is no constraint imposed on the system, then there always exists a linear quantum system that uniquely generates a given target Gaussian state by dissipation. However, of course in practice several constraints, which mainly stem from the structure of the system, have to be satisfied. The discussion at the end of Sect. 6.1.3 is one such example. In the literature, there have been further progress in this direction; for instance, [13] provides a condition for dissipation-induced generation of a given pure Gaussian state, under the condition that the system–field coupling has to be exactly local.

An alternative parameterization of linear quantum systems generating a pure Gaussian state at steady state was obtained in [14], based on a parameterization of completely passive linear quantum systems and the group of real symplectic matrices. This result was motivated by the introduction of the notion of quasi-balanceable linear quantum systems in the context of model reduction of linear quantum systems by balanced truncation [15].

Lastly, we remark that the dissipation-induced Gaussian state considered here is a system's internal state, such as an intracavity state in optics and a motional state of a nano-mechanical oscillator. That is, it is not the state of an external bosonic field. Thus, in order to perform an optical quantum information processing with the generated Gaussian state, it must be extracted to the outside; for this purpose, the method developed in [16] can be used. Note that if a desired Gaussian cluster state is generated and extracted to the outside, then universal quantum computation [2] can in principle be carried out, by employing some non-Gaussian operation such as the cubic-phase gate or photon counting on the extracted Gaussian state. On the other hand, if the goal is precision measurement for quantum metrology, we do not need to extract the dissipation-induced internal Gaussian state to the outside. For instance, a spin-squeezed state of a large atomic ensemble can be directly used for ultra-precise magnetometry [17].

6.2 Enhancing Continuous-Variable EPR Entanglement

Recall the non-degenerate parametric amplifier (NOPA) introduced in Sect. 1.5.2. It is a device that is used to generate EPR entanglement (two-mode squeezed states) between two of its output beams η_1 and η_2 in Fig. 1.3. Here, we will show that two identical NOPAs can be interconnected in a coherent feedback configuration to enhance the two-mode squeezing while achieving this in a power-efficient manner. This is inspired by earlier theoretic [18–20] and experimental results [20, 21] that have shown that single-mode squeezing (recall Sect. 1.5.3) can be improved and/or spectrally shaped by employing static feedback (using static optical devices) [18, 19, 21] or dynamic coherent feedback [20]. The exposition below largely follows [22].

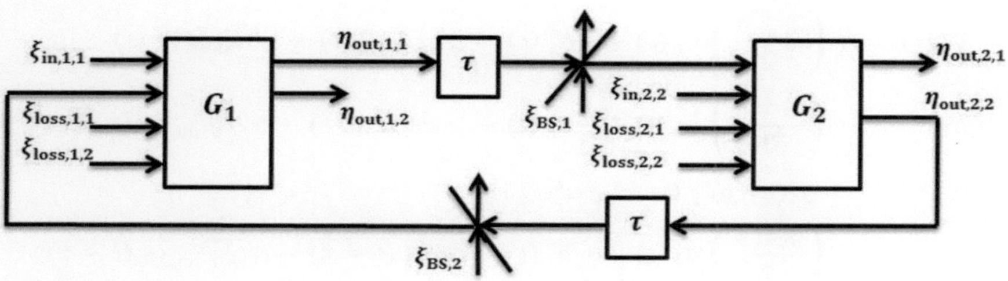

Fig. 6.2 Dual-NOPA (G_1 and G_2) coherent feedback network. The system contains transmission time delays τ and transmission losses represented by beam splitters with vacuum noise inputs $\xi_{\mathrm{BS},1}$ and $\xi_{\mathrm{BS},2}$. This figure does not show output fields associated with losses, which are not observable. Figure adapted from [22]. With permission of Springer

Consider the coherent feedback network shown in Fig. 6.2, consisting of two identical NOPAs labeled G_1 and G_2, Following Sect. 1.5.2, each NOPA G_j has two resonator modes a_j and b_j that are non-degenerate. The two modes are coupled by an interaction Hamiltonian $H_j = \frac{\iota}{2}\epsilon\left(a_j^* b_j^* - a_j b_j\right)$, where ϵ is the effective pump amplitude. Each NOPA has four input fields. The mode a_j is coupled to the incoming field $\xi_{\mathrm{in},j,1}$ and additional noise $\xi_{\mathrm{loss},j,1}$ via the coupling operators $L_{j,1} = \sqrt{\gamma}a_j$ and $L_{j,3} = \sqrt{\kappa}a_j$, respectively. In a similar fashion, the mode b_j is coupled to the input $\xi_{\mathrm{in},j,2}$ and noise $\xi_{\mathrm{loss},j,2}$ via the operators $L_{j,2} = \sqrt{\gamma}b_j$ and $L_{j,4} = \sqrt{\kappa}b_j$, respectively. The constant γ is the decay rate (coupling coefficient) of the mirror through which a mode interacts with its input, and κ is the coupling coefficient of a mode to the additional noise, assumed to be the same for a_j and b_j. The noises $\xi_{\mathrm{loss},j,1}$ and $\xi_{\mathrm{loss},j,2}$ represent channels through which photons created in modes a_j and b_j are lost during the interaction, respectively. The outputs corresponding to these two sources of loss are typically not observable (photons once lost are lost forever).

Due to the separation of G_1 and G_2, there are transmission losses and time delays present in the quantum network of Fig. 6.2. For each path, the transmission loss is effectively modeled by a beam splitter with (positive-valued) transmission rate α and reflection rate β satisfying $\alpha^2 + \beta^2 = 1$. The rates are determined by the distance traversed in a path [23]. The beam splitters have a port for additional vacuum noise inputs $\xi_{\mathrm{BS},1}$ and $\xi_{\mathrm{BS},2}$ (see Fig. 6.2) that represent channels which carry photons away as losses. For the NOPAs, a reference value of $\gamma_r = 7.2 \times 10^7$ Hz is used for the decay rate of the mirrors, following the values reported for the experiment in [21]. Since the pump amplitude ϵ and coupling coefficient γ are both adjustable quantities, we define $\epsilon = x\gamma_r$ Hz and $\gamma = \gamma_r/y$ Hz, where x and y are dimensionless parameters satisfying $0 < x, y \leq 1$ (in practice, y cannot be too small so that multiple resonance frequencies of the cavity are not excited). The dynamics of the dual-NOPA coherent feedback network is given by

$$
\begin{aligned}
\dot{a}_1(t) &= -\left(\frac{\gamma+\kappa}{2}\right) a_1(t) + \frac{\epsilon}{2} b_1^*(t) - \sqrt{\gamma}\xi_{\text{in},1,1}(t) - \sqrt{\kappa}\xi_{\text{loss},1,1}(t), \\
\dot{b}_1(t) &= -\left(\frac{\gamma+\kappa}{2}\right) b_1(t) + \frac{\epsilon}{2} a_1^*(t) - \alpha\gamma b_2(t-\tau) - \alpha\sqrt{\gamma}\xi_{\text{in},1,2}(t-\tau) \\
&\quad -\beta\sqrt{\gamma}\xi_{\text{BS},2}(t) - \sqrt{\kappa}\xi_{\text{loss},1,2}(t), \\
\dot{a}_2(t) &= -\left(\frac{\gamma+\kappa}{2}\right) a_2(t) + \frac{\epsilon}{2} b_2^*(t) - \alpha\gamma a_1(t-\tau) - \alpha\sqrt{\gamma}\xi_{\text{in},2,1}(t-\tau) \\
&\quad -\beta\sqrt{\gamma}\xi_{\text{BS},1}(t) - \sqrt{\kappa}\xi_{\text{loss},2,1}(t), \\
\dot{b}_2(t) &= -\left(\frac{\gamma+\kappa}{2}\right) b_2(t) + \frac{\epsilon}{2} a_2^*(t) - \sqrt{\gamma}\xi_{\text{in},2,2}(t) - \sqrt{\kappa}\xi_{\text{loss},2,2}(t),
\end{aligned} \tag{6.16}
$$

with outputs

$$
\begin{aligned}
\eta_{\text{out},1,2}(t) &= \sqrt{\gamma} b_1(t) + \alpha\sqrt{\gamma} b_2(t-\tau) + \alpha\xi_{\text{in},2,2}(t-\tau) + \beta\xi_{\text{BS},2}(t), \\
\eta_{\text{out},2,1}(t) &= \sqrt{\gamma} a_2(t) + \alpha\sqrt{\gamma} a_1(t-\tau) + \alpha\xi_{\text{in},1,1}(t-\tau) + \beta\xi_{\text{BS},1}(t).
\end{aligned} \tag{6.17}
$$

We are primarily interested in the entanglement between these two output Gaussian fields. In particular, we would like to investigate how transmission losses, amplification losses, and time delays impact the system's stability and the entanglement between the two output fields. Let Alice and Bob be two spatially separated interacting parties. The output field $\xi_{\text{out},1,2}$ can be on Alice's side, while the other output field $\xi_{\text{out},2,1}$ can be at Bob's location. Thus, Alice and Bob can share entanglement over a distance via the given quantum network. Define the quadratures

$$
\begin{aligned}
z &= (a_1^q, a_1^p, b_1^q, b_1^p, a_2^q, a_2^p, b_2^q, b_2^p)^\top, \\
\xi &= (\xi_{\text{in},1,1}^q, \xi_{\text{in},1,1}^p, \xi_{\text{in},2,2}^q, \xi_{\text{in},2,2}^p, \xi_{\text{loss},1,1}^q, \xi_{\text{loss},1,1}^p, \xi_{\text{loss},1,2}^q, \xi_{\text{loss},1,2}^p, \xi_{\text{loss},2,1}^q, \\
&\qquad \xi_{\text{loss},2,1}^p, \xi_{\text{loss},2,2}^q, \xi_{\text{loss},2,2}^p, \xi_{\text{BS},1}^q, \xi_{\text{BS},1}^p, \xi_{\text{BS},2}^q, \xi_{\text{BS},2}^p)^\top.
\end{aligned}
$$

Then, we have

$$
\begin{aligned}
\dot{z}(t) &= Az(t) + B\xi(t), \\
\eta_{\text{out},1,2}^q(t) + \eta_{\text{out},2,1}^q(t) &= C_1 z(t) + D_1\xi(t), \\
\eta_{\text{out},1,2}^p(t) - \eta_{\text{out},2,1}^p(t) &= C_2 z(t) + D_2\xi(t),
\end{aligned} \tag{6.18}
$$

where

$$A = \begin{bmatrix} -\frac{\gamma+\kappa}{2} & 0 & \frac{\epsilon}{2} & 0 & 0 & 0 & 0 & 0 \\ 0 & -\frac{\gamma+\kappa}{2} & 0 & -\frac{\epsilon}{2} & 0 & 0 & 0 & 0 \\ \frac{\epsilon}{2} & 0 & -\frac{\gamma+\kappa}{2} & 0 & 0 & 0 & -\alpha\gamma & 0 \\ 0 & -\frac{\epsilon}{2} & 0 & -\frac{\gamma+\kappa}{2} & 0 & 0 & 0 & -\alpha\gamma \\ -\alpha\gamma & 0 & 0 & 0 & -\frac{\gamma+\kappa}{2} & 0 & \frac{\epsilon}{2} & 0 \\ 0 & -\alpha\gamma & 0 & 0 & 0 & -\frac{\gamma+\kappa}{2} & 0 & -\frac{\epsilon}{2} \\ 0 & 0 & 0 & 0 & \frac{\epsilon}{2} & 0 & -\frac{\gamma+\kappa}{2} & 0 \\ 0 & 0 & 0 & 0 & 0 & -\frac{\epsilon}{2} & 0 & -\frac{\gamma+\kappa}{2} \end{bmatrix}, \tag{6.19}$$

$$B = \left[\begin{array}{cccc|c|cccc} \sqrt{\gamma} & 0 & 0 & 0 & & 0 & 0 & 0 & 0 \\ 0 & \sqrt{\gamma} & 0 & 0 & & 0 & 0 & 0 & 0 \\ 0 & 0 & \alpha\sqrt{\gamma} & 0 & & 0 & 0 & \beta\sqrt{\gamma} & 0 \\ 0 & 0 & 0 & \alpha\sqrt{\gamma} & & 0 & 0 & 0 & \beta\sqrt{\gamma} \\ \alpha\sqrt{\gamma} & 0 & 0 & 0 & \sqrt{\kappa}I_8 & \beta\sqrt{\gamma} & 0 & 0 & 0 \\ 0 & \alpha\sqrt{\gamma} & 0 & 0 & & 0 & \beta\sqrt{\gamma} & 0 & 0 \\ 0 & 0 & \sqrt{\gamma} & 0 & & 0 & 0 & 0 & 0 \\ 0 & 0 & 0 & \sqrt{\gamma} & & 0 & 0 & 0 & 0 \end{array}\right],$$

$$\begin{aligned} C_1 &= \sqrt{\gamma}\left[\alpha\ 0\ 1\ 0\ 1\ 0\ \alpha\ 0\right], \\ C_2 &= \sqrt{\gamma}\left[0\ -\alpha\ 0\ 1\ 0\ -1\ 0\ \alpha\right], \\ D_1 &= \left[\alpha\ 0\ \alpha\ 0\ 0_{1\times 8}\ \beta\ 0\ \beta\ 0\right], \\ D_2 &= \left[0\ -\alpha\ 0\ \alpha\ 0_{1\times 8}\ 0\ -\beta\ 0\ \beta\right]. \end{aligned}$$

Define $H_j[\imath\omega] = C_j\,(\imath\omega I - A)^{-1}\,B + D_j$ $(j = 1, 2)$. The two-mode squeezing spectra $V_\pm$ for the two output fields, introduced in Sect. 1.5.2, can be computed as in the case of a single NOPA using the formulas (1.33) and (1.34), with A, B, C_i, and D_i as defined above for $i = 1, 2$.

6.2.1 Stability Condition

The dual-NOPA coherent feedback system is required to be stable to operate properly. The next theorem provides a general stability condition for the system as a function of the parameters x, y, κ, and α.

Theorem 6.2 *The coherent feedback interconnection of two NOPAs is stable if and only if*

$$xy < -\alpha + \sqrt{\alpha^2 + (1 + y\frac{\kappa}{\gamma_r})^2} \tag{6.20}$$

with $0 < x, y \leq 1$.

Proof The determinant of the matrix A given in (6.19) is

$$\det(A-\lambda I)=\left\{\left[\left(\frac{\gamma+\kappa}{2}+\lambda\right)^2-\frac{\epsilon^2}{4}\right]^2-\frac{\alpha^2\epsilon^2\gamma^2}{4}\right\}^2.$$

Replacing ϵ, γ with $\epsilon = x\gamma_r$, $\gamma = \gamma_r/y$ and solving the inequality

$$\max(\Re\{\text{eig}(A)\}) = -\frac{\gamma+\kappa}{2}+\sqrt{\frac{\epsilon^2}{4}+\frac{\alpha\epsilon\gamma}{2}}<0$$

we obtain the theorem. □

Thus, for the same decay rate γ_r for the mirrors in a NOPA, in the ideal lossless scenario the dual-NOPA coherent feedback system has a lower stability threshold than a single NOPA, compare (6.20) with the condition $0 < xy < 1$ for a single NOPA.

6.2.2 The Ideal Lossless Case

In the lossless scenario with $\alpha = 1$ and $\kappa = 0$, from (1.33) and (1.34), we find

$$\begin{aligned}V_{\pm}(\imath\omega) = {} & \frac{2}{T(\imath\omega)^* T(\imath\omega)}(P(\imath\omega)^* P(\imath\omega)+Q(\imath\omega)^* Q(\imath\omega)\\ & +R(\imath\omega)^* R(\imath\omega)+S(\imath\omega)^* S(\imath\omega)),\end{aligned} \tag{6.21}$$

where

$$\begin{aligned}
P(\imath\omega) &= \alpha\,(2\imath\omega+\kappa)^4 - 2[\alpha\epsilon^2+\left(1+\alpha^2\right)\epsilon\gamma+\alpha\gamma^2]\,(2\imath\omega+\kappa)^2\\
&\quad -4\left(1-\alpha^2\right)\epsilon\gamma^2\,(2\imath\omega+\kappa)+\alpha\epsilon^4+2\alpha\epsilon^2\gamma^2+2\left(1+\alpha^2\right)\epsilon^3\gamma\\
&\quad -2\left(1+\alpha^2\right)\epsilon\gamma^3+\alpha\gamma^4,\\
Q(\imath\omega) &= 2\sqrt{\gamma\kappa}[-\alpha\,(2\imath\omega+\kappa)^3-(\alpha\gamma+\epsilon)\,(2\imath\omega+\kappa)^2\\
&\quad +\left(\alpha\epsilon^2+\alpha\gamma^2-2\left(1-\alpha^2\right)\epsilon\gamma\right)(2\imath\omega+\kappa)\\
&\quad +\epsilon^3+\alpha\gamma^3-\left(1+2\alpha^2\right)\epsilon\gamma^2+\alpha\epsilon^2\gamma],\\
R(\imath\omega) &= 2\sqrt{\gamma\kappa}[-(2\imath\omega+\kappa)^3-(3\gamma+\alpha\epsilon)\,(2\imath\omega+\kappa)^2+\left(\epsilon^2-3\gamma^2\right)(2\imath\omega+\kappa)\\
&\quad +\alpha\epsilon^3+\alpha\epsilon\gamma^2+\left(1+2\alpha^2\right)\epsilon^2\gamma-\gamma^3],\\
S(\imath\omega) &= \beta[(2\imath\omega+\kappa)^4+2\gamma\,(2\imath\omega+\kappa)^3-2\left(\epsilon^2+\alpha\epsilon\gamma\right)(2\imath\omega+\kappa)^2\\
&\quad -2\left(\gamma^3+\epsilon^2\gamma\right)(2\imath\omega+\kappa)+\epsilon^4+2\alpha\epsilon^3\gamma+2\alpha\epsilon\gamma^3-\gamma^4],\\
T(\imath\omega) &= \left[\epsilon^2-(2\imath\omega+\kappa+\gamma)^2\right]^2-4\alpha^2\epsilon^2\gamma^2.
\end{aligned}$$

Recall from Sect. 1.5.2 that we will use the two-mode squeezing spectra $V_{\pm}(0)$ at $\omega = 0$ as a figure of merit for entanglement. We have the following result.

Theorem 6.3 *In the absence of transmission and amplification losses, the two-mode squeezing spectra of the two outputs in the dual-NOPA coherent feedback system are*

$$V_{\pm}(0) = 2\left(\frac{1 - 2xy - x^2y^2}{1 + 2xy - x^2y^2}\right)^2. \tag{6.22}$$

For a fixed $0 < x \leq 1$, *the two-mode squeezing* $V_{\pm}(0)$ *at* $\omega = 0$ *decreases to 0 as* y *approaches the value* $\frac{\sqrt{2}-1}{x}$, *the threshold of stability, from below.*

Proof When $\alpha = 1$ and $\kappa = 0$, the condition of stability (6.20) becomes $xy < \sqrt{2}-1$. Based on (6.21), we can easily get (6.22), which is a monotonically decreasing function with respect to xy over the interval $(0, \sqrt{2} - 1)$, under which the stability condition of Theorem 6.2 is satisfied. It is straightforwardly verified that $V_{\pm}(0)$ goes to 0 as $y \uparrow \frac{\sqrt{2}-1}{x}$. □

Thus, the dual-NOPA coherent feedback displays a behavior that is analogous to that of a single NOPA operating below threshold, namely that the two-mode squeezing is maximized as the pump amplitude ϵ is increased to the threshold value of instability.

For the rest of this section, all two-mode squeezing spectra $V_{\pm}(\imath\omega)$ will be presented in log–log plots. The squeezing spectra are thus presented in dB unit, $V_{\pm}(\imath\omega)(\text{dB}) = 10\log_{10} V_{\pm}(\imath\omega)$ and $V_{+}(\imath\omega) + V_{-}(\imath\omega)(\text{dB}) = 10\log_{10}(V_{+}(\imath\omega) + V_{-}(\imath\omega))$. The more two-mode squeezing there is at a frequency ω, the more negative the plot of $V_{\pm}(\imath\omega)(\text{dB})$ at that frequency. Perfect two-mode squeezing at this frequency corresponds to $V_{\pm}(\imath\omega)(\text{dB}) = -\infty$. Also, since equations are expressed in a rotating frame at half of the common pump frequency ω_p of the NOPAs, 0 frequency corresponds to $\omega_p/2$, while the "low-frequency" region in these plots corresponds to positive frequencies around $\omega_p/2$.

Figure 6.3 shows the two-mode squeezing spectra $V_{\pm}(\imath\omega)$ (dB) and $V_{+}(\imath\omega) + V_{-}(\imath\omega)(\text{dB})$ versus the frequency ω for $x = 0.4$ and $y = 1$ and ignoring amplification and transmission losses (black solid lines). From (1.32), there is entanglement present when $V_{+}(\imath\omega) + V_{-}(\imath\omega)(\text{dB}) < 10\log_{10} 6\ \text{dB} = 6.0206$ dB. It can be seen that in this ideal lossless case, $V_{\pm}(\imath\omega)(\text{dB})$ is about -29 dB and $V_{+}(\imath\omega) + V_{-}(\imath\omega)(\text{dB})$ has a practically constant value of around -26 dB for frequencies ω roughly up to 10^5 rad/s.

6.2.3 Effect of Losses

The presence of transmission and amplification losses inevitably limits the two-mode squeezing that can be produced by the dual-NOPA network. Here, we will illustrate how the losses affect the two-mode squeezing while neglecting time delays in the network. The effect of time delays will be discussed later on.

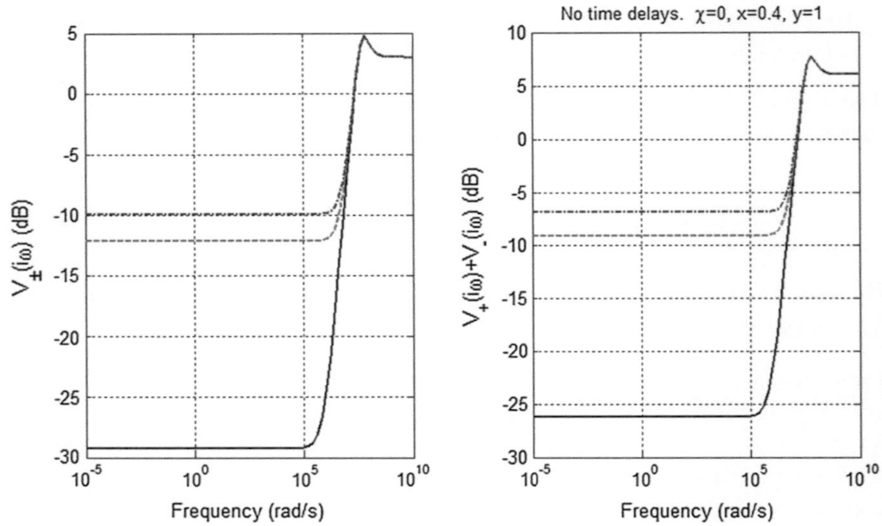

Fig. 6.3 Log–log plots of $V_{\pm}(\imath\omega)$ (*left*) and $V_+(\imath\omega) + V_-(\imath\omega)$ (*right*), when $\alpha = 1$ (*black solid line*), $\alpha = 0.97$ (*red dashed line*), $\alpha = 0.95$ (*blue dash-dot line*), $\kappa = 0$, $x = 0.4$, $y = 1$. Figure adapted from [22] (color figure online). With permission of Springer

Figure 6.4 shows the two-mode squeezing spectra when amplification losses are present, but there are no transmission losses and time delays. Based on [21], it is assumed that $\kappa \propto |\epsilon|$ with a reference value of $\kappa = \frac{3\times10^6}{\sqrt{2}}$ for $\epsilon = 0.6 \times \gamma_r$. Therefore, $\kappa = M\epsilon$, with $M = \frac{3\times10^6}{\sqrt{2}\times0.6\times\gamma_r}$. The figure shows how V and $V_{\pm}$ have higher values in the frequency range up to 10^5 rad/s compared to ideal case (black solid line), and this value steadily worsens as κ increases. Figure 6.5 illustrates the two-mode squeezing spectra $V_{\pm}(\imath\omega)$ when both transmission and amplification losses are present. It can be seen that entanglement becomes worse when both losses are present compared to when only one of them is present.

6.2.4 Comparison with Conventional Schemes

We will now compare the EPR entanglement generated at two spatially separated locations, say A (Alice) and B (Bob), by the dual-NOPA coherent feedback system to that of a single NOPA or two-cascaded NOPAs[1] placed midway between A and B, say, at C (Charlie), when there are no amplification losses in any of the NOPAs ($\kappa = 0$) and transmission delays are neglected (see Fig. 6.6a, b, and c1). The purpose is to assess how the presence of transmission losses as the sole imperfection in the

[1] See [24, 25] for experiments on the entanglement produced by cascading two or more NOPAs.

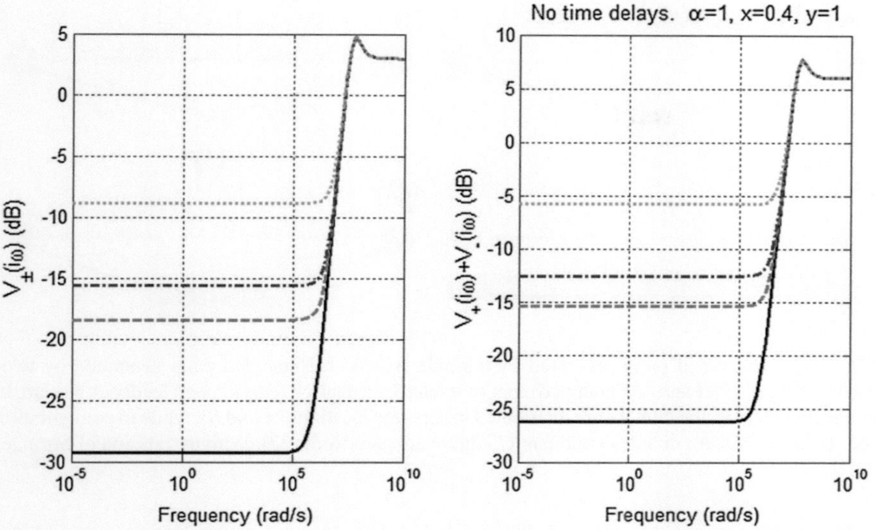

Fig. 6.4 Log–log plots of $V_{\pm}(\imath\omega)$ (*left*) and $V_{+}(\imath\omega)+V_{-}(\imath\omega)$ (*right*), when $\kappa = 0$ (*black solid line*), $\kappa = 0.1\left(\frac{3\times10^6}{0.6\times\sqrt{2}}\right)x$ (*red dashed line*), $\kappa = 0.2\left(\frac{3\times10^6}{0.6\times\sqrt{2}}\right)x$ (*blue dot-dash line*), $\kappa = \left(\frac{3\times10^6}{0.6\times\sqrt{2}}\right)x$ (*pink dotted line*), $\alpha = 1$, $x = 0.4$, $y = 1$. Figure adapted from [22] (color figure online). With permission of Springer

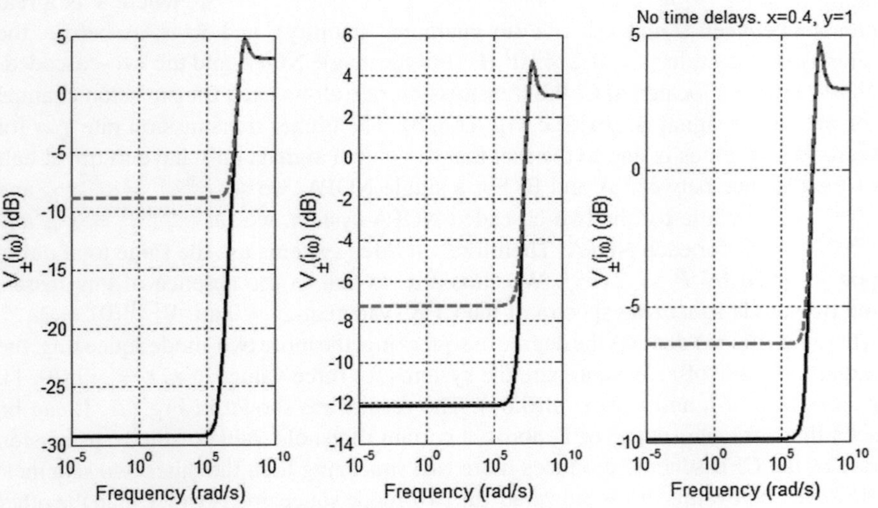

Fig. 6.5 Log–log plots of $V_{\pm}(\imath\omega)$ when $\alpha = 1$ (*left*), $\alpha = 0.97$ (*middle*), and $\alpha = 0.95$ (*right*), $\kappa = 0$ (*black solid line*), $\kappa = \left(\frac{3\times10^6}{0.6\times\sqrt{2}}\right)x$ (*red dashed line*), $x = 0.4$, $y = 1$. Figure adapted from [22] (color figure online). With permission of Springer

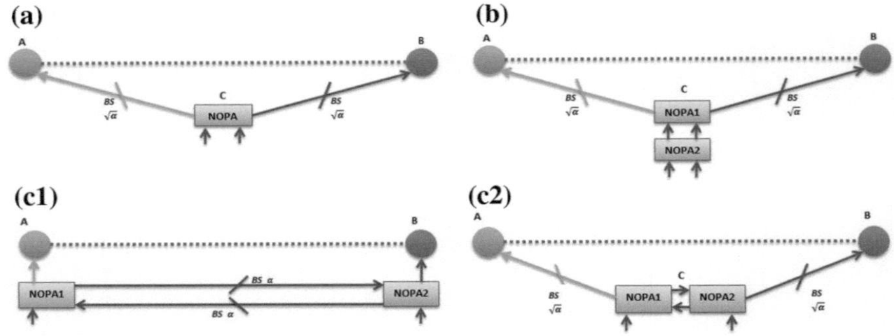

Fig. 6.6 **a** Entangled pairs generated by a single NOPA. **b** Entangled pairs generated by two-cascaded NOPAs. **c1** and **c2** Entangled pairs generated by a dual-NOPA coherent feedback system. In configuration (**c1**), the NOPAs are distributed in separate locations (*A* and *B*), while in configuration (**c2**) both NOPAs are in one location, at *C*. Figure adapted from [22]. With permission of Springer

system will impact the performance of these different schemes. For fairness of the comparison, it will be made for the situation where the total pump power consumed by all three systems is the same.

For the coherent feedback (CFB) system with the NOPAs placed at the endpoints (see Fig. 6.6c1), the transmission rate is α, and for each NOPA, we set $y = 1$, the pump power $\epsilon^{\text{CFB}} = x\gamma_r$, and decay rate of mirror $\gamma^{\text{CFB}} = \gamma_r$, where x is a real number between 0 and $\sqrt{2} - 1$ (to guarantee stability), and γ_r is, as before, the reference decay rate $\gamma_r = 7.2 \times 10^7$ Hz. For the single NOPA and the two-cascaded-NOPA systems located at C, the transmission rate along each transmission channel for the output signal is $\sqrt{\alpha}$ (see Fig. 6.6a, b). The higher transmission rate $\sqrt{\alpha}$ for these two schemes is due to the fact that the output signals only have to travel half of the distance between A and B. For a single NOPA, we set $\epsilon^{\text{single}} = \sqrt{2}x\gamma_r$ and $\gamma^{\text{single}} = \gamma_r$, while for the two-cascaded-NOPA system, we set $\epsilon^{\text{cascaded}} = x\gamma_r$ and $\gamma^{\text{cascaded}} = \gamma_r$ for each NOPA. Therefore, all three systems use the same total pump power, given by $P = 2x^2\gamma_r^2$. We choose x so that in the absence of any losses, the two-mode squeezing spectra of the CFB system at $\omega = 0$ are $V_{\pm}^{\text{CFB}}(0) = 2e^{-r}$, where $r \in [0, \infty)$ denotes the degree of squeezing; the more two-mode squeezing, the larger the value of r. We compare the systems for three values of r, $r = -\ln(0, 1)$, $r = -\ln(0.05)$, and $r = -\ln(0.01)$. The results are shown in Fig. 6.7. It can be seen that for each setting of r, above a certain threshold value of the transmission rate α, the CFB scheme produces more two squeezing than the other two schemes. However, below this threshold value, the two-mode squeezing is worse than the other two schemes. We conclude that when all three schemes consume the same amount of total pump power, the CFB scheme is advantageous over the two other schemes when transmission losses are not too large.

Rather than placing the two NOPAs at spatially separated locations A and B in a decentralized configuration, it is also possible to have both NOPAs located at C, as in configurations (a) and (b). This yields the centralized dual-NOPA configuration (c2)

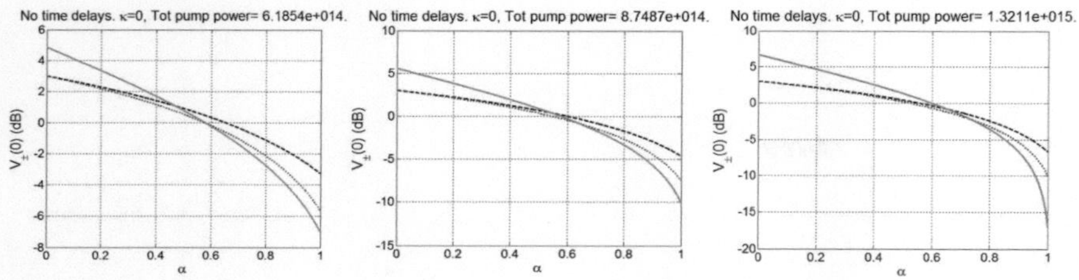

Fig. 6.7 Plots of $V_{\pm}(0)$ (dB) as a function of α, when all the systems consume the same total pump power ($P = 2x^2\gamma_r^2$) and $\kappa = 0$, with respect to a single NOPA (*black dashed line*), two-cascaded-NOPA system (*blue dotted line*), and a dual-NOPA coherent feedback system in configuration (c1) (*red solid line*). We set values of x so that $V^{\text{CFB}}(0) = 2e^{-r}$ with $r = -\ln(0.1)$ (*left*), $r = -\ln(0.05)$ (*middle*), and $r = -\ln(0.01)$ (*right*), when $\alpha = 1$ (color figure online)

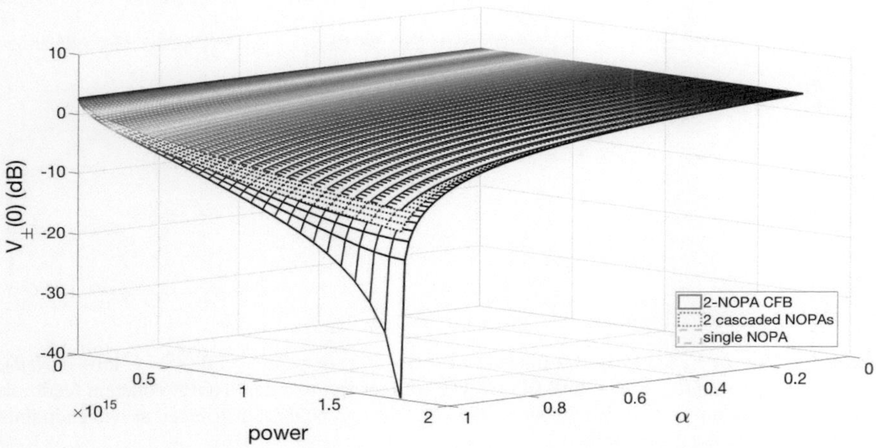

Fig. 6.8 Plots of $V_{\pm}(0)$ (dB) as a function of α and power, $P^{\text{CFB}} = P^{\text{cascaded}} = P^{\text{single}}$, when $\kappa = 0$, with respect to a single NOPA (*dashed line*), two-cascaded-NOPA system (*dotted line*), and a dual-NOPA coherent feedback system in configuration (c2) (*solid line*)

in Fig. 6.6. In all of the centralized configurations, one output beam will be sent from C to A and the other from C to B; thus, both beams travel a distance that is half of that traveled in configuration (c1), with a transmission rate of $\sqrt{\alpha}$. Figure 6.8 compares the entanglement generated from the three configurations (a), (b), and (c2), for the same values of α and total pump power. The figure shows that the coherent feedback network in the centralized configuration (c2) generates more two-mode squeezing than the other two configurations for all $0 \leq \alpha \leq 1$ and as the parameter x ranges in value from 0.01 to 0.414 (thus within the stability threshold).

Finally, we compare the dual-NOPA coherent feedback system in the centralized configuration (c1) and decentralized configuration (c2), when both configurations consume the same amount of pump power and ignoring amplification losses. Figure 6.9 shows the entanglement generated by the two configurations for various

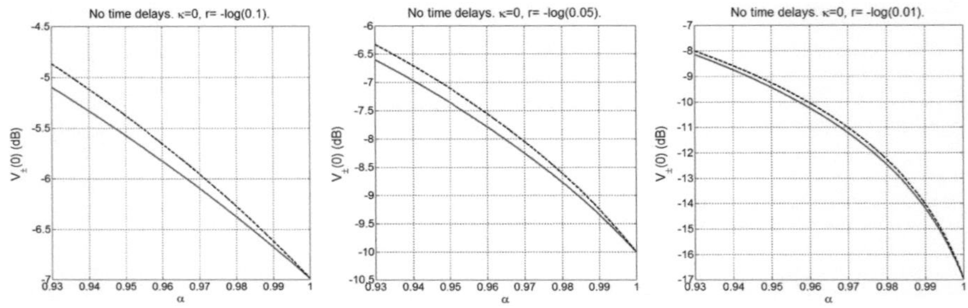

Fig. 6.9 Plots of $V_{\pm}(0)$ (dB) as a function of α, for $0.93 \leq \alpha \leq 1$ and $\kappa = 0, r = -\ln(0.1)$ (*left*), $r = -\ln(0.05)$ (*middle*), $r = -\ln(0.01)$ (*right*), with respect to a dual-NOPA coherent feedback system placed in the *middle* (*black dashed line*) and the two NOPAs distributed at two endpoints (*red solid line*). Reprinted from [22] (color figure online). With permission of Springer

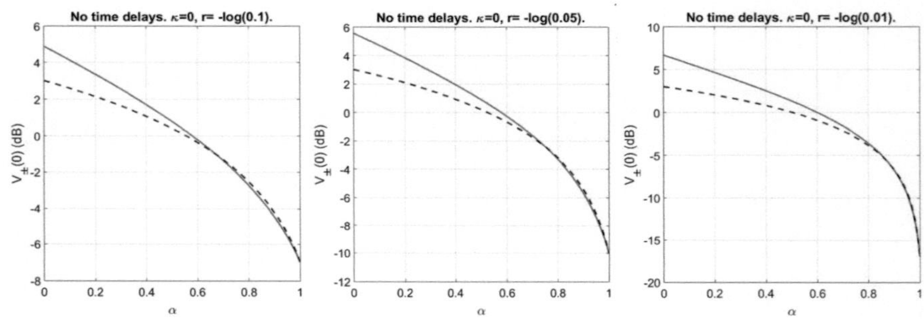

Fig. 6.10 Plots of $V_{\pm}(0)$ (dB) as a function of α, for $0 \leq \alpha \leq 1$, $\kappa = 0$, $r = -\ln(0.1)$ (*left*), $r = -\ln(0.05)$ (*middle*), $r = -\ln(0.01)$ (*right*), with respect to a dual-NOPA coherent feedback system placed in the *middle* (*black dashed line*) and the two NOPAs distributed at two endpoints (*red solid line*) (color figure online)

values of transmission rate α. It can be seen that configuration (c1) has a noticeably higher degree of EPR entanglement compared to configuration (c2) for higher transmission rates α, but this diminishes as r is increased. Figure 6.10 extends Fig. 6.9 to compare the two systems over the whole range of $0 \leq \alpha \leq 1$. It can be seen that the centralized configuration (c2) produces better entanglement for lower α.

To summarize, when the same total pump power is consumed, and amplification losses and time delays are neglected, the centralized coherent feedback configuration (c2) is superior to the single NOPA and two-cascaded-NOPA system for any value of the transmission rate α, while the distributed coherent feedback configuration (c1) beats the centralized configuration (c2) for higher transmission rates α. Thus, for lower transmission losses, one should employ a distributed coherent feedback network for entanglement generation; for higher transmission losses, one should switch from the distributed to a centralized coherent feedback scheme.

The results in this section indicate how feedback can, in certain situations, provide a performance advantage for entanglement generation, even when there are

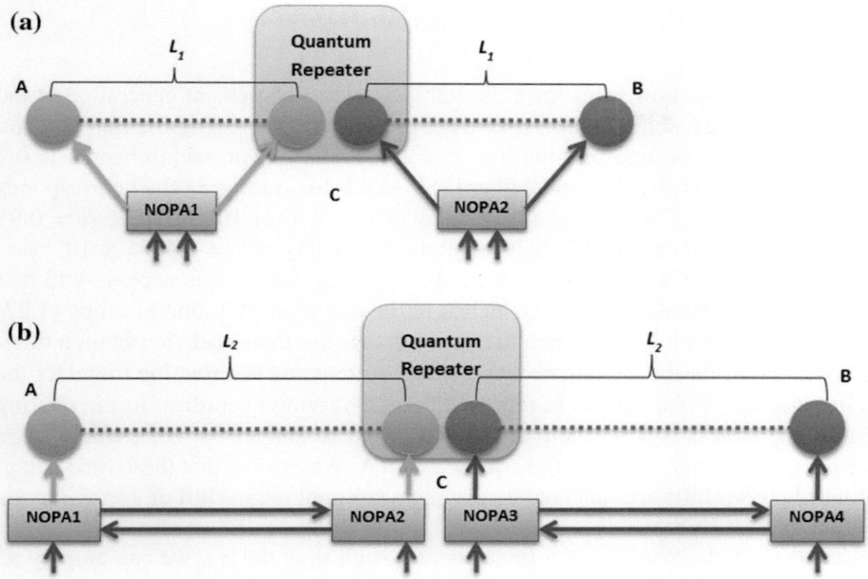

Fig. 6.11 **a** Quantum repeater structure with two single NOPAs. There are two endpoints (*A* and *B*) and a repeater station that has two nodes (*middle C*). Each node contains a quantum memory that can store a quantum state. NOPA1 and NOPA2 create two pairs of continuous-mode entangled photons between endpoints and repeater nodes (*blue pair* and *green pair*). The two outer quantum memories can be entangled via entanglement swapping. **b** Quantum repeater structure with two dual-NOPA systems. NOPA1 and NOPA2 and NOPA3 and NOPA4 are connected to get entangled pairs between two points *A* and *C* (*green nodes*) and *C* and *B* (*blue nodes*), respectively. A quantum repeater in the *middle* (*C*) is used to generate entanglement between *A* and *B*. Figure adapted from [22] (color figure online). With permission of Springer

losses present in the transmission channels. A potential application of the dual-NOPA scheme is to extend the distance for entanglement distribution in a power-efficient manner. As suggested in Fig. 6.11b, for the same given amount of total pump power, a dual-NOPA system including NOPA1 and NOPA2 could potentially transport the same amount of entanglement over a distance L2 that is longer than the distance L1 achievable by a single NOPA1 in Fig. 6.11a. Given the same distance between the two endpoints A and B in this figure, the coherent feedback scheme could thus potentially reduce the number of quantum repeaters required compared to the system consisting of a single NOPA between every two quantum repeaters.

6.2.5 Effect of Nonzero Transmission Delays

In this section, we now investigate the stability and entanglement generation of the coherent feedback configuration (c1) in the presence of non-negligible transmission time delays. We assume that the loss rate across the transmission channel is 0.2 dB/km, which is typical in optical fiber [26]. At this loss rate, $\alpha = 0.97$ corresponds to a distance of 1.3230 km and a time delay of $\tau = 4.41 \times 10^{-6}$ s, while $\alpha = 0.95$ corresponds to a distance of 2.2276 km and a time delay of $\tau = 7.4255 \times 10^{-6}$ s.

We note that to model interconnections with time delays, it is necessary to first return to the original model of the NOPA in the non-rotating frame given by (1.27) and (1.28). Interconnections including time delays are then made to obtain a time-varying QSDE for the interconnected model, before going to a rotating frame for the entire interconnected system to remove the time-varying quantities. In the rotating frame, the time delays will introduce addition phase shifts of $e^{\imath\omega_p\tau/2}$, where ω_p is the pump frequency, at the outputs of each NOPA. We assume that these time delay-induced phase shifts are canceled by placing a constant phase shift of $e^{-\imath\omega_p\tau/2}$ at the output of each NOPA.

To analyze the stability of a linear system with time delays, we can employ an eigenvalue-based approach as detailed in [27]. This approach has been implemented in the freely available DDE-BIFTOOL toolbox for MATLAB [28, 29], which is used to compute the eigenvalues of the coherent feedback network with time delays. The calculation by DDE-BIFTOOL shows that all roots of the characteristic equation for the coherent feedback system with time delays have negative real parts when $\alpha = 0.97, 0.95$, $x = 0.4$, $y = 1$, and $\kappa = \frac{3\times10^6}{0.6\sqrt{2}}x$. Therefore, for these parameter values, the dual-NOPA coherent feedback configuration (c1) is stable in the presence of the time delays.

The two-mode squeezing spectra $V_\pm(\imath\omega)$ in the presence of time delays is again given by the identities (1.33) and (1.34). Since the quantum Langevin equations of the network are linear, the non-rational transfer functions (due to the time delays) $H_1[s]$ and $H_2[s]$ appearing in (1.33) and (1.34) can be numerically computed in MATLAB, from which $V_\pm(\imath\omega)$ can then be computed. Figure 6.12 shows the EPR entanglement generated in the presence of time delays for a specific setting of the parameters of the dual-NOPA coherent feedback system in configuration (c1). It can be seen that the time delays have caused the frequency range (bandwidth) of entanglement to be reduced to less than 10^5 rad/s. Moreover, the degree of entanglement can also be observed to decrease, with a larger decrease for longer time delays. However, the latter effect is not due to the delays per se but is a consequence of the losses experienced by signals as they propagate through the transmission channels. Longer delays are associated with longer distances and hence with larger losses and a larger drop in the entanglement. Finally, in Fig. 6.12, we can also see erratic looking sharp peaks and dips in the squeezing spectra. We note that this is not a feature which is peculiar to the dual-NOPA scheme but is common in the frequency response of systems with internal time delays (see, e.g., [30, p. 182]).

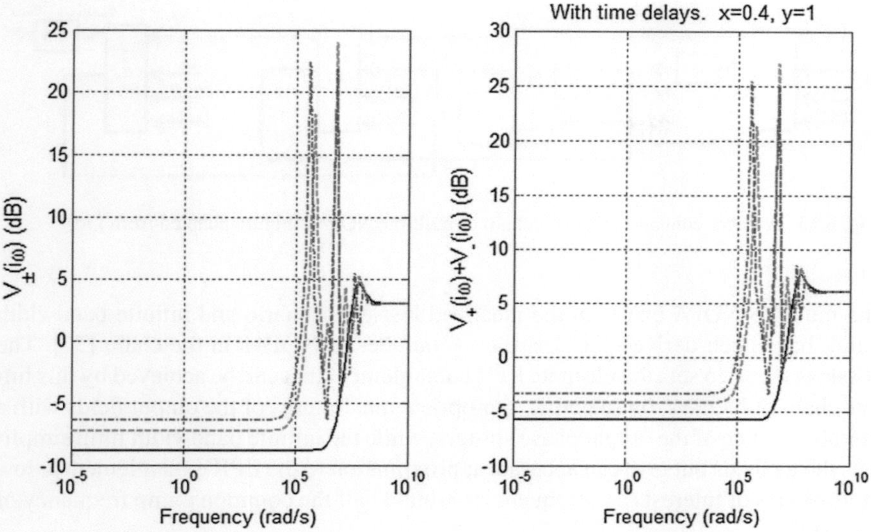

Fig. 6.12 Log–log plots of $V_{\pm}(\imath\omega)$ (*left*) and $V_{+}(\imath\omega) + V_{-}(\imath\omega)$ (*right*), when $\tau = 0$ (*black solid line*), $\tau = 4.41 \times 10^{-6}$ (*red dashed line*), $\tau = 7.4255 \times 10^{-6}$ (*blue dash-dot line*), $x = 0.4$, $y = 1$, $\kappa = \frac{3\times10^6}{0.6\sqrt{2}}x$. Decrease in entanglement (increase in dB value of $V_{\pm}(0)$) for higher delays is not due to the delays per se (see the text for details). Figure adapted from [22] (color figure online). With permission of Springer

6.2.5.1 Further Reading

The properties of the dual-NOPA coherent feedback scheme have been further analyzed in [31–33]. This includes studying the effect of phase shifts on the transmission channels on the performance of the scheme [31] and establishing some local optimal properties [33].

Related experiments have been reported that demonstrate how coherent feedback can enhance EPR entanglement in continuous-variable systems [34, 35]. However, these schemes consider the use of a static coherent feedback as in [18, 19, 21] to enhance EPR entanglement rather than interconnecting two NOPAs together in a coherent feedback network to improve entanglement.

The dual-NOPA scheme can be extended to a multiple-NOPA scheme involving multiple NOPAs $G_1, G_2, \ldots, G_N$, connected in linear coherent feedback chain equipped with adjustable phase shifters at two of its outputs, $\eta_{\text{out},1}$ and $\eta_{\text{out},2}$, as depicted in Fig. 6.13 and analyzed in [36, 37]. Similar conclusions to the dual-NOPA scheme emerge in the multiple-NOPA scenario, in that the same degree of two-mode squeezing between the outputs $\eta_{\text{out},1}$ and $\eta_{\text{out},2}$ can be achieved with much lower pump power and that the scheme produces a stronger degree of two-mode squeezing with more NOPAs in the chain for the same amount of total pump power and when only transmission losses are present [36]. Explicit formulae for EPR entanglement in

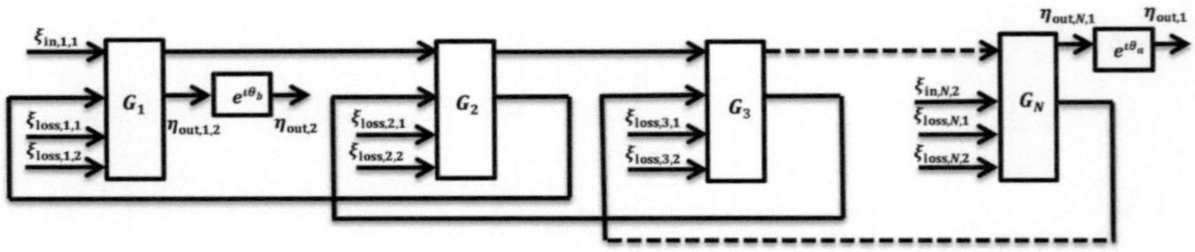

Fig. 6.13 A linear coherent feedback chain of multiple NOPAs. Figure adapted from [36]

the multiple-NOPA chain, in the idealized lossless scenario and infinite bandwidth limit, have been derived for an arbitrary number of NOPAs in the chain [37]. The lossless scenario sets the ultimate EPR entanglement that can be achieved by this linear chain of NOPAs, by selecting appropriate quadratures of the output fields with a suitable choice of the output phase shifters, while the infinite bandwidth limit simplifies the analysis but gives an accurate approximation to the EPR entanglement at low frequencies of interest (i.e., frequencies around half the common pump frequency of the NOPAs).

6.3 Force Sensing and Back-Action Evasion

In this section, we move to another very important research field in quantum information science, i.e., *quantum metrology*. The goal of this subject is, in a broad sense, to detect a very weak (quantum-level) signal such as a gravitational wave force [38–41] and an ultra-small deviation of frequency in an atomic clock [42], with the full use of quantum mechanical properties such as squeezing and entanglement. In particular, as mentioned in Sect. 1.5.4, the linear force sensing problem, which is a special subclass of quantum metrology, can be formulated within the framework of systems and control theory. More specifically, a linear open quantum system coupled to a probe field functions as a sensor to detect the signal, where the information about the signal can be extracted by measuring the probe output field. However, as immediately expected, this probe field unavoidably introduces a noise in the input port and eventually degrades the signal-to-noise ratio. In particular, the so-called *back-action (BA)* noise places the *standard quantum limit (SQL)* on the detection sensitivity. Fortunately, as has been proven in a number of papers, the SQL can be beaten by some *back-action evasion (BAE)* techniques, which achieve better detection sensitivity over SQL. The aim of this section is to describe a general force sensing problem within the framework of linear systems and control theory [43, 44] and demonstrate a coherent feedback-controlled system that achieves BAE.

6.3.1 Back-Action Evasion and the Standard Quantum Limit

First, let us recall the opto-mechanical oscillator described in Sect. 1.5.4, to see the meaning of SQL and why BAE is necessary. The simplified system dynamics, which is obtained by adiabatically eliminating the cavity mode, is given by (1.40); again, f is the force we want to detect, m is the mass, and ω_m is the resonance frequency of the oscillator. From this dynamics, we find that the measurement output $\eta^p = \sqrt{\lambda} q_1 - \xi^p$ brings some information about f. However, notably, η^p contains both of the probe input noise ξ^q and ξ^p; the unavoidable noise ξ^p is called the *shot noise*, while ξ^q implicitly appears in the measurement output η^p through the dynamics of q_1 and this is the BA noise. This fact can be explicitly seen in the Laplace domain:

$$\eta^p[s] = \frac{\sqrt{\lambda}}{m(s^2+\omega_m^2)}(\sqrt{\lambda}\xi^q[s] + \sqrt{2}f[s]) + \xi^p[s],$$

showing that η^p is indeed affected by both the noise quadratures. Now, instead of the force magnitude f, we focus on the change of oscillator's position induced by f; let us denote this change by g, and then in the Fourier domain $s = \imath\omega$, this is related to f as $f[\imath\omega] = -mL\omega^2 g[\imath\omega]$, where L is the optical path length of the cavity. Hence, under the assumption $\omega \gg \omega_m$, the normalized signal containing g is given by

$$\tilde{\eta}^p[\imath\omega] = \frac{\eta^p[\imath\omega]}{\sqrt{2\lambda}L} = g[\imath\omega] - \frac{\sqrt{\lambda}}{\sqrt{2}mL\omega^2}\xi^q[\imath\omega] + \frac{1}{\sqrt{2\lambda}L}\xi^p[\imath\omega].$$

This shows that the BA noise ξ^q becomes dominant in the low-frequency region $\omega \approx 0$, while the shot noise ξ^p sets the constant noise floor; also, we see that for precise detection, both the mass m and the optical path length L should be large, which is the reason why a gravitational interferometer of a kilometer-scale size was constructed in order to detect a gravitational wave [39, 45]. Now, the noise power of $\tilde{\eta}^p$ is bounded from below by the following SQL:

$$\begin{aligned} S(\imath\omega) &= \langle|\tilde{\eta}^p[\imath\omega] - g[\imath\omega]|^2\rangle = \frac{\lambda}{2m^2L^2\omega^4}\langle|\xi^q[\imath\omega]|^2\rangle + \frac{1}{2\lambda L^2}\langle|\xi^p[\imath\omega]|^2\rangle \\ &\geq 2\sqrt{\frac{\langle|\xi^q[\imath\omega]|^2\rangle\langle|\xi^p[\imath\omega]|^2\rangle}{4m^2L^4\omega^4}} \geq \frac{1}{mL^2\omega^2} = S_{\rm SQL}(\imath\omega). \end{aligned} \tag{6.23}$$

The last inequality comes from $\langle|\xi^q[\imath\omega]|^2\rangle\langle|\xi^p[\imath\omega]|^2\rangle \geq 1$, i.e., the Heisenberg uncertainty relation. (Note that rigorously, this should be defined in terms of the power spectral density.) From the above relation, it is clear that SQL appears because the measurement output η^p contains the BA noise ξ^q in addition to the unavoidable shot noise ξ^p. Thus, for high-precision detection of g, a special system configuration should be devised so that the BA noise ξ^q is evaded and η^p is free from ξ^q; this is the meaning of BAE. In fact, if BAE is achieved, then by injecting a ξ^p-squeezed probe light field, we can possibly reduce the noise power below the SQL and may have chance to detect g with better accuracy.

6.3.2 System Theoretical Characterization of BAE

Here, we consider a general linear quantum system given by (2.5):

$$dx(t) = Ax(t)dt + Bdw(t), \quad dy(t) = Cx(t)dt + Ddw(t), \tag{6.24}$$

and describe the notion of BAE in this general setting [43]. First, let us assume that the signal to be detected has some correlation with the measurement output $\tilde{y}$, which is a part of the quadrature components:

$$d\tilde{y} = M_1 dy = M_1 Cxdt + M_1 Ddw = M_1 Cxdt + d\tilde{w}. \tag{6.25}$$

Again, the signal is already incorporated in $\tilde{y}$. Here, M_1 is a $m \times 2m$ projection matrix representing which quadrature is to be measured, satisfying $M_1 \mathbb{J}_m M_1^\top = 0$ and $M_1 M_1^\top = I_m$; these conditions ensure that all the elements of $\tilde{y}(t)$ are classical signals commuting with one another as well as with those of $\tilde{y}(t')$ for all times t, t', i.e.,

$$[\tilde{y}_i(t), \tilde{y}_j(t')] = 0, \ \forall i, j, \ \forall t, t'.$$

Note that $\tilde{w} = M_1 Dw$ is a generalized version of shot noise, which must be present in the measurement output.

Now, the BA noise defined by $\tilde{w}' = M_2 Dw$, i.e., observables conjugate to $\tilde{w}$. Here, M_2 is also a $m \times 2m$ projection matrix and is chosen so that the matrix $[M_1^\top \ M_2^\top]$ is a symplectic and orthogonal matrix, which as a result leads to

$$\begin{aligned} &M_2 \mathbb{J}_m M_2^\top = 0, \ \ M_2 M_2^\top = I_m, \ \ M_1 \mathbb{J}_m M_2^\top = I_m, \\ &M_1 M_2^\top = 0, \ \ M_1^\top M_1 + M_2^\top M_2 = I_{2m}. \end{aligned} \tag{6.26}$$

Actually, these conditions ensure the CCR $[d\tilde{w}(t), \ d\tilde{w}'^\top(t')] = \imath \min(t, t')dt$. In particular, the last equality $M_1^\top M_1 + M_2^\top M_2 = I_{2m}$ leads to $Dw = M_1^\top \tilde{w} + M_2^\top \tilde{w}'$. Hence, the dynamical equation in (6.24) can be rewritten as

$$dx = Axdt + BD^{-1}(M_1^\top d\tilde{w} + M_2^\top d\tilde{w}'). \tag{6.27}$$

We now arrive at the situation of writing the BAE condition in this general setup; for this, we need that the output $\tilde{y}$ in (6.25) does not contain the BA noise $\tilde{w}'$. That is, BAE is achieved if and only if the transfer function from $\tilde{w}'$ to $\tilde{y}$ is always zero:

$$\textbf{BAE:} \ \ \Xi_{\tilde{w}' \to \tilde{y}}[s] = M_1 C(sI - A)^{-1} BD^{-1} M_2^\top = 0, \ \ \forall s. \tag{6.28}$$

Moreover, as is well known in systems and control theory, this condition means that there is no subsystem that is controllable w.r.t. $\tilde{w}'$ and observable with respect to $\tilde{y}$. This is equivalently represented by the condition

$$M_1 C A^k B D^{-1} M_2^\top = 0, \quad \forall k \geq 0. \tag{6.29}$$

Under this condition, the system Eqs. (6.25) and (6.27) are represented in a transformed coordinate by

$$d\begin{bmatrix} x_1' \\ x_2' \end{bmatrix} = \begin{bmatrix} A_{11} & 0 \\ A_{21} & A_{22} \end{bmatrix}\begin{bmatrix} x_1' \\ x_2' \end{bmatrix} dt + \begin{bmatrix} B_{11} \\ B_{21} \end{bmatrix} d\tilde{w} + \begin{bmatrix} 0 \\ B_{22} \end{bmatrix} d\tilde{w}',$$
$$d\tilde{y} = [\, C_1 \; 0\,]\begin{bmatrix} x_1' \\ x_2' \end{bmatrix} dt + d\tilde{w}.$$

This implies that the input signal $\tilde{w}'$ has no effect on the output $\tilde{y}$. Finally, note that even if the above BAE condition (6.28) or (6.29) is satisfied, this by itself does not necessarily mean that the signal sensitivity is improved; in particular, squeezing operation on the input field $\tilde{w}$, in addition to the BAE property, is necessary to realize such an improvement.

6.3.3 Coherent Feedback for BAE

In this subsection, based on the BAE theory formulated in the previous subsection, we demonstrate a coherent feedback (CFB) control approach for BAE of a mechanical oscillator driven by an unknown force.

First, we address a general framework for CFB configuration for the linear quantum system

$$dx(t) = Ax(t)dt + \mathbb{J}_n C^\top \mathbb{J}_m dw(t), \quad dy(t) = Cx(t)dt + dw(t), \tag{6.30}$$

which is a special case of (2.5) with $D = I$ (see Theorem 2.1). For CFB, at most all the components of the output field y can be used for feedback, and also, at most all the components of the input probe field w can be controlled. Here, we take the CFB control configuration depicted in Fig. 6.14, where the controller dynamics is given by

$$dx_K = A_K x_K dt + \mathbb{J}_{n'} C_1^\top \mathbb{J}_m dw_1 + \mathbb{J}_{n'} C_2^\top \mathbb{J}_m dw_2,$$
$$dy_1 = C_1 x_K dt + dw_1, \quad dy_2 = C_2 x_K dt + dw_2, \tag{6.31}$$

where n' is the number of degrees of freedom of the controller. Note that the controller has two input–output ports, to avoid the self-interaction of optical fields. Now, y is connected to the controller's input w_2, and the controller's output y_1 is connected to w,

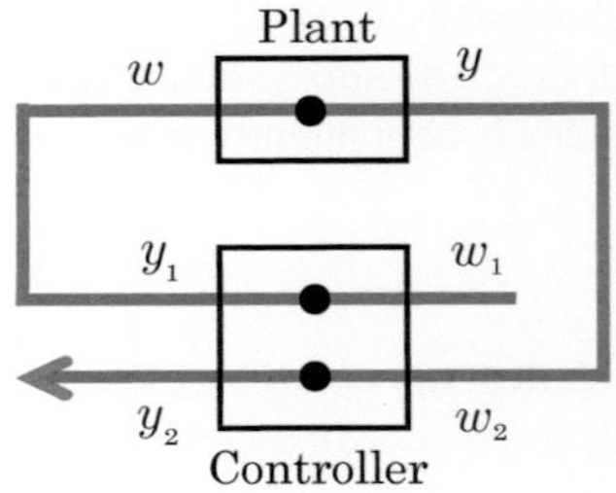

Fig. 6.14 A general configuration of the CFB control. Figure adapted from [43]

without involving any measurement process. That is, the CFB control is constructed by setting

$$w_2 = y, \quad w = y_1. \tag{6.32}$$

From this condition, the size of C_1 and C_2 has to be equal, although they are not necessarily of full rank. Combining (6.30), (6.31), and (6.32), we find that the dynamical equation of the closed-loop system is given by:

$$dx_e = A_e x_e dt + \mathbb{J}_{n+n'} C_e^\top \mathbb{J}_m dw_1, \ \ dy_2 = C_e x_e dt + dw_1, \tag{6.33}$$

where $x_e = [x^\top, x_K^\top]^\top$, $A_e = \mathbb{J}_{n+n'}(G_e + C_e^\top \mathbb{J}_m C_e/2)$, $C_e = [C, C_1 + C_2]$, and

$$G_e = \begin{bmatrix} G & C^\top \mathbb{J}_m C_1/2 - C^\top \mathbb{J}_m C_2/2 \\ \star & G_K + C_1^\top \mathbb{J}_m^\top C_2/2 + C_2^\top \mathbb{J}_m C_1/2 \end{bmatrix},$$

where $\star$ denotes the symmetric elements of G_e. Here, let us assume that a CFB controller satisfying $C_1 + C_2 = 0$ is implementable. Then, the closed-loop system (6.33) takes the following form:

$$\begin{aligned} dx_e &= \begin{bmatrix} A & \mathbb{J}_n C^\top \mathbb{J}_m C_1 \\ \mathbb{J}_{n'} C_1^\top \mathbb{J}_m^\top C & \mathbb{J}_{n'} G_K \end{bmatrix} x_e dt + \begin{bmatrix} \mathbb{J}_n C^\top \mathbb{J}_m \\ 0 \end{bmatrix} dw_1, \\ dy_2 &= [C \ \ 0] x_e dt + dw_1. \end{aligned} \tag{6.34}$$

It seems from the structure of this equation that the controller directly couples to the plant, but not to the external field. Hence, this type of plant–controller coupling is called a *direct interaction*. In this sense, (6.34) implies that direct interaction is realized within the CFB configuration.

Now, let us study the opto-mechanical oscillator from Sect. 6.3.1 as the plant of interest. To achieve highly precise detection of an unknown force f driving this system, we aim to apply the above-described CFB control configuration so that the closed-loop system (6.34) satisfies the BAE condition. The controller is a single-mode linear quantum system with variable $x_K = (q_3, p_3)^\top$, with two input fields $(\xi_1^q, \xi_1^p)^\top = dw_1/dt$ and $(\xi_2^q, \xi_2^p)^\top = dw_2/dt$. The controller's system matrices are assumed to satisfy

$$\mathbb{J}C_1^\top \mathbb{J}^\top C = \left[\begin{array}{cc|cc} & 0 & & 0 \\ 0 & & g & \end{array}\right], \quad \mathbb{J}G_K = \begin{bmatrix} & -\omega_m \\ \omega_m & \end{bmatrix},$$

which leads to

$$C_1 = -C_2 = \frac{g}{\sqrt{2\gamma}} \begin{bmatrix} 0 & 0 \\ 1 & 0 \end{bmatrix}, \quad G_K = \begin{bmatrix} -\omega_m & 0 \\ 0 & -\omega_m \end{bmatrix}. \tag{6.35}$$

We will discuss later about how to physically implement these system matrices. In this setting, the closed-loop system dynamics (6.34) is given by

$$dx_e = A_e x_e dt + B_e dw_1 + bf dt, \quad dy_2 = C_e x_e dt + dw_1, \tag{6.36}$$

where

$$A_e = \left[\begin{array}{cc|cc|cc} & 1/m & & 0 & & \\ -m\omega_m^2 & & \kappa & & & \\ \hline & 0 & -\gamma & & & 0 \\ \kappa & & & -\gamma & g & \\ \hline & & & 0 & & -\omega_m \\ & & g & & \omega_m & \end{array}\right], \quad B_e = C_e^\top,$$

$$C_e = \sqrt{2\gamma} \left[\begin{array}{cc|cc|cc} 0 & & 1 & & 0 & \\ & 0 & & 1 & & 0 \end{array}\right], \quad b = [0\ 1\ 0\ 0\ 0\ 0]^\top.$$

The output field is $(\eta_2^q, \eta_2^p)^\top = dy_2/dt$; clearly, η_2^q does not contain any information about f, and thus, we need to measure η_2^p, implying that the output signal is given by

$$\eta_2^p = cx_e + \xi_1^p = \sqrt{2\gamma}[0\ 0\ 0\ 1\ 0\ 0]x_e + \xi_1^p. \tag{6.37}$$

Here, we apply the BAE condition (6.29) and find $g = \kappa/\sqrt{m\omega_m}$. In fact, with this choice, the transfer function from the BA noise ξ_1^q to the output η_2^p is zero, so we have

$$\eta_2^p[s] = \frac{\sqrt{2\gamma}\kappa/m}{(s+\gamma)(s^2+\omega_m^2)} \xi_1^p[s] + \frac{s-\gamma}{s+\gamma} f[s].$$

Thus, by injecting a ξ_1^p-squeezed light field, which reduces the variance of ξ_1^p, we can perform a better detection of f below SQL.

Lastly, let us discuss how to physically implement the above CFB controller. The point is that the structure of C_1 (or C_2) in (6.35) represents a QND-type interaction between the controller and the field w_1; for instance, Ref. [2] provides an optical method to implement this interaction. On the other hand, the matrix G_K in (6.35) simply corresponds to an optical phase shifter. Therefore, the CFB controller designed here can be realized as a detuned optical cavity coupled to two input–output fields

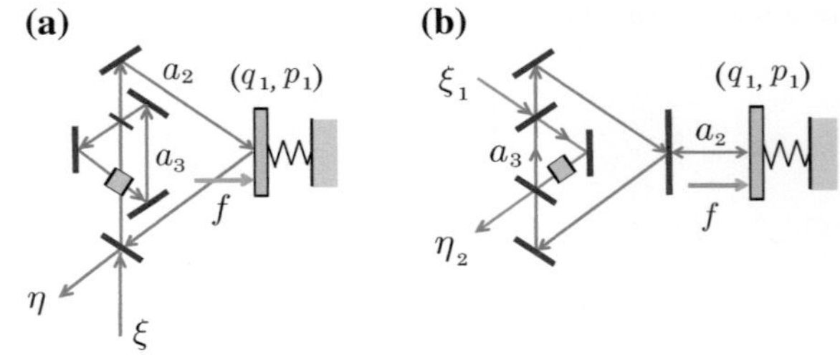

Fig. 6.15 **a** Direct interaction scheme achieving BAE for the opto-mechanical oscillator, proposed by Tsang and Caves [40]. **b** Equivalent realization via the CFB control. Figure adapted from [43]

via QND interactions, which is depicted in Fig. 6.15b. Note that the set of Eqs. (6.36) and (6.37) is exactly the same as that of the modified opto-mechanical oscillator proposed by Tsang and Caves [40], which is shown in Fig. 6.15a.

6.3.4 *Further Reading*

Synthesizing a system achieving BAE is an old problem, particularly in the field of gravitational wave detection (see for instance [38–41]). In contrast to their approach of looking at a specific optical interferometer for gravitational wave detection, the system theoretic characterization given in Sect. 6.3.2 can potentially yield a general approach for synthesizing a system having the BAE property. In fact, in [43] another coherent feedback controller realizing BAE for the optical interferometer was proposed, based on the general condition (6.28) or the equivalent (6.29).

In addition to BAE, in [43] system theoretic characterizations of quantum nondemolition (QND), interactions and decoherence-free subsystems (DFS) are given, which also provide a concrete coherent feedback controller achieving QND interaction and DFS. A crucial fact proven in [43] is that while we can find a coherent feedback controller achieving BAE, QND, or DFS for a given linear quantum system, it is generally not possible to construct a measurement-based feedback controller achieving these three goals. That is, there exists a no-go theorem showing the distinct superiority of coherent feedback over a measurement-based one, for those design goals in quantum information science. Again, it should be emphasized that this result is obtained based fully on a system theoretic approach.

6.4 Quantum Memory with Decoherence-Free Subsystem

A quantum memory is a device that can preserve a quantum state to realize tasks in computation and communication. Surprisingly, or not surprisingly, systems and control theory provides a unified methodology for constructing a linear quantum memory architecture [46]; more precisely, for linear quantum systems, we obtain a general procedure for perfect writing and storage of a single-photon or coherent state, with the full use of the controllability–observability properties and the notion of zero dynamics. This section is devoted to reviewing this theory, focusing on the case where the input state is a coherent state, for simplicity.

6.4.1 General Schematic of an Ideal Quantum Memory

Let us begin by describing the general and ideal schematic of a quantum memory. First of all, the system needs to contain a subsystem that preserves a quantum state for a long time, which is named here the *memory subsystem*. During the storage time, a memory subsystem must be decoupled from other subsystems as well as all dissipative environments; i.e., it is *decoherence-free (DF)* [47]. Note that in the storage stage, the memory subsystem must be disconnected with the transportation channel that transfers an input state or retrieves the memory state, while in the writing or reading process, we need to connect the memory subsystem with the transportation channel. Therefore, the system involves a switching mechanism that opens or closes the memory subsystem. Concrete examples are for instance an atomic ensemble with electromagnetic induced transparency (EIT) effect [48–50], a photonic crystal array [51], and an optical cavity [52].

The above basic function of a general quantum memory system is illustrated in Fig. 6.16. Let $a_{\rm M}$ denote the mode of the memory subsystem. $a_{\rm B}$ is the mode of the *buffer subsystem*, which mediates the optical field and the memory subsystem. As shown in the figure, these two subsystems can be coupled or decoupled. In the writing stage (a), we transport an input state of the optical field with mode ξ to the memory subsystem. Here, let us assume that with the use of a "nice procedure," the input state is perfectly transferred to the memory subsystem. Then, the two modes $a_{\rm B}$ and $a_{\rm M}$ are decoupled, and $a_{\rm M}$ becomes DF. In the storage stage (b), the memory subsystem preserves the state indefinitely. Finally, in the reading stage (c) at a later time, by coupling the modes again we can retrieve the state over the output channel with mode η.

Let us describe the scheme in a more detailed fashion, for the case of linear quantum systems. The system is given by a completely passive linear quantum system discussed in Sect. 2.4.1. That is, the vector of system variables $a = (a_1, \ldots, a_n)^\top$ obeys the following dynamical and output equations:

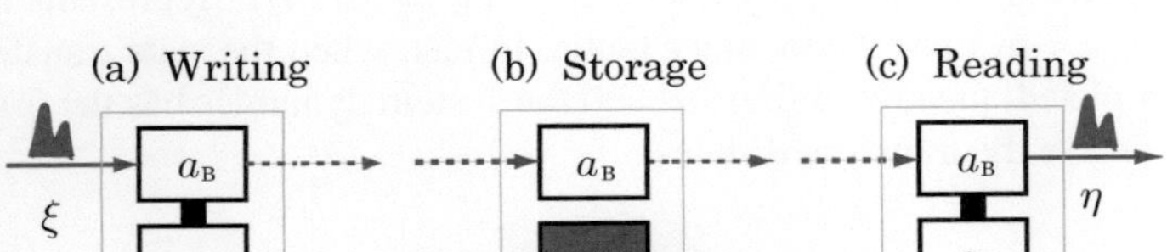

Fig. 6.16 Basic function of an ideal quantum memory. Figure adapted from [46]

$$\dot{a}(t) = Aa(t) - C^*\xi(t), \quad \eta(t) = Ca(t) + \xi(t). \tag{6.38}$$

Again, the system matrices are related to the Hamiltonian and the coupling operator in the form $H = a^*\Omega a$ and $L = Ca$. Also, recall that $A = -\imath\Omega - C^*C/2$. In particular, we assume that there is only one external channel, meaning that C is a $1 \times n$ complex row vector; hence, without loss of generality the scattering matrix S is now set to be $S = 1$. Now, let $a_{\rm B} = (a_1, \ldots, a_m)^\top$ and $a_{\rm M} = (a_{m+1}, \ldots, a_n)^\top$ be the modes of the buffer and memory subsystems, respectively. Then, in the writing stage, our task is to devise a procedure (as mentioned above) such that the state of ξ is perfectly transferred to that of $a_{\rm M}$. Once this process is completed, the system is switched to the storage mode, where (6.38) takes the form

$$\frac{d}{dt}\begin{bmatrix} a_{\rm B}(t) \\ a_{\rm M}(t) \end{bmatrix} = \begin{bmatrix} A_{\rm B} & O \\ O & O \end{bmatrix}\begin{bmatrix} a_{\rm B}(t) \\ a_{\rm M}(t) \end{bmatrix} - \begin{bmatrix} C_{\rm B}^* \\ O \end{bmatrix}\xi(t),$$
$$\eta(t) = C_{\rm B}a_{\rm B}(t) + \xi(t). \tag{6.39}$$

Hence, $a_{\rm M}$ is decoupled from both the input field ξ and the output field η. This implies that $a_{\rm M}$ is actually DF and its state can be stored. Importantly, (6.39) shows that in the storage stage, the memory subsystem is uncontrollable and unobservable with respect to the input and output channels (see Sect. 1.2.2). Thanks to this system theoretic characterization of the memory subsystem, a convenient necessary and sufficient condition for such a DF subsystem to exist has been developed in [53].

As a typical setup, we show an atomic ensemble system composed of N Λ-type atoms confined in an optical cavity [50]. Each atom contains two ground states $|s\rangle$ and $|g\rangle$, and an excited state $|e\rangle$. The transition between $|g\rangle$ and $|e\rangle$ occurs due to the coupling of atoms and the cavity mode a_1, with strength g. Also, a classical field with controllable Rabi frequency ω_R induces the transition between $|s\rangle$ and $|e\rangle$. Now, define the collective lowering operators $\sigma_{ge} = \sum_{k=1}^N (|g\rangle\langle e|)^{(k)}$ and $\sigma_{gs} = \sum_{k=1}^N (|g\rangle\langle s|)^{(k)}$ (see 1.42); in the large ensemble limit $N \gg 1$, these operators can be well approximated by annihilation operators as $a_2 = \sigma_{ge}/\sqrt{N}$ and $a_3 = \sigma_{gs}/\sqrt{N}$. The ideal dynamics of $a = (a_1, a_2, a_3)^\top$ is then given by

$$\frac{da(t)}{dt} = \begin{bmatrix} -\kappa & ig & 0 \\ ig & 0 & i\omega_R \\ 0 & i\omega_R & 0 \end{bmatrix} a(t) - \begin{bmatrix} \sqrt{2\kappa} \\ 0 \\ 0 \end{bmatrix}\xi(t), \quad \eta(t) = \sqrt{2\kappa}a_1 + \xi(t),$$

where κ denotes the cavity decay rate. Hence, $a_{\rm B} = (a_1, a_2)^\top$ represents the buffer mode and $a_{\rm M} = a_3$ does the memory mode. In fact, when the state transfer from ξ to a_3 is completed, then by setting $\omega_R = 0$ the system dynamics has the form (6.39) and can preserve the transferred state.

6.4.2 The Zero Dynamics Principle

As mentioned above, our task is to develop a procedure that perfectly transfers the input state of ξ to the state of $a_{\rm M}$. Now, note that the input state should be brought over a time temporal pulsed field rather than a continuous-wave one (for instance an optical field whose amplitude changes in time according to a sinusoidal function), because obviously the subsystem can absorb only the finite amount of energy. Then, the problem is to find an appropriate pulse shape for achieving perfect state transfer. As shown below, the zero dynamics principle gives us a clear solution to this problem.

First, let us assume that the initial state is given by the product of coherent states, $|\phi\rangle_S \otimes |f\rangle_F$, with $f(t)$ the amplitude of the field coherent state, corresponding to the pulse function to be determined. Here, the subscripts S and F are used to emphasize system and field states, respectively. Then, noting that ${}_S\langle\phi|{}_F\langle f|\xi(t)|\phi\rangle_S|f\rangle_F = f(t)$, we see that the vector of mean amplitudes, $m(t) = (\langle a_1(t)\rangle, \ldots, \langle a_n(t)\rangle)^\top$ with $\langle a_i(t)\rangle = {}_S\langle\phi|{}_F\langle f|a_i(t)|\phi\rangle_S|f\rangle_F$, follows

$$\dot{m}(t) = Am(t) - C^* f(t), \quad \tilde{f}(t) = Cm(t) + f(t), \tag{6.40}$$

where $\tilde{f}(t) = \langle\eta(t)\rangle$. Now, for the completely passive linear quantum system (6.38), the following *energy balance identity* [54] holds:

$$\int_{t_0}^{t} \eta^*(s)\eta(s)ds + a^*(t)a(t) = \int_{t_0}^{t} \xi^*(s)\xi(s)ds + a^*(t_0)a(t_0). \tag{6.41}$$

Thus, the total energy of the system and the field is preserved for all time. Actually, combined with (6.40), this equation leads to

$$\int_{t_0}^{t} |\tilde{f}(s)|^2 ds + \langle a^*(t)a(t)\rangle = \int_{t_0}^{t} |f(s)|^2 ds + \langle a^*(t_0)a(t_0)\rangle,$$

where the mean is taken for the state $|\phi, f\rangle$. Now, we assume that the system initial state is vacuum, $|\phi\rangle_S = |0\rangle_S$, which yields $\langle a^*(t_0)a(t_0)\rangle = 0$. Then, in order to achieve perfect transportation of the energy from the input to the system, we need that the output field must be vacuum, or equivalently, the amplitude of the output field is always "zero" because in this case the complete passivity means that the output is also a coherent state; this is the zero dynamics principle. Therefore, $\tilde{f}(t) = 0$ for all $t \in [t_0, t_1]$ with t_1, the stopping time of the writing process. Then, the condition $\tilde{f}(t) = Cm(t) + f(t) = 0$ leads to $C^*Cm(t) + C^*f(t) = 0$, and we thus have

$$\dot{m}(t) = (A + C^*C)m(t) = \Big(-\imath\Omega + \frac{1}{2}C^*C\Big)m(t) = -A^*m(t). \tag{6.42}$$

This corresponds to zero dynamics, meaning that if the system variable obeys the dynamics (6.42), the output is always vacuum. Now, the solution of (6.42) is given by $m(t) = e^{-A^*(t-t_1)}m_1$, with m_1 a fixed vector. Thus, the condition $Cm(t) + f(t) = 0$ yields

$$f(t) = -Cm(t) = -m(t)^\top C^\top = -m_1^\top e^{-A^\#(t-t_1)} C^\top.$$

Note that the input is transferred in the period $t \le t_1$ (writing stage) and $f(t) = 0$ for $t_1 \le t$ (storage and reading stages). As a result, we have the following proper expression:

$$f(t) = -m_1^\top e^{-A^\#(t-t_1)} C^\top \Theta(t_1 - t), \tag{6.43}$$

where $\Theta(t)$ is the Heaviside step function taking the value 1 for $t \ge 0$ and 0 for $t < 0$. Note that the real parts of all the eigenvalues of $-A^\#$ are strictly positive, because A is Hurwitz. Thus, $f(t)$ grows exponentially, and hence, it is called a *rising exponential function*.

6.4.3 Perfect State Transfer

First, we discuss the concrete procedure to encode (classical) information on the coherent field state with pulse function (6.43), which is perfectly transferred to the system; this procedure determines the parameter vector m_1 contained in (6.43). Note that a coherent field state $|f\rangle_F$ with temporal pulse shape $f(t)$ can be expressed as

$$|f\rangle_F = e^{B^*(f)-B(f)}|\Omega\rangle_F = \exp\Big[\int_{-\infty}^{\infty}\Big(f(t)\xi^*(t) - f^*(t)\xi(t)\Big)dt\Big]|\Omega\rangle_F,$$

where $B(f) = \int_{-\infty}^{\infty} f^*(t)\xi(t)dt$. The power $\int_{-\infty}^{\infty}|f(t)|^2 dt$ is assumed to be finite. Now, assume that $f(t)$ is characterized, in terms of the orthonormal functions $\{\gamma_k(t)\}_{k=1,\ldots,n}$, by

$$f(t) = \sum_{k=1}^{n} \alpha_k \gamma_k(t), \tag{6.44}$$

where $\alpha_k \in \mathbb{C}$ represents the classical information to be stored in the memory. Then, the power of $|f\rangle_F$ is given by $\int_{-\infty}^{\infty}|f(t)|^2 dt = \sum_k |\alpha_k|^2$, and the coherent field state is described by

$$|f\rangle_F = \exp\Big[\sum_k \{\alpha_k B^*(\gamma_k) - \alpha_k^* B(\gamma_k)\}\Big]|\Omega\rangle_F. \tag{6.45}$$

Next, let us turn our attention to the writing process, which actually transfers the classical information $\{\alpha_k\}$ to the memory subsystem. For this purpose, first recall that the system dynamics (6.38) has the solution

$$a(t) = e^{A(t-t_0)}a(t_0) - e^{At}\int_{t_0}^{t} e^{-As}C^*\xi(s)ds.$$

Since A is Hurwitz, we can take the limit $t_0 \to -\infty$, which yields

$$a^{\#}(t_1) = -e^{A^{\#}t_1}\int_{-\infty}^{t_1} e^{-A^{\#}s}C^{\top}\xi^*(s)ds,$$

where again t_1 is the stopping time of the writing process. We now introduce the following vector of rising exponential functions:

$$\nu(t) = -e^{-A^{\#}(t-t_1)}C^{\top}\Theta(t_1 - t). \tag{6.46}$$

Then, $a^{\#}(t_1)$ can be represented by

$$a^{\#}(t_1) = \int_{-\infty}^{\infty} \nu(t)\xi^*(t)dt = \big(I_S \otimes B^*(\nu_1), \ldots, I_S \otimes B^*(\nu_n)\big)^{\top}. \tag{6.47}$$

This is a vector of creation operators $a_k^*(t_1)$ satisfying the canonical commutation relation $aa^* - (a^{\#}a^{\top})^{\top} = I$, which means that $\{\nu_i(s)\}$ are orthonormal; $\int_{-\infty}^{\infty}\nu_i^*(t)\nu_j(t)dt = \delta_{ij}$. Recall now that at the initial time $t_0 \to -\infty$, the field input state is the coherent field state (6.45) and the system initial state is the ground state $|0,\ldots,0\rangle_S$. Then, after the interaction, the whole state at $t = t_1$ is given by

$$\begin{aligned}|\Psi(t_1)\rangle &= U(-\infty, t_1)|0,\ldots,0\rangle_S \otimes e^{\sum_k[\alpha_k B^*(\gamma_k)-\alpha_k^* B(\gamma_k)]}|\Omega\rangle_F \\ &= U(-\infty, t_1)e^{\sum_k[\alpha_k B^*(\gamma_k)-\alpha_k^* B(\gamma_k)]}U^*(-\infty, t_1)|0,\ldots,0\rangle_S|\Omega\rangle_F.\end{aligned}$$

Thus, by setting $\gamma_k(t) = \nu_k(t)$ and noting that (6.47) can be expressed as $a_k^*(t_1) = U^*(-\infty, t_1)a_k^*(-\infty)U(-\infty, t_1) = I_S \otimes B^*(\nu_k)$, we obtain

$$\begin{aligned}|\Psi(t_1)\rangle &= e^{\sum_k[\alpha_k a_k^*(-\infty)-\alpha_k^* a_k(-\infty)]}|0,\ldots,0\rangle_S|\Omega\rangle_F \\ &= \prod_k e^{\alpha_k a_k^*(-\infty)-\alpha_k^* a_k(-\infty)}|0,\ldots,0\rangle_S|\Omega\rangle_F \\ &= |\alpha_1,\ldots,\alpha_n\rangle_S|\Omega\rangle_F.\end{aligned}$$

Hence, the system state is transformed to the product of coherent states $|\alpha_k\rangle$. That is, the coherent field state with classical information $\{\alpha_k\}$ encoded in the rising exponential pulse functions (6.46) can be perfectly transferred to the system.

Now, we describe the ideal memory schematic in a more detailed way. Recall that the system can tune the memory subsystem to the DF mode in the storage stage or to the non-DF mode in the other two stages. Hence, after the perfect state transfer in the writing process has been finished at time $t = t_1$, the system structure is instantaneously modified so that the system dynamics takes the form (6.39):

$$\frac{d}{dt}\begin{bmatrix} a_{\rm B} \\ a_{\rm M}\end{bmatrix} = \begin{bmatrix} A_{\rm B} & O \\ O & O\end{bmatrix}\begin{bmatrix} a_{\rm B} \\ a_{\rm M}\end{bmatrix} - \begin{bmatrix} C_{\rm B}^* \\ O\end{bmatrix}\xi(t), \quad \eta(t) = C_{\rm B}a_{\rm B}(t) + \xi(t).$$

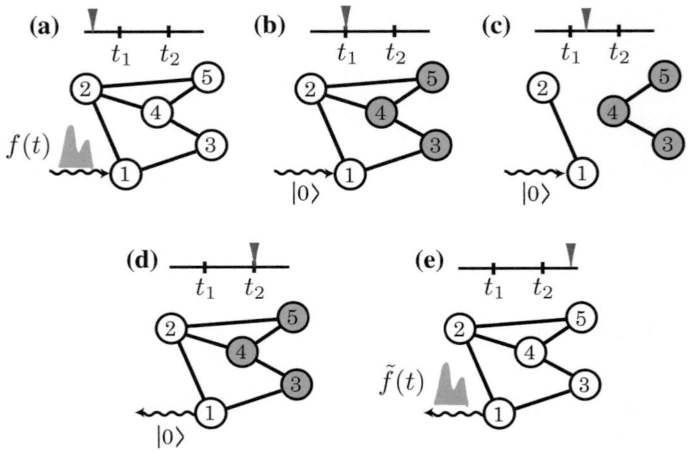

Fig. 6.17 The perfect memory procedure in a 5-node linear network. Figure adapted from [46]

Clearly, $a_{\rm M}$ constitutes a DF subsystem. Now, for simplicity, we assume $n = 5$ and consider the specific case $a_{\rm B} = [a_1, a_2]^\top$ (the buffer mode) and $a_{\rm M} = [a_3, a_4, a_5]^\top$ (the memory mode). Hence, the whole state $|\alpha_1, \ldots, \alpha_5\rangle$ cannot be stored, but only its (3, 4, 5) components can be. This means that to achieve the perfect state transfer and storage, the input field must be prepared to a specific coherent field state $|f(t)\rangle_F$ with $\alpha_1 = \alpha_2 = 0$ and sent over the rising exponential pulse function (6.46); more precisely, the input pulse shape should be synthesized by first multiplying the classical information $(\alpha_3, \alpha_4, \alpha_5)$ with the basis functions $(\nu_3(t), \nu_4(t), \nu_5(t))$ and then generating $f(t) = \alpha_3\nu_3(t) + \alpha_4\nu_4(t) + \alpha_5\nu_5(t)$. In fact, as described before, the whole state when the writing process is finished is then given by

$$|\Psi(t_1)\rangle = |0, 0\rangle \otimes |\alpha_3, \alpha_4, \alpha_5\rangle \otimes |\Omega\rangle_F,$$

and consequently, by decoupling the buffer and memory subsystems, the state $|\alpha_3, \alpha_4, \alpha_5\rangle$ can be preserved. The above writing and storage processes together with the reading process are illustrated in Fig. 6.17, and now, those can be explicitly described as follows. (a) The coherent field state with $\alpha_1 = \alpha_2 = 0$ is transferred over the pulse shape $f(t)$. (b) At time $t = t_1$, the field state has been perfectly transferred to the memory subsystem. Then, by switching the system architecture, the memory subsystem is decoupled. (c) In the storage state in the time period $[t_1, t_2]$, the transferred state is preserved. (d) At $t = t_2$, the memory subsystem is coupled again to the buffer subsystem and accordingly the output field. (e) The output field with pulse shape $\tilde{f}(t)$, which contains the perfect copy of the information $\{\alpha_3, \alpha_4, \alpha_5\}$, can be retrieved (see [46] for the detailed time evolution of this retrieved state).

6.4.4 Further Reading

In this section, based on the zero dynamics principle, we derived the rising exponential function (6.43) as the desired input pulse realizing perfect state transfer to a completely passive linear quantum system. This fact can be derived by another system theoretic approach as follows. Let $\tilde{f}[s] = G[s]f[s]$ be the input–output relation of (6.40) and z_j be a zero of the transfer function $G[s]$, i.e., $G[z_j] = 0$ holds. Then, it is known in linear systems and control theory that an input of the form $f(t) = e^{z_j t}$ can yield an output $\tilde{f}(t) = G[z_j]e^{z_j t} = 0$ under appropriate conditions. Now, we can proceed to prove (6.43) as a linear combination of those functions $\{e^{z_j t}\}$. That is, the condition for perfect state transfer can be completely characterized by the zero of the system's transfer function; this is a special case of the general zero dynamics principle. Importantly, as mentioned before, this fact can be extended beyond coherent states to include single-photon fields, which is an important class of non-Gaussian states which is of much interest in quantum information science (see [46]). Moreover, we can generalize the theory of quantum memory discussed in this section to general completely passive linear quantum systems with multiple inputs and outputs, in which case the notion of zeros is extended to transmission zeros or particularly blocking zeros [55].

The rising exponential function was found in some specific settings as the desired pulse function achieving complete absorption of the state by a quantum system [56–58], and recently, we find several experimental proposals to generate a simple (monotonous) rising exponential wave packet of a single-photon field [59–62]. However, the pulse function (6.43) has a complicated non-monotonous form in the general setup (see [46]), and producing such a complicated pulse function in experiments is challenging. In this case, an alternative approach proposed in [49, 50, 80] can be used for the purpose of high-efficiency state transfer. In this approach, an additional open-loop control input is applied on the system so that an input single-photon field with a given pulse shape is transferred to the system more efficiently, in the sense that the energy of the internal system (set to the ground state at the initial time) is maximized.

6.5 Robust Quantum Amplification via Coherent Feedback

Let us begin with a classical amplifier, which is doubtless a key component incorporated in almost all current electrical devices. The basic function of an autonomous amplifier is to transform an input signal u to $y = Gu$ with gain $G > 1$. However, usually the system parameters contained in G are fragile, and thus, the output y can change easily and eventually become distorted. Fortunately, this serious issue was finally resolved in 1927 by Black [63, 64]. The central idea is in the use of feedback as shown in Fig. 6.18; that is, an autonomous amplifier (plant) is combined with a passive system (controller) in a feedback interconnection. Then, the output of the closed-loop system is given by

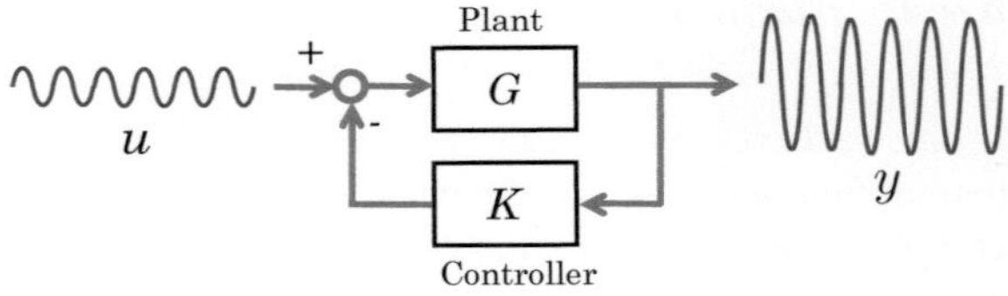

Fig. 6.18 Classical feedback amplification scheme: G is the gain of an autonomous amplifier, and K is the gain of a passive controller. Reprinted with permission from [65]. © (2016) by the American Physical Society

$$y = G^{(\text{fb})}u, \quad G^{(\text{fb})} = \frac{G}{1 + GK},$$

where $K < 1$ is the attenuation level of the passive controller. Then, a large gain $G \gg 1$ leads to $G^{(\text{fb})} \approx 1/K > 1$. That is, the gain of the controlled system is determined only by the controller's parameters. This is a very important result, because a passive device such as a resistor is very robust and its parameters contained in K can be made to be very stable. Thus, thanks to the feedback, the controlled system functions as a robust amplifier.

Now, a pertinent quantum version of the above-described commercial classical amplifier is the quantum *phase-preserving amplifier (PPA)*. This amplifier plays a central role in, for instance, quantum communication and weak-signal detection [66–70], but as in the classical case, it is always fragile with respect to the system parameters. In [65], a general theory was provided to resolve this issue; that is, a coherent feedback with a completely passive linear quantum system makes the amplifier significantly robust. This section is devoted to describing this theory, with a concrete optical example.

6.5.1 The Phase-Preserving Amplifier

The general idea of a PPA is simple. That is, it is a linear transformation of a bosonic mode b_1 to the output

$$\tilde{b}_1 = g_1 b_1 + g_2 b_2^*, \tag{6.48}$$

where b_2 is an auxiliary mode. Here, the point is that b_2 is necessary in order to satisfy the CCR of the output mode, $[\tilde{b}_1(t), \tilde{b}_1(t')^*] = \delta(t - t')$. As a consequence, the coefficients have to satisfy $|g_1|^2 - |g_2|^2 = 1$, and hence, the output $\tilde{b}_1$ is an amplified mode of b_1 with gain $|g_1| > 1$. Usually, b_2 is set to be a vacuum state; hence, we have $\langle \tilde{b}_1 \rangle = g_1 \langle b_1 \rangle$, implying that the phase of b_1 is preserved. Note that contrary to this, a squeezing operation does not preserve the phase of the input; hence, it is sometimes referred to as phase-sensitive amplification.

A typical physical realization of a PPA is provided by a NOPA as discussed in Sect. 1.5.2. Recall that this is an optical cavity system with two internal orthogonally polarized fields a_1 and a_2. The cavity contains a $\chi^{(2)}$ optical crystal that facilitates interaction between a_1 and a_2 via the quadratic Hamiltonian $H = \imath\epsilon(a_1^* a_2^* - a_1 a_2)$, where $\epsilon \in \mathbb{R}$ is assumed for simplicity (note that in Sect. 1.5.2, we discussed the NOPA with coefficient $\epsilon/2$). Also, the cavity couples to two external traveling fields ξ_1 and ξ_2; the dynamics is given by Sect. 1.29:

$$\frac{da_1}{dt} = \Big(-\frac{\gamma}{2} - i\Delta_1 \Big) a_1 + \epsilon a_2^* - \sqrt{\gamma}\xi_1,$$
$$\frac{da_2^*}{dt} = \Big(-\frac{\gamma}{2} + i\Delta_2 \Big) a_2^* + \epsilon a_1 - \sqrt{\gamma}\xi_2^*, \tag{6.49}$$

where Δ_1 *and* Δ_2 are detuning parameters. Also, for simplicity, optical losses are ignored. The output equations (boundary conditions) are given by (1.30):

$$\eta_1 = \sqrt{\gamma} a_1 + \xi_1, \quad \eta_2^* = \sqrt{\gamma} a_2^* + \xi_2^*. \tag{6.50}$$

In the standard scenario of quantum amplification, we only consider the first output mode η_1 and ignore the second one η_2, although in the scenario of entanglement generation via NOPA both the output modes are of course kept, as discussed in Sects. 1.5.2 and 6.2. Hence, let us focus on η_1, the Laplace transform of which is given by

$$\eta_1[s] = g_1[s]\xi_1[s] + g_2[s]\xi_2^*[s],$$
$$g_1[s] = \frac{(s - \gamma/2 + \imath\Delta_1)(s + \gamma/2 - \imath\Delta_2) - \epsilon^2}{D[s]}, \quad g_2[s] = \frac{-\gamma\epsilon}{D[s]},$$
$$D[s] = \big(s + \frac{\gamma}{2} + \imath\Delta_1\big)\big(s + \frac{\gamma}{2} - \imath\Delta_2\big) - \epsilon^2.$$

The stability condition is obtained by examining the characteristic equation $D[s] = 0$; particularly, when $\Delta_1 = \Delta_2 = 0$, it is $\gamma^2/4 - \epsilon^2 > 0$. The amplification process is described in the Fourier domain $s = \imath\omega$ with ω, the frequency; that is, we consider the linear transformation at steady state, $\eta_1[\imath\omega] = g_1[\imath\omega]\xi_1[\imath\omega] + g_2[\imath\omega]\xi_2^*[\imath\omega]$. Note that g_1 and g_2 satisfy $|g_1[\imath\omega]|^2 - |g_2[\imath\omega]|^2 = 1$ for all ω. In particular, when $\Delta_1 = \Delta_2 = 0$, the amplification gain at the resonance frequency of a_1 and a_2, $\omega = 0$ (in the rotating frame, $\omega = 0$ corresponds to $\omega_p/2$ with ω_p, the pump frequency) is given by

$$|g_1(0)| = \frac{\gamma^2 + 4\epsilon^2}{|\gamma^2 - 4\epsilon^2|}.$$

Hence, it takes a large value by setting $\epsilon \approx \gamma/2$ while keeping $|\epsilon| < \gamma/2$.

The above example can be generalized to any PPA; it is modeled as a linear system with two inputs and two outputs as follows:

$$\begin{bmatrix} \eta_1[s] \\ \eta_2^*[s] \end{bmatrix} = G[s] \begin{bmatrix} \xi_1[s] \\ \xi_2^*[s] \end{bmatrix}, \quad G[s] = \begin{bmatrix} G_{11}[s] & G_{12}[s] \\ G_{21}[s] & G_{22}[s] \end{bmatrix}, \tag{6.51}$$

where $\xi_1[s]$ is the Laplace transform of ξ_1, etc. The transfer function matrix $G[s]$ at $s = \imath\omega$ (called the scattering matrix in this field) satisfies

$$|G_{11}[\imath\omega]|^2 - |G_{12}[\imath\omega]|^2 = |G_{22}[\imath\omega]|^2 - |G_{21}[\imath\omega]|^2 = 1,$$
$$G_{21}[\imath\omega]G_{11}^*[\imath\omega] - G_{22}[\imath\omega]G_{12}^*[\imath\omega] = 0, \quad \forall\omega. \tag{6.52}$$

Thus, $|G_{11}[\imath\omega]|$ represents the amplification gain.

6.5.2 *Coherent Feedback Control for a Quantum Amplifier*

As will be shown later, and as in the classical case, an autonomous amplifier is fragile against parameter fluctuation. Hence, we need feedback control. Here, we take a coherent feedback strategy, because a measurement-based control inevitably introduces extra classical noise. In particular, the controller is given by a completely passive linear quantum system with two inputs ξ_3, ξ_4 and two outputs η_3, η_4; note that a single-input and single-output passive system has gain equal to 1, and therefore, it does not work as an attenuator. The controller is given in the Laplace domain as follows:

$$\begin{bmatrix} \eta_3^*[s] \\ \eta_4^*[s] \end{bmatrix} = K[s] \begin{bmatrix} \xi_3^*[s] \\ \xi_4^*[s] \end{bmatrix}, \quad K[s] = \begin{bmatrix} K_{11}[s] & K_{12}[s] \\ K_{21}[s] & K_{22}[s] \end{bmatrix}, \tag{6.53}$$

where the creation operator representation is taken to make the notation simple. Note that the transfer function matrix $K[s]$ is unitary in the Fourier domain; i.e., $K[\imath\omega]^* K[\imath\omega] = I$ holds for all ω [71]. We then consider the following feedback connection, as shown in Fig. 6.19:

$$\eta_2 = \xi_3, \quad \xi_2 = \eta_4, \tag{6.54}$$

which is of course equivalent to $\eta_2^* = \xi_3^*$ and $\xi_2^* = \eta_4^*$. By combining (6.51), (6.53), and (6.54), we find that the controlled system, with inputs ξ_1, ξ_4^* and outputs η_1, η_3^*, has the following input–output relation in the Laplace domain:

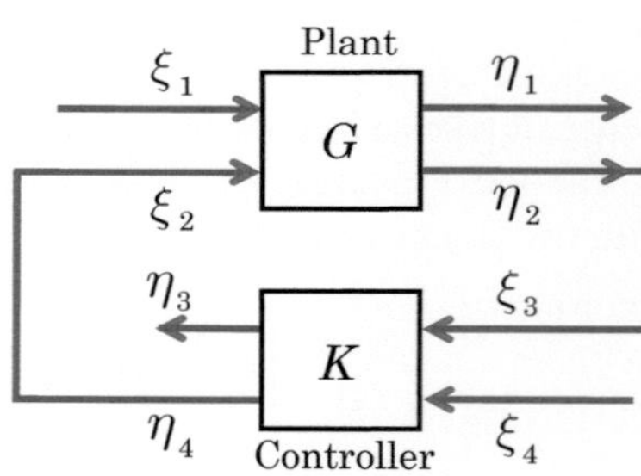

Fig. 6.19 Coherent feedback configuration composed of the autonomous amplifier (plant) G and the controller K. Reprinted with permission from [65]. © (2016) by the American Physical Society

$$\begin{bmatrix} \eta_1[s] \\ \eta^*[s] \end{bmatrix} = \begin{bmatrix} G_{11}^{(\mathrm{fb})}[s] \; G_{12}^{(\mathrm{fb})}[s] \\ G_{21}^{(\mathrm{fb})}[s] \; G_{22}^{(\mathrm{fb})}[s] \end{bmatrix} \begin{bmatrix} \xi_1[s] \\ \xi_4^*[s] \end{bmatrix},$$

where

$$\begin{aligned} G_{11}^{(\mathrm{fb})} &= [G_{11} - K_{21}(G_{11}G_{22} - G_{12}G_{21})]/(1 - K_{21}G_{22}), \\ G_{12}^{(\mathrm{fb})} &= (G_{12}K_{22})/(1 - K_{21}G_{22}), \\ G_{21}^{(\mathrm{fb})} &= (G_{21}K_{11})/(1 - K_{21}G_{22}), \\ G_{22}^{(\mathrm{fb})} &= [K_{12} + G_{22}(K_{11}K_{22} - K_{12}K_{21})]/(1 - K_{21}G_{22}). \end{aligned}$$

The matrix entries satisfy a condition corresponding to (6.52), i.e., $|G_{11}^{(\mathrm{fb})}[\imath\omega]|^2 - |G_{12}^{(\mathrm{fb})}[\imath\omega]|^2 = 1 \; \forall\omega$.

Let us now discuss the effect of feedback control on the amplified signal in the Fourier domain:

$$\eta_1[\imath\omega] = G_{11}^{(\mathrm{fb})}[\imath\omega]\xi_1[\imath\omega] + G_{12}^{(\mathrm{fb})}[\imath\omega]\xi_4^*[\imath\omega].$$

Our focus is on the amplification gain $|G_{11}^{(\mathrm{fb})}[\imath\omega]|$ especially when the original gain $|G_{11}[\imath\omega]|$ is large. The point is the use of (6.52); first, from $|G_{21}||G_{11}| = |G_{22}||G_{12}|$ together with the other two equations, we have $|G_{11}| = |G_{22}|$ and $|G_{12}| = |G_{21}|$ (for the moment, we omit the variable $\imath\omega$). Moreover, $G_{11}G_{22} - G_{12}G_{21} = G_{22}/G_{11}^*$ holds. Then, in the limit $|G_{11}| \to \infty$, it follows that

$$\frac{|G_{11}G_{22} - G_{12}G_{21}|}{|G_{11}|} = \frac{|G_{22}|}{|G_{11}|^2} = \frac{1}{|G_{11}|} \to 0.$$

This implies that $(G_{11}G_{22} - G_{12}G_{21})/|G_{11}|$ converges to zero in this limit. As a consequence, we have

$$\begin{aligned} |G_{11}^{(\mathrm{fb})}| &= \left| \frac{G_{11}/|G_{11}| - K_{21}(G_{11}G_{22} - G_{12}G_{21})/|G_{11}|}{1/|G_{11}| - K_{21}G_{22}/|G_{11}|} \right| \\ &\to \left| \frac{G_{11}/|G_{11}|}{-K_{21}G_{22}/|G_{11}|} \right| = \frac{1}{|K_{21}|}. \end{aligned}$$

Hence, in the frequency range where $|G_{11}[\imath\omega]| \gg 1$ holds, the amplification gain of the controlled system is given by $|G_{11}^{(\mathrm{fb})}[\imath\omega]| \approx 1/|K_{21}[\imath\omega]| > 1$. Therefore, because the completely passive controller is much more robust compared to the original amplifier, even if G_{11} changes while still maintaining a large value, the controlled system functions as a very robust amplifier with gain $1/|K_{21}|$. Hence, a pertinent quantum counterpart of the classical feedback amplification theory is now constructed.

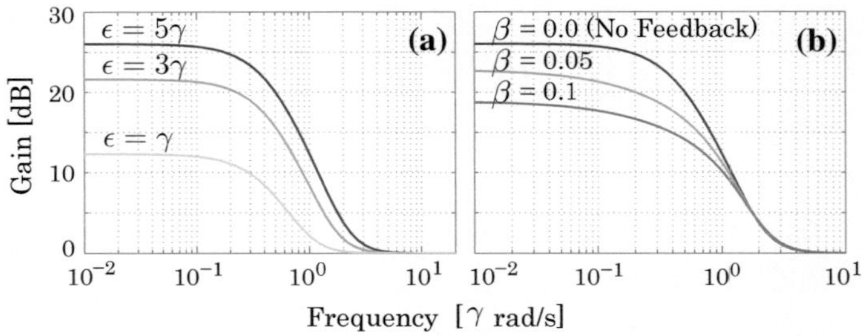

Fig. 6.20 **a** Gain profile of the specially detuned NOPA without feedback. **b** Gain profile of the feedback-controlled system with parameters $\epsilon = 5\gamma$ and various β. Reprinted with permission from [65]. © (2016) by the American Physical Society

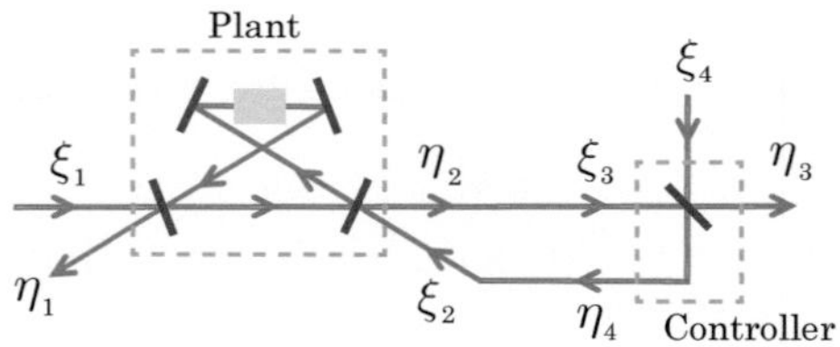

Fig. 6.21 Coherent feedback of NOPA (plant) implemented with a beam splitter (controller). Reprinted with permission from [65]. © (2016) by the American Physical Society

6.5.3 Example: Non-degenerate Optical Parametric Amplifier

Let us consider again the NOPA given by (6.49) as an example. In particular, we here consider the non-standard parameter choice $\Delta_1 = \Delta_2 = \epsilon$, which circumvents the so-called *gain–bandwidth constraint* as shown below. The transfer function matrix in this case is then given by

$$G[s] = \frac{1}{(s+\gamma/2)^2}\begin{bmatrix} s^2 - \gamma^2/4 + \imath\gamma\epsilon & -\gamma\epsilon \\ -\gamma\epsilon & s^2 - \gamma^2/4 - \imath\gamma\epsilon \end{bmatrix}.$$

The maximum gain is $|G_{11}[0]| = \sqrt{1 + 16\epsilon^2/\gamma^2}$, which becomes larger by increasing the pump strength ϵ. Figure 6.20a depicts the three cases corresponding to $\epsilon = \gamma, 3\gamma, 5\gamma$. Remarkably, the bandwidth (i.e., the frequency range where nearly flat amplification gain is realized) does not depend on the gain, which is nearly $\gamma/10$; that is, the amplification is free from the gain–bandwidth constraint. Note that because the pole of the transfer function matrix is fixed to $-\gamma/2$, the system is always stable irrespective of ϵ; this is a clear contrast to the standard NOPA that requires $|\epsilon| < \gamma/2$ for stability. As a consequence, in a proper parameter regime such that the linearized model (6.49) is valid, there is no upper limit on ϵ and accordingly on the gain $|G_{11}[0]|$ as well.

Now, let us consider coherent feedback control of this amplifier. Here, as illustrated in Fig. 6.21, we take a beam splitter with transmissivity α and reflectivity β as a controller, in which case the transfer function matrix is constant:

$$K[s] = \begin{bmatrix} \alpha & \beta \\ \beta & -\alpha \end{bmatrix}, \quad \alpha, \beta \in \mathbb{R}.$$

Thus, $K_{21} = \beta$ represents the attenuation level. The amplification gain of the controlled system is then

$$G_{11}^{(\mathrm{fb})}[s] = \frac{(1-\beta)s^2 + \beta\gamma s - (1+\beta)\gamma^2/4 + \imath\gamma\epsilon}{(1-\beta)s^2 + \gamma s + (1+\beta)\gamma^2/4 + \imath\beta\gamma\epsilon}.$$

As expected from the general theory discussed above, we find that the limit $|G_{11}[\imath\omega]| \to \infty$ (i.e., $\epsilon \to \infty$) leads to $|G_{11}^{(\mathrm{fb})}[\imath\omega]| \to 1/\beta$ in a certain frequency bandwidth. Also, by examining the denominator of $G_{11}^{(\mathrm{fb})}[s]$, we find that the controlled system is stable when $|\epsilon| < (\gamma/2\beta)\sqrt{(1+\beta)(1-\beta)}$. This yields $|\beta| < \gamma/2|\epsilon|$ when $\beta^2 \ll 1$; hence, let us choose $\epsilon = 5\gamma$, meaning $|\beta| < 0.1$. Figure 6.20b shows the gain $|G_{11}^{(\mathrm{fb})}[\imath\omega]|$ for the cases $\beta = 0.1$, $\beta = 0.05$, and $\beta = 0$. From this, we observe that the gain of the controlled NOPA is reduced from the value of the autonomous one; as shown later, by sacrificing this gain, the controlled NOPA obtains a fine robustness property against parameter fluctuation. Note also that the controlled NOPA has the gain–bandwidth constraint. This is in fact a general feature of a controlled amplifier: The gain and bandwidth of the controlled amplifier can be easily modified by simply tuning the controller.

Finally, let us discuss the robustness property of the controlled amplifier, which is in fact the main advantage of using feedback. To see this, let us consider the following imperfect case. First, the system parameters are fragile; the pump strength ϵ can change as $\epsilon = (1 + 0.1e_0)\epsilon_0$ where ϵ_0 is the nominal value; similarly, the detunings Δ_1 and Δ_2 can slightly deviate from ϵ, which is modeled by $\Delta_1 = (1 + 0.001e_1)\epsilon$ and $\Delta_2 = (1+0.001e_2)\epsilon$. Here, (e_0, e_1, e_2) are independent random variables drawn from the uniform distribution on $[-1, 1]$. The blue lines in Fig. 6.22 are 50 sample values of the autonomous gain $|G_{11}[\imath\omega]|$ in the case $\epsilon_0 = 5\gamma$. This shows that the parameter fluctuations above lead to significant fragility of the amplification gain. Hence, this is the situation where we should apply the feedback. The red lines in

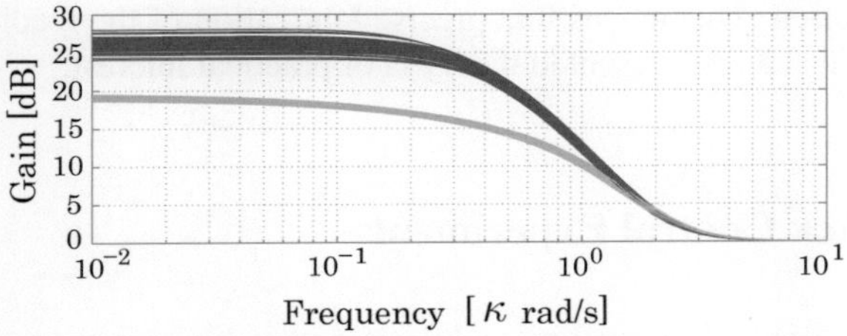

Fig. 6.22 The *upper blue lines* represent the gain profile of the autonomous NOPA, $|G_{11}[\imath\omega]|$, while the *lower red lines* correspond to the controlled case, $|G_{11}^{(\mathrm{fb})}[\imath\omega]|$ with $\beta = 0.1$. In both cases, $\epsilon_0 = 5\gamma$. Reprinted with permission from [65]. © (2016) by the American Physical Society (color figure online)

Fig. 6.22 are 50 sample values of the controlled gain $|G_{11}^{(\mathrm{fb})}[\imath\omega]|$ with $\beta = 0.1$. This clearly demonstrates that gain fluctuation in the amplifier is greatly suppressed by feedback.

6.5.4 Added Noise

In this section, we have discussed only the amplification of the mean, but not of the higher-order moments, especially the variance. In the general form (6.48), certainly the mean $\langle \tilde{b}_1 \rangle = g_1 \langle b_1 \rangle$ is amplified, but note that the noise variance defined by

$$\langle |\Delta b|^2 \rangle := \frac{1}{2} \langle \Delta b \Delta b^* + \Delta b^* \Delta b \rangle, \quad \Delta b = b - \langle b \rangle$$

must also be amplified as follows. If b_2 is the vacuum mode, (6.48) leads to $\langle |\Delta \tilde{b}_1|^2 \rangle = |g_1|^2 \langle |\Delta b_1|^2 \rangle + |g_2|^2/2$. This directly implies the degradation of the signal-to-noise ratio as follows:

$$\widetilde{(\mathrm{S/N})} = \frac{|\langle \tilde{b}_1 \rangle|^2}{\langle |\Delta \tilde{b}_1|^2 \rangle} = \frac{|\langle b_1 \rangle|^2}{\langle |\Delta b_1|^2 \rangle + \mathcal{A}} < \frac{|\langle b_1 \rangle|^2}{\langle |\Delta b_1|^2 \rangle} = (\mathrm{S/N}).$$

Hence, the *added noise*

$$\mathcal{A} := \frac{|g_2|^2}{2|g_1|^2} = \frac{|g_1|^2 - 1}{2|g_1|^2}$$

quantifies how well the signal information is amplified [72, 73]. In particular, in the large amplification limit $|g_1| \to \infty$, we find $\mathcal{A} \to 1/2$, and this is called the *quantum noise limit*.

In [65], it was proven that in general, the controlled amplifier functions as a PPA achieving the quantum noise limit $\mathcal{A} \to 1/2$ in the large amplification limit, even if several imperfections are present in the feedback loop. This effect was demonstrated in [65] in the numerical simulation for the same NOPA example studied in the previous subsection. This means that precise fabrication of the feedback control is not necessary; hence, the presented theory is of practical interest.

6.6 Feedback Control Experiments

We have seen in the previous sections and chapters that there are two types of quantum feedback schemes: coherent feedback and measurement-based quantum feedback. Here, we describe two actual experiments that demonstrate the effectiveness of those feedback schemes in different physical setups. The first one is a fully optical system that contains a coherent feedback architecture, and the second is a large atomic ensemble controlled by measurement-based feedback. In both cases, the control goal is to create an enhanced (optical or atomic) squeezed state.

6.6.1 Coherent Feedback for Optical Squeezing Enhancement

As discussed before, a squeezed state is a genuine non-classical state such that the fluctuation is reduced below the vacuum noise variance in one of the quadrature observables q and p. Coherent feedback has the ability to further suppress this fluctuation compared to the case without feedback, as theoretically proven in [18, 19] in the quantum optical setting. Figure 6.23a illustrates the proposed coherent feedback configuration composed of an optical parametric oscillator (OPO; see Sect. 1.5.3) as the plant and a beam splitter (BS) as the controller; the output squeezed state of OPO is then enhanced by the feedback and produced at the output of BS. This theoretical proposal was experimentally demonstrated in [21], and this is what we discuss in detail in this subsection.

We begin by describing the theoretical model of this feedback method, which is depicted in Fig. 6.23b. First, let us recall from Sect. 1.5.3 that the dynamics of OPO is given, in a rotating frame, by

$$\frac{da}{dt} = -\frac{\gamma + \kappa}{2}a + \frac{\epsilon}{2}a^* - \sqrt{\gamma}\xi_1 - \sqrt{\kappa}\xi_{\text{loss},1}, \quad \eta_1 = \sqrt{\gamma}a + \xi_1, \tag{6.55}$$

where $\gamma = cT_1/\ell$ and $\kappa = cL_1/\ell$ with ℓ the optical path length in the OPO, T_1 the transmissivity of the coupling BS, L_1 the transmissivity of a fictitious BS (which models the optical loss), and c the speed of light. In the quadrature form, the Fourier input–output relation is given by

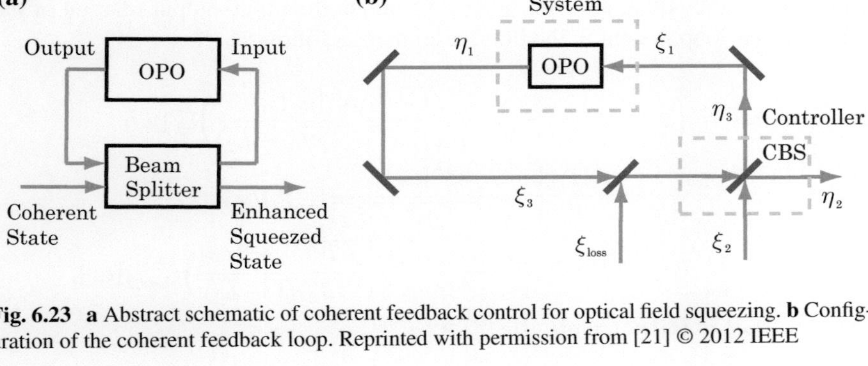

Fig. 6.23 **a** Abstract schematic of coherent feedback control for optical field squeezing. **b** Configuration of the coherent feedback loop. Reprinted with permission from [21] © 2012 IEEE

$$\eta_1^q[\imath\omega] = \big(G[\imath\omega] + g[\imath\omega]\big)\xi_1^q[\imath\omega] + \big(\bar{G}[\imath\omega] + \bar{g}[\imath\omega]\big)\xi_{\text{loss},1}^q[\imath\omega],$$
$$\eta_1^p[\imath\omega] = \big(G[\imath\omega] - g[\imath\omega]\big)\xi_1^p[\imath\omega] + \big(\bar{G}[\imath\omega] - \bar{g}[\imath\omega]\big)\xi_{\text{loss},1}^p[\imath\omega],$$

where

$$G[\imath\omega] = \frac{\gamma^2 - (\kappa - 2\imath\omega)^2 + \epsilon^2}{(\gamma + \kappa - 2\imath\omega)^2 - \epsilon^2}, \quad \bar{G}[\imath\omega] = \frac{2\sqrt{\gamma\kappa}(\gamma + \kappa - 2\imath\omega)}{(\gamma + \kappa - 2\imath\omega)^2 - \epsilon^2},$$
$$g[\imath\omega] = \frac{2\epsilon\gamma}{(\gamma + \kappa - 2\imath\omega)^2 - \epsilon^2},$$

and $\bar{g}[\imath\omega] = \sqrt{\kappa/\gamma}g[\imath\omega]$. The variance of $\eta_1^q[\imath\omega]$ and $\eta_1^p[\imath\omega]$ is simply given by the power spectrum $S_1^q(\imath\omega) = \langle|\eta_1^q[\imath\omega]|^2\rangle$ and $S_1^p(\imath\omega) = \langle|\eta_1^p[\imath\omega]|^2\rangle$. In particular, when the input field is a vacuum state, we have $S_1^q(\imath\omega) = |G[\imath\omega] + g[\imath\omega]|^2 + |\bar{G}[\imath\omega] + \bar{g}[\imath\omega]|^2$ and $S_1^p(\imath\omega) = |G[\imath\omega] - g[\imath\omega]|^2 + |\bar{G}[\imath\omega] - \bar{g}[\imath\omega]|^2$. If $\epsilon = 0$, then $S_1^q(\imath\omega) = S_1^p(\imath\omega) = 1, \forall\omega$, which is the vacuum noise variance.

Next, we describe the feedback configuration. The controller is called the *control BS* (CBS); its transmissivity T_2 is tunable. The coherent input field $\xi_2(t)$ is sent to the CBS. Next, one of the outputs of the CBS, $\eta_3(t)$, is sent to the OPO. The output of the OPO, $\eta_1(t)$, is then sent back to the CBS. This closed-loop system outputs the enhanced squeezed field $\eta_2(t)$. The input and output fields of the CBS are connected by

$$\eta_2 = \sqrt{1 - T_2}\xi_2 + \sqrt{T_2}\big[\sqrt{1 - L_2}\xi_3 + \sqrt{L_2}\xi_{\text{loss},2}\big],$$
$$\eta_3 = -\sqrt{1 - T_2}\big[\sqrt{1 - L_2}\xi_3 + \sqrt{L_2}\xi_{\text{loss},2}\big] + \sqrt{T_2}\xi_2,$$

where $\xi_{\text{loss},2}$ is a vacuum field entering through a fictitious BS with reflectivity L_2; this models the losses in the CF loop. Now, let $\tau_a = l_a/c$ ($\tau_b = l_b/c$) be the time delay resulting from the optical path length l_a (l_b) from (to) the CBS to (from) the OPO. Then, the feedback interconnection is represented by $\xi_1(t) = \eta_3(t - \tau_a)e^{i\omega_0\tau_a}$ and $\xi_3(t) = \eta_1(t - \tau_b)e^{i\omega_0\tau_b}$, where ω_0 is the resonance frequency of the OPO. Combining these equations with (6.55), we obtain the input–output relation of the whole closed-loop system in the Fourier domain as follows:

$$\begin{aligned}\eta_2^{\pm}[\imath\omega] &= \left(\sqrt{1 - T_2} + \frac{T_2\sqrt{1 - L_2}\alpha^{\pm}[\imath\omega]}{1 + \alpha^{\pm}[\imath\omega]\sqrt{(1 - T_2)(1 - L_2)}}\right)\xi_2^{\pm}[\imath\omega] \\ &\quad + \frac{\sqrt{T_2(1 - L_2)}\beta^{\pm}[\imath\omega]}{1 + \alpha^{\pm}[\imath\omega]\sqrt{(1 - T_2)(1 - L_2)}}\xi_{\text{loss},1}^{\pm}[\imath\omega] \\ &\quad + \left(\sqrt{T_2L_2} - \frac{\sqrt{T_2(1 - L_2)(1 - T_2)L_2}\alpha^{\pm}[\imath\omega]}{1 + \alpha^{\pm}[\imath\omega]\sqrt{(1 - T_2)(1 - L_2)}}\right)\xi_{\text{loss},2}^{\pm}[\imath\omega],\end{aligned}$$
$$\alpha^{\pm}[\imath\omega] = (G[\imath\omega] \pm g[\imath\omega])e^{\imath(\omega+\omega_0)(\tau_a+\tau_b)},$$
$$\beta^{\pm}[\imath\omega] = (\bar{G}[\imath\omega] \pm \bar{g}[\imath\omega])e^{\imath(\omega+\omega_0)\tau_b},$$

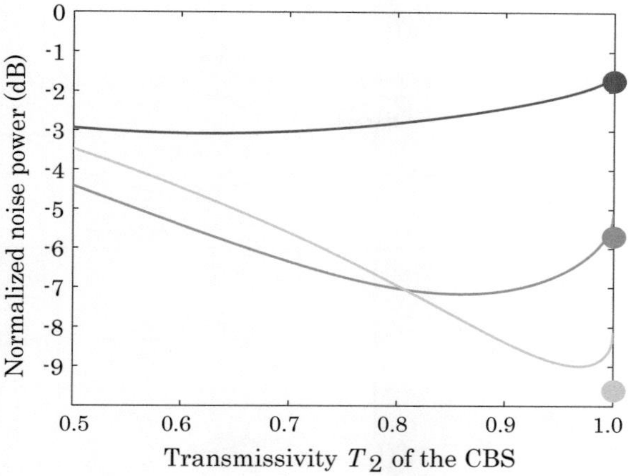

Fig. 6.24 The transmissivity T_2 of the CBS versus the squeezing levels for various normalized pumping strength $x = \epsilon/\gamma$. The *blue*, *red*, and *green lines* correspond to $x = 0.1, 0.35$, and 0.6, respectively. The *circles indicate* the values at $T_2 = 1$ and $L_2 = 0$, corresponding to the uncontrolled OPO. Reprinted with permission from [21] © 2012 IEEE (color figure online)

where $+$ and $-$ correspond to q and p, respectively. Here, the input is assumed to be a vacuum state (this is indeed the case in the experiment); then, the power spectrum of the output field can be calculated as $S_2^{\pm}(\imath\omega) = \langle|\eta_2^{\pm}[\imath\omega]|^2\rangle$. Also, we assume that the coherent feedback loop is on resonance, i.e., $e^{\imath\omega_0(\tau_a+\tau_b)} = -1$.

Now, we conduct a numerical simulation with the following parameters, to examine the performance of the feedback control for squeezing enhancement: $T_1 = 0.12$, $L_1 = 5.0 \times 10^{-3}$, $L_2 = 5.0 \times 10^{-2}$, $l = 0.5$ m, and $l_a = l_b = 0.25$ m. Figure 6.24 depicts the normalized noise power $10\log_{10}(\langle|\eta_2^q[\imath\omega]|^2\rangle/\langle|\xi_2^q[\imath\omega]|^2\rangle) = 10\log_{10} S_2^q(\imath\omega)$ dB versus T_2, at frequency $\omega/2\pi = 1$ MHz, for several normalized pumping strength $x = \epsilon/\gamma$. Note that the horizontal axis at 0 dB indicates the vacuum noise variance. The circles represent the values at $T_2 = 1$ with $L_2 = 0$, i.e., the squeezing levels of the uncontrolled OPO. This figure shows that if the pumping strength is weak ($x = 0.1$ or $x = 0.35$), there exists an optimal point of T_2 that lowers the variance than that in the uncontrolled case; that is, the coherent feedback certainly enhances the squeezing. However, if a large pump field is applied ($x = 0.6$), the autonomous OPO without control is the best system that minimizes the variance; the CFB does not enhance the squeezing at all. This is simply because, in general, a highly squeezed optical field is fragile to optical loss, leading to a trade-off between the squeezing level and the optical loss. Thus, for a strongly pumped OPO, the feedback introduces further optical loss rather than enhancing the squeezing. This is a clear limitation of coherent feedback control for the purpose of squeezing enhancement.

Finally, let us see the experimental demonstration of the above. The setup is illustrated in Fig. 6.25 (see [21] for the detailed configuration). A notable point is that the CBS is constructed so that it is involved in a Mach–Zehnder (MZ) interferometer, which allows us to tune the controller parameter T_2. In this experiment, the parameters are chosen as $x = 0.106$, $T_1 = 0.20$, $L_1 = 9.0 \times 10^{-3}$, $L_2 = 0.12$, $l = 0.5$ m, and $l_a = l_b = 0.25$ m. Figure 6.26 shows the squeezing and anti-squeezing levels in the

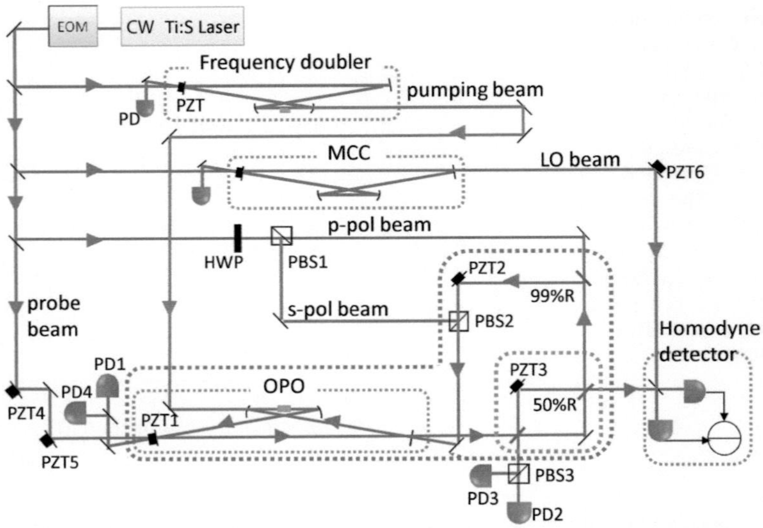

Fig. 6.25 Experimental configuration. MCC: mode cleaning cavity, PD: photodetector, PZT: piezo-electric transducer, PBS: polarized beam splitter, HWP: half-wave plate, and LO: local oscillator. The *blue dashed line* indicates the CFB loop. The *green dashed line* indicates a Mach–Zehnder interferometer, corresponding to the CBS. Reprinted with permission from [21] © 2012 IEEE (color figure online)

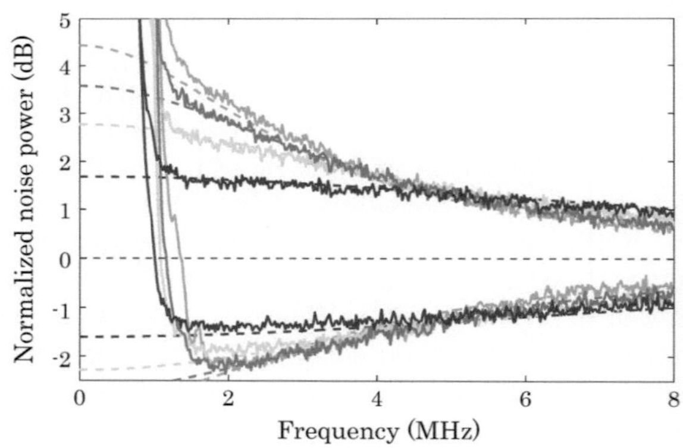

Fig. 6.26 Frequency dependence of the squeezing and anti-squeezing levels. The *blue lines* represent those without feedback ($T_2 = 1$), while the *green*, *red*, and *pink lines* correspond to the case with feedback under the condition $T_2 = 0.7$, 0.8, and 0.9, respectively. Dark noise (variance of the number of electrons in the detector) is subtracted. *Dashed lines* indicate theoretical values. Reprinted with permission from [21] © 2012 IEEE (color figure online)

frequency domain, with several T_2; the blue solid lines indicate those without the coherent feedback (i.e., $T_2 = 1$), while the green, red, and pink solid lines correspond to the feedback case with $T_2 = 0.7$, 0.8, and 0.9, respectively. The large noise found at lower frequencies comes from the laser noises and modulation signals that are used for cavity locking mechanisms. Nonetheless, at higher frequency regime, the experimental result agrees well with the theoretical values shown with dashed lines. Note that while better squeezing enhancement is realized if we take a smaller value of T_2, the effective squeezing bandwidth becomes narrower. This additional constraint imposed on the feedback control is mainly because the OPO and the feedback loop suffer from the time delays.

6.6.2 Measurement-Based Feedback for Spin Squeezing in Atomic Ensemble

Next, we consider a particular experimental demonstration of measurement-based feedback control as reported in [74]. As before, let us begin with the theoretical background.

The system is a large atomic ensemble system as described in Sect. 1.5.5, which is subjected to the Faraday interaction $L = \sqrt{M} J_z$ and the spin rotation implemented by the magnetic field $H = bJ_y$. Let us recall that this setup realizes QND measurement of J_z, through the dynamics (1.44) and the measurement output (1.45). In particular, we here assume the large ensemble limit $J = N/2 \gg 1$ and that the x-component is much larger than the others, leading to $J_x \approx J$ and the CCR relation $[J_y, J_z] = \imath J_x \approx \imath J$; further, assume that the initial state is the *spin-coherent state*, which is exactly the same as the optical coherent state in the two-dimensional tangent space at $(x, y, z) = (J, 0, 0)$ in the Bloch-like sphere depicted in Fig. 6.27a. Under these assumptions, the state is always Gaussian as long as it is confined in the tangent space. Now, we construct a quantum filter as described in Sect. 4.2.1, to estimate $j_t(J_z) = U(t)^* J_z U(t)$ using the measurement output (1.45), which in Itō form is given by $dy_\mathrm{m} = 2\sqrt{M} j_t(J_z)dt + d\mathcal{A} + d\mathcal{A}^*$. Then, as seen in (4.21) and (4.22), the conditional mean and variance of $j_t(J_z)$ change in time as

$$d\langle J_z\rangle_c = Jb(t)dt + 2\sqrt{M}\langle \Delta J_z^2\rangle_c dw, \quad d\langle \Delta J_z^2\rangle_c = -4M\langle \Delta J_z^2\rangle_c^2 dt, \qquad (6.56)$$

where $\langle J_z\rangle_c = \pi_t(J_z)$ is the least mean square estimate of $j_t(J_z)$. Also, note that $\langle \Delta J_z^3\rangle_c = 0$, because of the Gaussianity of the state. The above equations show that the variance of $j_t(J_z)$ deterministically decreases, while the mean fluctuates probabilistically; then, due to the Heisenberg uncertainty relation $\langle \Delta J_y^2\rangle_c\langle \Delta J_z^2\rangle_c \geq J^2/4$, the variance of J_y must increase. Hence, a squeezed state is generated, with its mean probabilistically fluctuating, which is illustrated in Fig. 6.27b. Note that in the general setting without the Gaussian approximation, the state finally converges to one of the eigenstate of J_z if $b(t) = 0$, as discussed in Sect. 4.2.4. This state is called a *spin-squeezed state*, in analogy to the optical squeezed state.

Now, note that without any control (i.e., $b(t) = 0$), ensemble averaging of the spin-squeezed states over all the atoms leads to $\langle \Delta J_z^2\rangle \geq J/2$, because of the fluctuation of the mean. Thus, the unconditional state is not anymore a spin-squeezed state, as

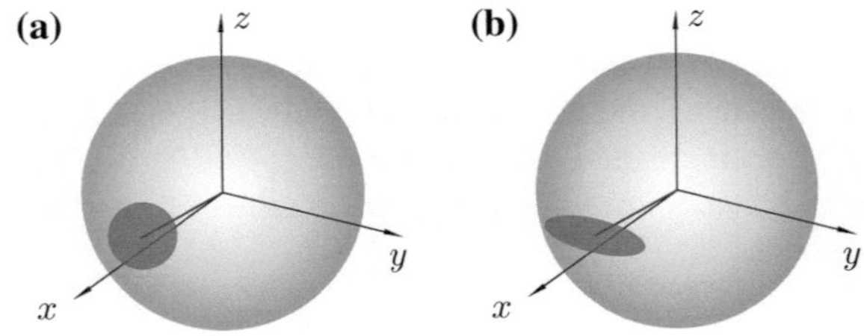

Fig. 6.27 **a** Spin-coherent state and **b** spin-squeezed state

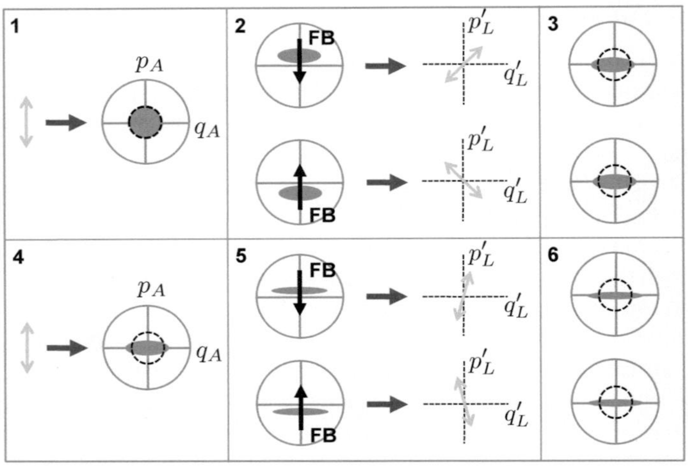

Fig. 6.28 Measurement-based feedback protocol for spin squeezing

discussed at the end of Sect. 4.3.1. Measurement-based feedback control is applicable to resolve this issue and eventually generate a spin-squeezed state in a deterministic way, as follows. The point is that now we have a real-time optimal estimate of $j_t(J_z)$, namely $\langle J_z \rangle_c = \pi_t(J_z)$, by updating the quantum filter (6.56). Hence, the control input signal $b(t)$ can be set to a function of $\langle J_z \rangle_c$; in particular, let us here take a simple choice $b(t) = -\bar{b}\langle J_z \rangle_c$ with $\bar{b} > 0$. Then, (6.56) yields $d\mathbf{E}[\langle J_z \rangle_c]/dt = -J\bar{b}\mathbf{E}[\langle J_z \rangle_c]$, leading to $\mathbf{E}[\langle J_z \rangle_c] \to 0$ in the long time limit. Then, because decrease of $\langle \Delta J_z^2 \rangle_c$ is still maintained under the control due to (6.56), the spin-squeezed state with mean zero is deterministically generated.

The above measurement-based feedback strategy was experimentally demonstrated in [74]. Let us now see the idea behind this experiment. Figure 6.28 illustrates the control procedure for deterministically generating a spin-squeezed state. In the figure, we denote $(q_A, p_A) = (J_y/\sqrt{J}, J_z/\sqrt{J})$, and (q'_L, p'_L) are the pair of normalized (q, p) components of the polarization of the output light field η. The procedure is to send a polarized beam into the atomic ensemble followed by applying a magnetic field depending on the measurement result, and in the experiment, this action was performed *twice*; that is, rather than the above-described continuous-time monitoring and feedback, what was demonstrated in the experiment is actually discrete-time feedback control. The detailed control procedure is as follows (the steps below correspond to the numbering in Fig. 6.28).

1. Initialization: The atomic ensemble and the light field are prepared to a spin-coherent state (red circle) and a vertical polarized state (green arrow), respectively.
2. After the atom–field Faraday interaction, the polarization rotates depending on the probabilistic shift of the atom state, which is now squeezed. The angle of polarization can be determined by measuring q'_L by a polarization beam splitter; this can be used to estimate the shift of the spin-squeezed state, and this shift is compensated by the feedback control implemented by a magnetic field.
3. As a result, a spin-squeezed state is deterministically generated.
4. Second initialization: The atom state is prepared in the spin-squeezed state generated through the Steps 1–2–3.
5. The same action as in the Step 2 is applied. Due to the conditional state evolution, the atom state is further squeezed. Also, the rotation angle of the polarization is reduced compared to the case in the Step 2.
6. As a result, a further squeezed state is generated.

As implied from the above procedure, the variance of the polarization angle of the output field depends on the variance of p_A. Therefore, the squeezing level of the atomic state can be evaluated by using the variance of the output $q'_L = \kappa p_A + q_L$, where q_L represents the normalized q-component of the polarization of the input light field ξ and κ is the interaction strength proportional to $\sqrt{M}$. (Note that this equation corresponds to $dy_{\rm m} = 2\sqrt{M} j_t(J_z)dt + d\mathcal{A} + d\mathcal{A}^*$.) In particular, the squeezing level can be defined by $\xi^2_{\rm sq} = \langle(\Delta p_A^{(2)})^2\rangle/\langle(\Delta p_A^{(1)})^2\rangle$, where $p_A^{(i)}$ denotes the initial p-component of the atomic state in the ith experiment. Then, from $\langle(\Delta q'_L{}^{(i)})^2\rangle = \kappa^2\langle(\Delta p_A{}^{(i)})^2\rangle + \langle(\Delta q_L^{(i)})^2\rangle = \kappa^2\langle(\Delta p_A{}^{(i)})^2\rangle + 1/2$. Therefore, the squeezing level of the atom is represented as

$$\xi^2_{\rm sq} = \frac{\langle(\Delta p_A{}^{(2)})^2\rangle}{\langle(\Delta p_A^{(1)})^2\rangle} = \frac{\langle(\Delta q'_L{}^{(2)})^2\rangle - 1/2}{\langle(\Delta q'_L{}^{(1)})^2\rangle - 1/2}.$$

In the experiment, it was observed that the best squeezed level (with respect to the feedback gain) is $\xi^2_{\rm sq} \approx 0.912$, i.e., about 0.8 dB spin squeezing was realized.

6.6.3 Further Reading

There are several experimental demonstrations of quantum feedback control beyond linear quantum systems; for instance, measurement-based feedback has been experimentally implemented to stabilize photon number states in a microwave cavity [75] and Rabi oscillations in a superconducting qubit [76]. As for coherent feedback, we find its application to stabilizing an entangled state of two superconducting qubits [77]. Here, in accordance with the theme of this monograph, we briefly highlight the experiments reported in [78, 79], which both experimentally demonstrate feedback control for linear quantum systems.

The first one [78] experimentally implemented a coherent feedback network for an electromechanical device, aimed at realizing efficient control and readout of the mechanical state. While these two operations cannot be simultaneously and efficiently achieved in the standard setup without a feedback architecture, the developed coherent feedback allows very fast switching between the two. Such very fast manipulation using a feedback controller, which is comparable to the system's decay rate, would be quite difficult to implement with a measurement-based feedback controller. Thus, the experiment illustrates one of the major advantages of coherent feedback control in an actual physical setup.

The next experiment, [79], demonstrates the use of measurement-based feedback for cooling a mechanical oscillator to achieve its motional ground state and showed a significant improvement over the conventional approach. There are two important system components to accomplish this goal. One is an ultra-precise position sensor that can resolve the zero-point fluctuation of a nano-mechanical oscillator, and the other is an actuator powered by radiation pressure force. This illustrates the requirements of having a very precise sensor and actuator in order to attain a real advantage of measurement-based feedback control in the quantum regime. As described in this work, this is now within reach of current state-of-the-art experiments in the field of opto-mechanics.

References

1. S.L. Braunstein, P. van Loock, Quantum information with continuous variables. Rev. Mod. Phys. **77**, 513 (2005)
2. A. Furusawa, P. van Loock, *Quantum Teleportation and Entanglement: A Hybrid Approach to Optical Quantum Information Processing* (Wiley-VCH, Berlin, 2011)
3. N.C. Menicucci, P. van Loock, M. Gu, C. Weedbrook, T.C. Ralph, M.A. Nielsen, Universal quantum computation with continuous-variable cluster states. Phys. Rev. Lett. **97**, 110501 (2006)
4. N.C. Menicucci, S.T. Flammia, P. van Loock, Graphical calculus for Gaussian pure states. Phys. Rev. A **83**, 042335 (2011)
5. J.F. Poyatos, J.I. Cirac, P. Zoller, Quantum reservoir engineering with laser cooled trapped ions. Phys. Rev. Lett. **77**, 4728 (1996)
6. N. Yamamoto, Parametrization of the feedback Hamiltonian realizing a pure steady state. Phys. Rev. A **72**, 024104 (2005)
7. B. Kraus, H.P. Buchler, S. Diehl, A. Kantian, A. Micheli, P. Zoller, Preparation of entangled state by quantum Markov processes. Phys. Rev. A. **78**, 042307 (2008)
8. F. Verstraete, M.M. Wolf, J.I. Cirac, Quantum computation and quantum-state engineering driven by dissipation. Nat. Phys. **5**, 633–636 (2009)
9. K. Vollbrecht, C.A. Muschik, J.I. Cirac, Entanglement distillation by dissipation and continuous quantum repeaters. Phys. Rev. Lett. **107**, 120502 (2011)
10. K. Koga, N. Yamamoto, Dissipation-induced pure Gaussian state. Phys. Rev. A **85**, 022103 (2012), Reprinted, with permission, © (2012) by the American Physical Society

11. N. Yamamoto, Pure Gaussian state generation via dissipation: a quantum stochastic differential equation approach. Philos Trans. R. Soc. A **370**, 5324–5337 (2012)
12. R. Simon, E.C.G. Sudarshan, N. Mukunda, Gaussian pure states in quantum mechanics and the symplectic group. Phys. Rev. A **37**, 3028–3038 (1988)
13. Y. Ikeda, N. Yamamoto, Deterministic generation of Gaussian pure states in a quasilocal dissipative system. Phys. Rev. A **87**, 033802 (2013)
14. O. Techakesari, H.I. Nurdin, On the quasi-balanceable class of linear quantum stochastic systems. Syst. Control Lett. **78**, 25–31 (2015)
15. H.I. Nurdin, Structures and transformations for model reduction of linear quantum stochastic systems. IEEE Trans. Autom. Control **59**(9), 2413–2425 (2014)
16. T. Tufarelli, A. Ferraro, A. Serafini, S. Bose, M.S. Kim, Coherently opening a high-Q cavity. Phys. Rev. Lett. **112**, 133605 (2014)
17. J. Ma, X. Wang, C.P. Sun, F. Nori, Quantum spin squeezing. Phys. Rep. **509**, 89–165 (2011)
18. M. Yanagisawa, H. Kimura, Transfer function approach to quantum control - part ii: control concepts and applications. IEEE Trans. Autom. Control **48**(12), 2121–2132 (2003)
19. J.E. Gough, S. Wildfeuer, Enhancement of field squeezing using coherent feedback. Phys. Rev. A **80**, 042107 (2009)
20. O. Crisafulli, N. Tezak, D.B.S. Soh, M.A. Armen, H. Mabuchi, Squeezed light in an optical parametric amplifier oscillator network with coherent feedback quantum control. Opt. Express **21**, 18371–18386 (2013)
21. S. Iida, M. Yukawa, H. Yonezawa, N. Yamamoto, A. Furusawa, Experimental demonstration of coherent feedback control on optical field squeezing. IEEE Trans. Autom. Control **57**(8), 2045–2050. Reprinted, with permission, © 2012 IEEE (2012)
22. Z. Shi, H.I. Nurdin, Coherent feedback enabled distributed generation of entanglement between propagating Gaussian fields. Quant. Inf. Process. **14**, 337–359 (2015). © 2014 Springer. Reprinted, with permission of Springer
23. C.C. Gerry, P.L. Knight, *Introductory Quantum Optics* (Cambridge University Press, Cambridge, 2005)
24. W.P. He, F.L. Li, Generation of broadband entangled light through cascading nondegenerate optical parametric amplifiers. Phys. Rev. A **76**, 012328 (2007)
25. Z. Yan, X. Jia, X. Su, Z. Duan, C. Xie, K. Peng, Cascaded entanglement enhancement. Phys. Rev. A **85**, 040305(R) (2012)
26. B.C. Jacobs, T.B. Pittman, J.D. Franson, Quantum relays and noise suppression using linear optics. Phys. Rev. A **66**, 052307 (2002)
27. W. Michiels, S.I. Niculescu, *Stability and Stabilization of Time-Delay Systems: An Eigenvalue-Based Approach.* Advances in Design and Control (Society of Industrial and Applied Mathematics, 2007)
28. K. Engelborghs, T. Luzyanina, G. Samaey, DDE-BIFTOOL vol. 2.00: A Matlab package for bifurcation analysis of delay differential equations. Department of Computer Science, Katholieke Universiteit Leuven, Belgium, Technical Report TW-330 (2001)
29. K. Engelborghs, T. Luzyanina, D. Roose, Numerical bifurcation analysis of delay differential equations using DDE-BIFTOOL. ACM Trans. Math. Softw. **28**(1), 1–21 (2002)
30. M. Di Loreto, M. Dao, L. Jaulin, J.-F. Lafay, J.J. Loiseau, Applied interval computation: a new approach for time-delays systems analysis, in *Applications of Time Delay Systems*, ed. J. Chiasson, J.J. Loiseau (Springer, Berlin 2007)
31. Z. Shi, H.I. Nurdin, Effect of phase shifts on EPR entanglement generated on two propagating Gaussian fields via coherent feedback, in *Proceedings of the 53rd IEEE Conference on Decision and Control (CDC) (Dec 15–17, 2014)*, pp. 5813–5818
32. Z. Shi, H.I. Nurdin, Optimization of distributed entanglement generated between two Gaussian fields by the modified steepest descent method, in *Proceedings of the 2015 American Control Conference (ACC) (Jul 1–3, 2015)*, pp. 2697–2702
33. Z. Shi, H.I. Nurdin, Local optimality of a coherent feedback scheme for distributed entanglement generation: the idealized infinite bandwidth limit, in *Proceedings of the 54th IEEE Conference on Decision and Control (CDC)* (Dec 15–18, 2015), pp. 7755–7760

34. Y. Zhou, X. Jia, F. Li, J. Yu, C. Xie, K. Peng, Quantum coherent feedback control for generation system of optical entangled state. Sci. Rep. **5**, 11132 (2015)
35. D. Wang, C. Xia, Q. Wang, Y. Wu, F. Liu, Y. Zhang, M. Xiao, Feedback-optimized extraordinary optical transmission of continuous-variable entangled states. Phys. Rev. B **91**, 121406(R) (2015)
36. Z. Shi, H.I. Nurdin, Entanglement in a linear coherent feedback chain of nondegenerate optical parametric amplifiers. Quant. Inf. Comput. **15**(13–14), 1141–1164 (2015)
37. Z. Shi, H.I. Nurdin, Formulae for entanglement in a linear coherent feedback network of multiple nondegenerate optical parametric amplifiers: the infinite bandwidth case, in *Proceedings of the 2016 American Control Conference (ACC) (Jul 6–8, 2016)*, pp. 4769–4774
38. V.B. Braginsky, F.Y. Khalili, *Quantum Measurement* (Cambridge University Press, Cambridge, 1992)
39. C.M. Caves, K.S. Thorne, R.W.P. Drever, V.D. Sandberg, M. Zimmermann, On the measurement of a weak classical force coupled to a quantum mechanical oscillator. I. Issues of principle. Rev. Mod. Phys. **52**, 341–392 (1980)
40. M. Tsang, C.M. Caves, Coherent quantum-noise cancellation for optomechanical sensors. Phys. Rev. Lett. **105**, 123601 (2010)
41. H. Miao, *Exploring Macroscopic Quantum Mechanics in Optomechanical Devices* (Springer, Berlin, 2012)
42. E.M. Komar, M. Kessler, L. Bishof, A.S. Jiang, J. Sorensen, Ye, M.D. Lukin, A quantum network of clocks. Nat. Phys. **10**, 582–587 (2014)
43. N. Yamamoto, Coherent versus measurement feedback: linear systems theory for quantum information. Phys. Rev. X **4**, 041029 (2014)
44. Y. Yokotera, N. Yamamoto, Geometric control theory for quantum back-action evasion. EPJ Quantum Technol. **3**(15), 1–22 (2016)
45. B.P. Abbott et al., Observation of gravitational waves from a binary black hole merger. Phys. Rev. Lett. **116**, 061102 (2016)
46. N. Yamamoto, M.R. James, Zero dynamics principle for perfect quantum memory in linear networks. New J. Phys. **16**, 073032 (2014)
47. D.A. Lidar, K.B. Whaley, Decoherence-free subspaces and subsystems, in *Irreversible Quantum Dynamics*. Lecture Notes in Physics, vol. 622 (2003), p. 83
48. D.F. Phillips, A. Fleischhauer, A. Mair, R.L. Walsworth, M.D. Lukin, Storage of light in atomic vapor. Phys. Rev. Lett. **86**, 783 (2001)
49. I. Novikov, A.V. Gorshkov, D.F. Phillips, A.S. Sorensen, M.D. Lukin, R.L. Walsworth, Optimal control of light pulse storage and retrieval. Phys. Rev. Lett. **98**, 243602 (2007)
50. A.V. Gorshkov, A. Andre, M.D. Lukin, A.S. Sorensen, Photon storage in lambda-type optically dense atomic media I. Cavity model. Phys. Rev. A **76**, 033804 (2007)
51. Q. Xu, P. Dong, M. Lipson, Breaking the delay-bandwidth limit in a photonic structure. Nat. Phys. **3**, 406–410 (2007)
52. J. Yoshikawa, K. Makino, S. Kurata, P. van Loock, P,A. Furusawa, Creation, storage, and on-demand release of optical quantum states with a negative Wigner function. Phys. Rev. X **3**, 041028 (2013)
53. N. Yamamoto, Decoherence-free linear quantum subsystems. IEEE Trans. Autom. Control **59**(7), 1845–1857 (2014)
54. M. Hush, A.R.R. Carvalho, M. Hedges, M.R. James, Analysis of the operation of gradient echo memories using a quantum input-output model. New J. Phys. **15**, 085020 (2013)
55. N. Yamamoto, H.I. Nurdin, M.R. James, Quantum state transfer for multi-input linear quantum systems, in *Proceedings of the 55th IEEE Conference on Decision and Control (CDC) (Dec 15–17, 2014)* (2016)
56. C.A. Muschik, K. Hammerer, E.S. Polzik, J.I. Cirac, Efficient quantum memory and entanglement between light and an atomic ensemble using magnetic field. Phys. Rev. A **73**, 062329 (2006)
57. Q.Y. He, M.D. Reid, E. Giacobino, J. Cviklinski, P.D. Drummond, Dynamical oscillator-cavity model for quantum memories. Phys. Rev. A **79**, 022310 (2009)

58. Y. Wang, J. Minar, G. Hetet, V. Scarani, Quantum memory with a single two-level atom in a half cavity. Phys. Rev. A **85**, 013823 (2012)
59. S.A. Aljunid, G. Maslennikov, Y. Wang, H.L. Dao, V. Scarani, C. Kurtsiefer, Excitation of a single atom with exponentially rising light pulses. Phys. Rev. Lett. **111**, 103001 (2013)
60. M. Bader, S. Heugel, A.L. Chekhov, M. Sondermann, G. Leuchs, Efficient coupling to an optical resonator by exploiting time-reversal symmetry. New J. Phys. **15**, 123008 (2013)
61. G.K. Gulati, B. Srivathsan, B. Chng, A. Cere, D. Matsukevich, C. Kurtsiefer, Generation of an exponentially rising single-photon field from parametric conversion in atoms. Phys. Rev. A **90**, 033819 (2014)
62. H. Ogawa, H. Ohdan, K. Miyata, M. Taguchi, K. Makino, H. Yonezawa, J. Yoshikawa, A. Furusawa, Real-time quadrature measurement of a single-photon wavepacket with continuous temporal-mode-matching. Phys. Rev. Lett. **116**, 233602 (2016)
63. H.S. Black, Inventing the negative feedback amplifier. IEEE Spectr. **14**, 55 (1977)
64. H.S. Black, Stabilized feedback amplifiers. Proc. IEEE **72**, 716–722 (1984)
65. N. Yamamoto, Quantum feedback amplification. Phys. Rev. Appl. **5**, 044012 (2016), Reprinted, with permission, © (2016) by the American Physical Society
66. A.A. Clerk, M.H. Devoret, S.M. Girvin, F. Marquardt, R.J. Schoelkopf, Introduction to quantum noise, measurement, and amplification. Rev. Mod. Phys. **82**, 1155 (2010)
67. N. Bergeal, F. Schackert, M. Metcalfe, R. Vijay, V. Manucharyan, L. Frunzio, D.E. Prober, R.J. Schoelkopf, S.M. Girvin, M.H. Devoret, Phase-preserving amplification near the quantum limit with a Josephson ring modulator. Nature **465**, 64–68 (2010)
68. H.M. Chrzanowski, N. Walk, S.M. Assad, J. Janousek, S. Hosseini, T.C. Ralph, T. Symul, P.K. Lam, Measurement-based noiseless linear amplification for quantum communication. Nat. Photonics **8**, 333–338 (2014)
69. F. Hudelist, J. Kong, C. Liu, J. Jing, Z.Y. Ou, W. Zhang, Quantum metrology with parametric amplifier-based photon correlation interferometers. Nat. Commun. **5**, 3049 (2014)
70. A. Metelmann, A.A. Clerk, Quantum-limited amplification via reservoir engineering. Phys. Rev. Lett. **112**, 133904 (2014)
71. J. Gough, R. Gohm, M. Yanagisawa, Linear quantum feedback networks. Phys. Rev. A **78**, 061204 (2008)
72. H.A. Haus, J.A. Mullen, Quantum noise in linear amplifiers. Phys. Rev. **128**, 2407–2413 (1962)
73. C.M. Caves, Quantum limits on noise in linear amplifiers. Phys. Rev. D **26**, 1817–1839 (1982)
74. R. Inoue, S. Tanaka, R. Namiki, T. Sagawa, Y. Takahashi, Unconditional quantum-noise supression via measurement-based quantum feedback. Phys. Rev. Lett. **110**, 163602 (2013)
75. C. Sayrin, I. Dotsenko, X. Zhou, B. Peaudecerf, T. Rybarczyk, S. Gleyzes, P. Rouchon, M. Mirrahimi, H. Amini, M. Brune, J.-M. Raimond, S. Haroche, Real-time quantum feedback prepares and stabilizes photon number states. Nature **477**, 73–77 (2011)
76. R. Vijay, C. Macklin, D.H. Slichter, S.J. Weber, K.W. Murch, R. Naik, A.N. Korotkov, I. Siddiqi, Stabilizing Rabi oscillations in a superconducting qubit using quantum feedback. Nature **490**, 77–80 (2012)
77. S. Shankar, M. Hatridge, Z. Leghtas, K.M. Sliwa, A. Narla, U. Vool, S.M. Girvin, L. Frunzio, M. Mirrahimi, M.H. Devoret, Autonomously stabilized entanglement between two superconducting quantum bits. Nature **504**, 419–422 (2013)
78. J. Kerckhoff, R.W. Andrews, H.S. Ku, W.F. Kindel, K. Cicak, R.W. Simmonds, K.W. Lehnert, Tunable coupling to a mechanical oscillator circuit using a coherent feedback network. Phys. Rev. X **3**, 021013 (2013)
79. D.J. Wilson, V. Sudhir, N. Piro, R. Schilling, A. Ghadimi, T.J. Kippenberg, Measurement-based control of a mechanical oscillator at its thermal decoherence rate. Nature **524**, 325–329 (2015)
80. H. Nakao, N. Yamamoto, Optimal control for perfect state transfer in linear quantum memory, J. Phys. B: At. Mol. Opt. Phys. **50**, 065501 (2017)

Index

H.I. Nurdin and N. Yamamoto, *Linear Dynamical Quantum Systems*,
Communications and Control Engineering, DOI 10.1007/978-3-319-55201-9

MIX
Papier aus verantwortungsvollen Quellen
Paper from responsible sources
FSC® C105338

If you have any concerns about our products,
you can contact us on
ProductSafety@springernature.com

In case Publisher is established outside the EU,
the EU authorized representative is:
Springer Nature Customer Service Center GmbH
Europaplatz 3, 69115 Heidelberg, Germany

Printed by Libri Plureos GmbH
in Hamburg, Germany